2007
SUBMARINE
ALMANAC

2007

ALMANAC

SUBSIM.COM

ARTICLES, STORIES AND ART FROM SUBVETS, GAME DEVELOPERS, PLAYERS AND HISTORIANS

Neal Stevens, Editor

DEEP DOMAIN

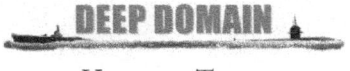

Houston, Texas

2007 Submarine Almanac
Copyright © 2007 by Neal Stevens
Subsim ~ www.subsim.com

All rights reserved. No part of this book shall be reproduced, stored in a retrieval system, or transmitted by any means, electronic, mechanical, photocopying, recording, or otherwise, without written permission from the publisher. Brief excerpts are permitted as the subject of reviews of this book.

In Memoriam: Erich Topp (U-552) © 2006 by Theodore P. Savas
The Legend of Odin; Things They Don't Tell You at
New London © 2003 by Bob 'Dex' Armstrong
Lyrics to *Billy Bones* by Skip Henderson used with permission
Front Cover art by Spencer Burnham, with U-boat by Mark Davies,
"Subsim ring" by Craig Dinkleman
Back Cover art by Spencer Burnham

Second Edition
ISBN 978-0-6151-5381-0
Printed in the United States of America

To my sister, Lisa

Published by the Deep Domain
ww.deepdomain.net
Houston, Texas USA
2006

Foreword

In every generation since the historic transition from sail to steam and from wooden hulls to steel, no type of ship or sailor has played a more vital, versatile role in peace and war than the naval submarine and her crew. From harbor defense to fleet escort, from marauder of enemy seaborne logistics to reliable sinker of surface combatants dozens of times their own size, subs and submariners have "been there, done that" to the max. As platforms for relentless strategic deterrence, for amazing intelligence-gathering and undersea salvage capabilities, for countless covert insertions/extractions of SEALs and other commando teams, whose stories must go forever untold, subs have a proven track record for delivering the goods, and their present-day and future utility remain indispensable.

Whether their vessels were (or are) powered by human muscle, or gasoline engines or diesel and batteries, or by nuclear reactors, or by several different kinds of air-independent technologies, submariners have always been a breed apart. For raw courage and grit, for long separations from family, for extremely rough living conditions in crowded and claustrophobic spaces deep under the waves, no other branch of military service compares. Weeks of repetitive, uneventful watchstanding can change without warning into a frenzy of well-coordinated thought and action where the lives of every soul aboard, and sometimes the fate of humanity, are instantly at stake. For instance, if an emergency action message comes through to a boomer, if a fast-attack suddenly detects a hostile contact approaching, if Tomahawk launch mission orders arrive unexpectedly on a guided-missile sub — a submarine must be ever-vigilant for conflict. Even in peacetime any one of a myriad potentially deadly mechanical casualties may occur. A sub is always at war with its natural elements: the sea. But the career of a submariner is not without its unique rewards. The degree of bonding and camaraderie that results from so many perpetual challenges is hard for laymen to comprehend. The heavy burdens of command can be especially elusive.

Human nature being what it is, many people do want to understand and participate vicariously in the exciting adventures that can only be experienced by those who go down into the sea in ships designed to dive instead of sink. The best way of learning and sharing, of course, is to clamber through the hatch and go for a ride. As a civilian author and commentator on undersea warfare and national defense, I was very privileged to twice be invited by the U.S. Navy to travel in nuclear subs — one trip was an unforgettable four-day tiger cruise aboard *USS Miami*, and the other was an impressive media underway on the littoral-combat reconfigured *USS Ohio*. But with America at war against terror on a global scale, and this nation's Submarine Force stretched thin by an extremely demanding operational tempo, such first-hand opportunities are scarce. a submarine enthusiast to do?

Visiting a museum ship, such as *USS Cavalla* in Galveston, *U-505* in Chicago, or *USS Nautilus* in Groton, is certainly a good way to start. And chatting with submarine veterans is always informative and fun; these great guys will have sea stories that leave you chuckling uncontrollably or literally make your hair stand on end. For decades, memoirs, novels, and movies have been excellent methods — part education, part entertainment — for savoring the submarine experience. Yet as realistic and scary as these depictions can be, their plots are predetermined. Given the form of the medium used, whether an old war movie you might have watched on a black-and-white TV as a kid, or an e-book you plan to read at the beach with a PDA on your next vacation, your involvement is limited to being somewhat passive, once removed.

Alas, with such pre-scripted content, no matter how well it might keep you on the edge of your seat with heart-pounding excitement, you just aren't completely *there*. It's not *you* making those impossible tradeoffs, those split-second decisions against wily, ruthless enemy sub and destroyer captains or antisubmarine aircrews all hell-bent on your demise. It's not *you* peering up through the periscope, ferociously giving the order to fire.

Then along came personal computers and installable (not arcade) videogames. The two grew up together, symbiotically, more capable PCs and laptops allowing more complex and textured first-person and multiplayer games. Actual real-time submarine simulations became first an achievable dream, next an affordable pastime, and eventually an entire industry as games about various sorts of subs and missions of different eras competed and multiplied. Aided by the worldwide connectivity of the Internet, players wanting to meet other players, trade tactical tips and technical help, exchange product reviews, even plan group get-togethers on one continent or another, were able to conveniently do so. Most of all, at long last, whatever particular game you were playing, you were there.

And so was Neal Stevens' Subsim.com, now celebrating a wonderful benchmark, its tenth anniversary on the Web. By a funny coincidence, my full-time career as a submarine writer began back in '97, with my very first sale, a non-fiction article published in a professional submarine journal. In what to me, and Neal, now seems like fated synchronicity regarding our eventual crossing of paths, to try to improve my tactics I began to drill on Jane's 688(I). Let me tell you, I discovered fast that submariners genuinely mean what they say, how while they and their buddies are immersed in a combat scenario the whole thing really feels as if it's happening for keeps. I'll never forget the sensation of stark terror when I had a Soviet torpedo screaming along its merry way and pinging faster and faster as it closed on my ship — and my crew.

With practice, I found that countermeasures and well-executed evasive maneuvers could save my backside and avoid creating a hundred-plus widows and orphans, which would have all been my fault, my personal responsibility as imaginary CO. And man, did I ever sweat; the spot on the couch where I sat while I trained for hours on end would be soaked. As I write this, a Dell Inspiron 6400 with 2GB of RAM is on order. Sonalysts' "Dangerous Waters" already sits here on my desk. I look forward with relish to playing the one on the other.

Yes, the games have gotten better and better during this memorable past decade. (The experiences available are just starting to rival those possible in the computer-controlled team attack simulators at the New London Naval Submarine Base.) The users of Subsim's many resources have grown vastly more numerous and interconnected, too. It is accurate to say that Subsim holds pride of place as the center of the online universe for submarine and other naval simulation computer games and the international community of diverse folks who love participating in them.

I am extremely honored that Neal invited me to write this foreword for the *2007 Submarine Almanac*. You're in for a delicious treat with a compendium you will turn to again and again and be proud to display on your bookshelf. Here's wishing Neal Stevens, Subsim, and all the website's devoted players and contributors a prosperous, safe, and enjoyable second decade of even greater gaming to come!

Joe Buff
Dutchess County, NY
November 24, 2006

Introduction

What do you get when you cross a love for computers, submarines, and journalism?

The first Internet website appeared on August 6th, 1991. Five and a half years later, on January 26th, 1997, a guy who had always liked submarine subject matter put together a small collection of sub game reviews and some pictures of the *USS Cavalla* to form a simple homepage called *Neal's Subsim Reviews*. That guy was me, and still is.

Ten years have passed since I FTP'ed a batch of Notepad html files and hit F5 on my browser. That first Subsim website resided on a meager 1MB of server space provided by my Internet service provider, Computron. It grew to include more reviews, articles, a patches section, and a public discussion forum. The homepage soon outgrew the ISP's allocated space and was migrated to the local college server, which was replaced in turn with shared hosting accounts, and eventually took over two EV1 servers. We got big. Game developers began taking note of us, and Subsim members formed the beta teams for several top-notch games, including Silent Hunter II and Dangerous Waters.

I first came up with the idea for this book about a year ago. I was thinking, man, Subsim has been around nearly ten years. In Internet time, that is like AT&T and Western Electric. Homepages and websites come and go. Most of the websites Subsim shared the Net with in 1997 have long since gone to the World Wide Web graveyard.

Websites and online communities are still relatively new to society. Before Sir "Tim" Berners-Lee revolutionized the way we share information, finding other people with uncommon interests like submarines was generally achieved through print magazines and newspaper advertisements. If you liked building model ships or stamp collecting, you had to search out fellow aficionados and join a club. They might provide a newsletter and meetings, but the way you personally interacted with other enthusiasts and shared information was limited.

Now, with the Internet and online forums such as Subsim's Radio Room, submarine enthusiasts from any part of Planet Earth can pull up a chair in cyberspace and discuss sub simulations, sub history, and naval topics. Subsim became more than a website, it became a community.

For ten years, Subsim has been a hub for sub fanatics eager to learn about new games coming out, speculate on features, and share tips and tactics.

The good fellowship, the help and assistance in playing the games, and the humor and wit of the people who populate the Radio Room forums are the core of Subsim. We have held good-natured mock trials for fellow players who spread game rumors, expressed viewpoints, and debated an endless number of topics, both trivial and serious. Subsim members have raised money for cancer victims, venerated the Red Triangle, developed mods for games, and held numerous meetings, including three that drew members from around the world. We have talked about buying a surplus Russian Kilo sub and outfitting it as a cruise ship/clubhouse for vacationing sub enthusiasts, but that's still in the planning stage.

The people of Subsim are always full of insight and ideas, and that gave me the idea to make this Almanac — like the website — a platform for subsim developers, submarine veterans, and naval enthusiasts.

So, it is in celebration of you, the naval enthusiast, the guy who played Silent Service on a Commodore computer, who reads books about Mush Morton, builds models of Gunter Prien's U-boat, and knows all the character's names in *Das Boot*; the fellow who understands how a torpedo data computer works, and who is genetically inclined to grow fangs and howl at the moon when the term "scripted campaign" is uttered…

… this tip of the cap and this book is for you.

Good hunting!
Neal Stevens
Houston, Texas
December 1, 2006

CONTENTS

Silent Hunter III Captain's Log - Florin Boitor	15
You Know You've Been With Subsim a Long Time When… Craig "Torplexed" Dinkleman	24
The Legend of Odin - Bob 'Dex' Armstrong	25
Just Another Cargo Ship - Mariano Sciaroni	29
The *Kraka* Story - Jonathan Beck Jorgensen	32
Tesseraction Games and Enigma: Rising Tide - Kelly Asay	38
Back to the Future - Bill Nichols	47
The Lucky Lighter - Jason Lobo	59
Civilian Submariner - Donald Ross, Ph.D.	75
Chief Mac and the Contact - Dave Stoops	85
Ron Martini – Profile of a Web Pioneer	87
Cavalla Makes its Mark in Naval History - Zeke Zellmer	90
The Flanders U Boat Flotilla 1915 – 1918 - D. Swetnam, MA	95
Silent Hunter II Play Test Report - Neal Stevens	112

Blood & Honor - Gerard Cuomo	120
Things They Don't Tell You at New London - Bob 'Dex' Armstrong	137
Sub Club Inter-Continental Meetings – Laura Sands	140
The Dreadnought Era - David Millichope	145
Subsim Jim – Valerie Stevens	169
In Memoriam: Erich Topp (*U-552*) - Theodore P. Savas	172
The History of Subsims - Brian H. Danielson	188
Submarine Silencer - Donald Ross, Ph.D.	198
Life aboard a U.S. Nuclear Submarine - Tim Grab	212
Submarines as Time Machines - Paul Farace	232
Ourselves Alone; The Lost Patrols of *U-49* - Clifford J. Hurgin, Jr	242
Subsim Roll Call	350
Origin of the *Laconia* Order - Dr. Maurer Maurer and Lawrence J. Paszek	356
Contributors	370

I first heard about Subsim.com way back in 1998 when I was looking for tips for the Commander's Edition of Silent Hunter. The first article I can remember reading was an interview with Mike Jones, and the first file I downloaded was the 1.1 patch for 688(I). I remember seeing the first screenshots for SHII back in 1998, and being glad that we'd recently updated our computer.

Daryl "Subnuts" Carpenter

2007 SUBMARINE Almanac

"It would be hard to trust a boat that was not designed to sink."
Mark Breece

Silent Hunter III Captain's Log

Florin Boitor, Executive Producer

"What do you know about Silent Hunter?"

The first time I was asked this question was in the summer of 2002. The answer was quite simple: nothing. But, that's why Google is great. One hour later I was already aware of what the Silent Hunter brand represents. Of course, the main source of information was Subsim.com. At that moment I was the production manager at Ubisoft Romania and I was supposed to give an answer if we should start development on this brand. Not that we had too many choices at that moment, but let's say we could have tried to avoid it if we figured it was an inappropriate project for us. Fortunately, I realized not only was it a perfect project for us at Ubisoft Romania, but I also decided I would do my best to be the producer for this project.

The discussions about the game started that summer and I have to admit: I was very surprised to learn that WWII subs didn't have windows. But not everybody at Ubisoft Romania was that ignorant concerning the subject. There were several guys that knew and played SH1 and SH2 and, even more, were WWII buffs. By September 2002 I also had decent knowledge about the subject. What I was even more interested in was what had happened with previous installments. Were they successful? What kind of people play these games? Well, Subsim had all the answers in this department and I soon realized what the legacy of the brand was at that moment.

"Gran Turismo on the Sea"- Jawohl!

The first document we got for SH3 was a wish list from the previous developers. On top of this list was an assumption: "User controls Pacific Theater U.S. submarines only." Well, this wasn't our choice. I mean, it was probably a logical decision to alternate the theaters, but not for us. We were really fresh and when we started the discussions internally, our assumptions were: the game should be named SH3 and it should be with submarines. There were even discussions about a *WWI version!* Well, we soon realized that the "fattest" subject is still U-boats in WWII. We can't miss this one. But, there were also other reasons. First, we were far more familiar with the European Theater of operations, and second, we really started to be fascinated by the U-boat story in WWII. And I guess that another factor was that somehow we identified ourselves with the German side because we were allies in WWII until a certain point. This isn't something to underestimate. I guess it is very important for a simulation development team, especially at the first, to work on a project about "our guys." Of course, if the initial project proved successful, sequels can alternate sides and theaters of operations. Simulation fans know many examples of this.

With that decision in mind, I made a trip to San Francisco in November 2002 for some meetings about SH3. First stop: *USS Pampanito*. I was really impressed and I realized how lucky we were to develop a game on such a fabulous theme as WWII submarines.

The meetings weren't exactly what I expected. I soon realized that our excitement about the project was really not matched by the guys from the Ubi SF office. Probably our decision to continue with U-boats was a reason. They had their arguments, but I was probably really convincing because they gave them up quite easily. I guess they realized how passionate we were on the subject. In game development, if there is no passion about the subject, you better be prepared to double your budget if you want to realize something decent. Especially when you want to develop a simulation title. However, were we talking about a pure simulation? Not at that meeting. What was discussed was mainly how to develop an action simulation game with submarines. Well, "Gran Turismo on the sea" was the closest match. The idea was to attract as many gamers as possible from outside the traditional sim fans pool because the conclusion was that "they (sim fans) will buy it anyway." As outrageous as that may sound, this was in fact a good decision. If SH3 is probably the best-looking simulator until now, it is mainly because of this idea to make a beautiful and fun game.

The conclusion of this meeting was that we should start the development on a prototype in two steps. The first step should be a non-playable technical demo showing how great the game will look. The second step should be a playable prototype that should validate the gameplay and kick off production.

Happy Times - The Prototype

At that time it was a special moment in Ubisoft Romania. In 2002 we celebrated ten years of Ubisoft development in Romania. During this period, we developed several titles on all possible platforms and a lot of effort and R&D was put in developing an original game development framework called GDS. However, we were missing that single title that would put us on the map. Our team consisted of several programmers, designers, and graphic artists with huge experience in game development, yet we were very frustrated that our work wasn't recognized. You can imagine that when I managed to convince these people that this was our title, there was great excitement. Well, don't imagine people cheering and in tears because they finally had an opportunity to demonstrate that they are very good. No, I am talking about some guys in their mid-thirties, very rational, who realized that it was indeed the most promising project they had ever worked on.

With that positive feeling we started working on the technical demo prototype. As mentioned, this was supposed to be a non-playable demo validating the GDS engine, mainly the graphical part. We started researching the Internet, finding movies and all the possible games showing water in order to have a reference on what we should achieve. Finally, we stopped at only two. *Das Boot* is a no-brainer of course, but the other reference was a movie circulating on the Internet at that time. It was a movie claiming that the entire footage was real-time capture from a PC adventure-strategy game in development about the naval war in the Pacific during WWII and supposed to be released in 2003. While we soon realized that the movie was doctored, at least in the sound department, we were, nevertheless, really impressed. It was obvious that with this competition (indirectly because it wasn't a naval simulation but only shared the naval theme) in development and possibly to be released before SH3, we could say good-bye to the naval games crown if we didn't match them. Well, that sounds funny now because that game is yet to be released and SH3 was released more than one year ago as I'm typing this in July 2006. We knew something was strange with that movie.

Anyway, it was a very good motivation factor for us and we decided to build three scripted scenarios that would run with our engine. Our goal was to create the most impressive real-time naval scenes known at that moment to the general public. Without any doubt I can say now that the period from the prototype kick off in January 2003 until May 2003 was our "Happy Times" during the entire SH3 development period. It was a kernel of great developers and we had a lot of fun working insane hours. During that time we had unforgettable moments like our first render of the sea, first ship, first explosions, first U-boat, etc. We were really feeling we were creating a world. Well, kind of a "Waterworld" because there was no land at that time.

The fact is that before May 1 we managed to develop an application running these real-time scripted scenarios: two attacks against cargo ships (day and night) and one U-boat submerged under destroyer attack.

We were really proud and eager to show off our work. But, to whom? May is a particular time of the year when all publishers go crazy because of E3. There is no way in the world you will get the slightest attention unless your game is in the show. SH3 wasn't even announced in development at that time. First reactions were: "We are still unclear on if this movie was created from your own game engine in real-time." So, we decided to wait for reactions until after E3 settled down.

During the entire SH3 development I was obsessed with providing the team with any kind of materials that could help us in development. We bought dozens of books about U-boats, plans, movies, reel footage, documentaries, games, but the ultimate experience when you develop a game about U-boats is actually seeing one. So, off we went, four guys from Romania on a trip to Germany to see as many U-boats as possible. We drove 5000 km in five days to see: the *Das Boot* Museum at the Bavaria Studios in Munchen, *U-1* at the Deutches Museum in Munchen, *U-995* in Laboe, *U-2540* in Bremerhaven, Deutsches Marinemuseum in Wihelmshaven, and of course, the U-boot Archive in Cuxhaven. We took a lot of pictures that proved very useful, especially for the graphic and design development. We were ready now to start the real development and play with U-boats.

Unfortunately, after E3 and during the summer of 2003, it became obvious for us that SH3 wouldn't get the attention we had hoped. It was the year of Splinter Cell and Prince of Persia, two titles that promoted Ubisoft to the big guys' league. SH3 was never supposed to match these types of titles in terms of sales and importance. The bad part was that the some people involved were losing interest in working for a title that was PC only and had no chance of selling millions of copies like other titles. The good part was that we weren't under pressure anymore. SH3 wasn't supposed to be that kind of game "that can make or break" Ubisoft. So, as long as we could stay on budget and on time, we could do what we wanted. And, speaking of budget, it became obvious that proposals like obtaining the *Das Boot* license, at least for the soundtrack, wasn't going to happen.

Under those circumstances we continued our development in order to build a playable version that should be presented to get the final okay before production. Our first playable version was presented in October 2003 and it was based on the sinking of *Bismarck* scenario. It was our feeling that involving *Bismarck* in SH3 would make it sexier for people who weren't familiar with submarine simulations. We were right! Actually, that first playable version had very few submarine simulation gameplay elements and was in fact a spectacular action demo. Everybody was impressed at how the game looked (externally) and sounded and not so much by how it played. I can't describe in detail this approval process, but I can tell you that it went very well.

Except for one thing: the interior scenes weren't at the same level as the external scenes, especially the crew look and animation because at that time our experience with characters and animation was very limited. This sparked a controversy inside our team that wasn't dramatic, but definitely influenced the team spirit until the end of the project. From that moment it was obvious that while SH3 would become better and better, some guys' morale was going downhill.

The last important event in 2003 was the meeting with Mr. Jurgen Oesten. The meeting was arranged by Ted Savas, who helped us to find a U-boat commander that we wanted to interview for our game. It was really the most important experience I had during this project. We had two meetings in his house on the outskirts of Hamburg, his residence since 1936 when he was in the Kriegsmarine. It was a privilege for me to hear the stories of this gentleman. We gathered about six hours of videotaped material with him. Some parts were released on the official website.

It's official!

In January 2004 the game was announced "officially in development" and the release date was set for September 2004.

To be honest, when we started full production in January 2004, we didn't know at that time how the game would be played. While the basic game mechanics started to take shape, other things like the interface, AI, crew management, missions, etc. were very fuzzy. Initially the idea was to control the game mainly through the crewmembers like in Star Trek: Bridge Commander. But the more advanced we got in the development, the more we realized it wasn't enough. This became obvious when we had the AI for ships and airplanes in place. It was an intense period with a lot of development, meetings, and questions to be answered. On top of that, we were told we should present the game at E3 2004 in a playable state. The official website launch was also in that period. It was the time when some people first cracked under pressure and left the team.

Well, that was pretty much the picture when the Subsim community started to become very interested in our title and a lot of speculation was made. So, I decided to contact Neal Stevens and ask him to post a message from the dev team. That is because we decided that no matter what, we won't post on forums. Our reasons were quite simple: we didn't have time and we didn't want to spark forum frenzy amongst our team. It happened on many other development teams and many times it is a source of misunderstanding between community and developers. Reading forums was okay and even encouraged, but posting as a developer wasn't. Fortunately, submarine simulation fans are a great community and they not only understood this situation, they also started to post pertinent suggestions and systematize the materials that appeared about the game. At a certain moment during development, the FAQ list held by the Subsim forums was the most comprehensive document about SH3's announced features, and we were sometimes checking it to see where we were.

E3 2004: "To be or not to be a simulation?"

Presenting the game at E3 is an important step in any game's development cycle. It is an opportunity to gather reactions from several points of view: customers, marketing, reviewers, and other game developers. Each category is important because it can help you identify before release what your strengths and weaknesses are, and if you have time, you can react accordingly before the game is shipped.

Usually, the developers build some special levels in order to show as much gameplay as possible, but in a very short period of time. For us it was a big challenge because submarine simulators aren't well suited for these kinds of presentations. I mean, it can take several hours in real time between seeing a ship and actually firing a torpedo. That is why we came up with this idea of "hiding" the U-boat behind an iceberg, and just around the corner having a British taskforce. Of course, it sounds stupid, but it was the best solution to quickly show a torpedo attack, a depth charge attack, and an air attack. With this mission presented in an Ubisoft booth and a Microsoft booth, and a movie rolling on the Ubisoft big screen, SH3 was a *very* big success at E3 2004. It was very rewarding after all the efforts we had made to see that the game was really appreciated. SH3 was awarded best simulation of the show by Gamespot and nominated several times in the same category.

But of course, our most important meeting was with Neal Stevens from Subsim. We showed him everything we had available at that moment, and it was obvious he liked what he saw. What he didn't like were, in fact, the missing parts. And those were a dynamic campaign and harbors. He didn't tell us directly, he is a Texan gentleman after all, but we felt it, and later when we came back to Bucharest, we were certain that we weren't going to please the hardcore community without those features included.

On our way back to Bucharest we arranged a visit to *U-505* at the Museum of Science and Industry in Chicago. The U-boat was under heavy restoration and had just been relocated to a new indoor exhibition. The conditions were quite difficult for a documentary visit, but we managed to have a look. *U-505* is by far the most well-preserved U-boat on display so we took this opportunity to take some rare pictures. The TDC interface in the game is based on them. We have to thank again Ted Savas and Keith Gill (the curator of *U-505*) for arranging this visit.

Dynamic Campaign

After E3 the conclusion was simple. We have a great game with a lot of potential, especially among "casual" WWII fans, but it looked like the true Silent Hunter fans would be disappointed. Again! Even though we never promised a dynamic campaign! When we started to develop this project, one of the first things that everybody (in Ubi) agreed on, was to develop an action-oriented simulation that should cut traveling times and try to have instant action scenarios in order to attract more casual fans. All this without losing the SH community. Community reaction after the game was revealed at E3 was mixed. Everybody agreed on the fact that the game was looking great and appreciated some of the new features we proposed (especially cameras and crew management). However, the reaction on the campaign system we proposed (branched campaign with scripted missions containing dynamic elements) was tougher then we expected. Initially we estimated that this was only a reaction from the real hardcore community, so we designed a poll to check the expectations from a larger audience. With more then 2000 respondents the results were quite surprising for us, showing that the large majority was really interested in taking their time playing the game. They wanted to spend several hours on a mission, they wanted to plan, and not engage the enemy immediately, and of course they wanted to roam around the oceans and hunt for the prey themselves, no matter how much time it took.

To be honest this wasn't our only problem. SH3 was supposed to be released in September and we were late. Very, very late! There was a lot of tension inside the dev team. We really wanted to have this feature, but it was obvious we needed more time. And time is the only asset a developer won't get very easily. Sometimes it is easier to get a million dollars rather than a single month delay. But the unthinkable happened. At the end of June we were informed that the game release would be postponed to Q1 2005, and it should include a dynamic campaign. It was truly a great decision and this kind of decision is instrumental in making a great game. Kudos to the Silent Hunter community and Ubisoft.

Usually when this kind of announcement is made (game release date is pushed back), the communities are very unhappy. But, the Silent Hunter community is really one of a kind, and they were cheering. Last year I had the opportunity to meet some of them at the Subsim 2005 meeting in Holland and I understood better why their reaction was so positive all the time. I guess in fact the community isn't based only on the game itself. The game is only one of the vectors that gather them.

At that time we knew that the game was looking very good graphically and attracting a lot of buzz after E3 so the development started to be dedicated almost exclusively to the Dynamic Campaign and the simulation features.

A dynamic campaign is a very tricky feature. We had already been in development for eighteen months and we didn't have too many things prepared for this feature. For one month we had a lot of meetings and debates about this feature. Finally we agreed on what the dynamic campaign would be: A huge scripted mission! Yep, that was it. In SH3 the dynamic campaign is in fact a huge mission developed on the entire world map, with several layers of ships, several time periods, hundreds of spawn points and hundreds of mini-missions around critical areas like Scapa Flow, Gibraltar, etc. Sounds simple and like cheating, but don't jump to conclusions. It works great and in order to be easy to design, a lot of difficult implementations were done at the editor level. And it looks like it is great for mods, too!

Our next encounter with the community was in August at the Game Convention in Leipzig, Germany. We put in a lot of effort to show that we cared about the E3 reactions. That is why we included harbors in our presentation. This time it was Subsim's Drebbel waiting for us with questions and observations. We think that after Leipzig the SH community was a little bit more relaxed, knowing we are really considering their reactions. The Dynamic Campaign was still a mystery. The show was a success again, and we had a lot of positive feedback from the German press and communities. Germany was, in fact, our main target audience.

After Leipzig we continued to work like hell. The deadline was approaching and it was already the second one. The last six months were no fun at all. Everybody was tired after daily extra hours and weekends at work. In this period the best game developer was "AXE"[1]. And he helped us a lot: Goodbye W98/Me compatibility, goodbye playable Enigma Machine, so long MilchCows, see you later fellow captains from Wolf Packs etc. Clean, minimum bugs, I tell you, he is the best game developer ever.

Land!

In January 2005 we had our first version of the Dynamic Campaign. With all the stress it was a major breakthrough. As mentioned, we decided to bet on programming and put a lot of the development into the programming part. So, besides the major changes to the mission editor, a kind of engine running the dynamic campaign was developed. This took a lot more work than we expected, and for the implementation almost all the programmers were involved. But what a beauty! When we ran the first simulation of the world naval traffic in the editor, we knew we had hit the jackpot with this feature. I am serious; if anybody is really crazy out there, the editor and campaign engine are powerful enough to exactly reproduce the naval traffic anytime, anywhere. You just have to put enough waypoints and cruising speeds. Of course, we weren't that crazy. But we were crazy enough to reproduce all the major traffic lanes for convoys during WWII to constitute a dynamic campaign.

With all the major features (that survived) in place, there was one more battle to fight before release: Debug against the clock. Or around the clock. With people already leaving for other projects, it wasn't a fair fight. During the last weeks the remains of the team felt more like they were fighting for survival and not for winning. We probably lost this battle, but I like to believe that we won the war.

In the Harbor

On February 28, 2005 Silent Hunter III was declared Gold, and a go for manufacture was given. Two weeks later the game was released in the U.S., Germany, and UK.

---

[1] "AXE" is a fictional character haunting game development in critical times before release. It generally represents the willingness of the dev team, but speaks only through the producer's voice.

Honestly, we were very surprised to see how the game was received. It was better than expected. I have to admit that given the difficult conditions to finish the game, we almost lost our faith that the game would have reviews better than 75% and the sales would go beyond breaking even. We felt like returning home from a long voyage at sea and nobody was waiting for us. However, yet again we underestimated the Silent Hunter community. One of the first reviews was from Subsim: 100%! And others followed: 88, 89, 90, 94, 91, etc. And two more 100s from GameSpy and Computer Gaming World! And when the first sale reports came, it was obvious we had done it.

But what was really touching was the reaction from the community. Even with all the flaws of the release version, they were still really happy. We just managed to create something that is a catalyst for people that share this submarine passion around the world.

Well, this is the brief story of the SH3 development. I guess it is a very common story in game development. What is probably remarkable is the fact that it was a success story with a naval simulation title in the year 2005. And this is really encouraging for other titles like SH4. But somebody else will write that story.

....Aircraft shadows tend to make you a little bit jumpy.

The Legend of Odin

Bob "Dex" Armstrong

My daughter said, "Dad, it looks like all you did was have fun." I guess it looks that way to folks who never did what we did for a living. Most people have no idea what life was like inside one of those steel monsters. People always ask, "When you were underwater, could you see out?" They have the idea that submarine duty is like riding a glass bottom boat in Tarpon Springs, Florida. We just enjoyed life and watched fish go scooting by.

Walt Disney caused folks to think like that. In his rendition of the Jules Verne version of submarine service, his sub had a big glass window. Folks sat in big, overstuffed red velvet chairs, smoked imported tobacco, drank sherry, and watched the crew go out some magic hatch and play grab-ass all over the ocean floor. That, boys and girls, is pure, unadulterated bullshit. Strictly 20,000 Leagues of Grade A horse manure.

You can't see out... It is hot... It stinks. You're cooped up in less moving around room than you have in your garage. You share your living space with very active, one-inch long, multi-legged wildlife and eighty two-legged critters.

Without stupid activity, life could become unacceptably boring. There were times when life was so uneventful, you could actually hear your toenails growing.

So we did nutty stuff. We spent hours thinking up stupid stuff to do. It was either that, or a trip to the loony bin. When you lived in the North Atlantic, the only circus that came to town was the one you created in your head. We had to manufacture any fun we had.

Only boat sailors will think this is funny. Why? Because they did it. If any submariner tells you he never pulled this one, he's lying.

When you got some JG or fresh 'out of the cabbage patch' lieutenant standing the diving watch, you waited. You waited until he had trimmed the boat. Then by twos and threes, you made your way to the forward room. You waited some more. Then all of you moved by ones... twos... until all of you were in the after room. The boat would take on weird angles. The diving officer compensated. The trim manifold operator laughed as he responded to instructions.

"Pump 500 lbs. aft... No, forward... Wait... Make that after trim... Forward trim... Belay my last... Make that zero bubble! More dive on the stern planes... What the hell's going on? What's happening? Boat's really acting weird!"

It never took long for the COB to get a handle on what was going on. There was another outbreak of crew lunacy on *Requin*. Most possibly the best. At the very least, the most memorable.

If you visit the *Requin* in Pittsburgh, Pennsylvania, she's sitting out in the river in front of Three Rivers Stadium. If you go through the boat, you will find a little aluminum fish dangling over the control room chart table, hanging down on a bead chain with the legend 'ODIN' die-stamped in the aluminum.

They've got tour guides. Non-qual wannabe fellows who make up answers for John Q. Public to cover what they have not the slightest clue about. There are as many stories about that little fish as there are tour guides.

Here is the straight dope. I was there. I was one of the idiots involved in it and had a front row seat in the "I will shoot the next Viking" major ass chewing.

Stuart was the primary instigator. A major player and father of that aluminum fish. I am not ratting on a fellow shipmate. Far from it. At reunions, Stuart is a celebrity. He starred in a video, signs autographs and I am told, will contract to father children for anyone wishing to have a certified diesel boat maniac in their family tree. Knowing Stu, it would probably fall out of the tree and land on its head. Stuart deserves the credit line on this one.

It was winter. Up north, cold as a witch's tit. We had rigged in all the brass monkeys. Before we singled up and took in the brow, we got this film, *The Vikings*. Great flick. Some other boat in SUBRON SIX gave it up, as I recall, because we got orders that didn't allow time for a movie run.

We showed it the first time on the second day out. Good movie. We then saw it six or seven times in a row. Weird story...if you haven't seen it, rent the video. Kirk Douglas, Tony Curtis, Ernest Borgnine, and I think Curtis' wife at the time. Some good-lookin' blonde.

The Vikings were a ratty-ass looking bunch. They did a lot of drinking. Fondled a lot of blonde, blue-eyed women and went to sea on a regular basis. It sounded familiar....

One night, someone announced that we, the crew of the *Requin*, had to be the spiritual descendents of the Vikings. WHAM! In that instant, we all became Vikings. Everyone spoke in Scandinavian-Minnesotan-Inger Stevens dialect.

"Ja Sven, you see da cheef? He's da beeg fella wit da beeg moudt!"

Everybody got into it. The skipper became Ragnar. The exec, Einar. We turned our foul weather jackets inside out so the brown, hairy looking fake fur stuff was on the outside. We made cardboard horns and stapled them to both sides of our watch caps. When we passed each other going fore and aft, we banged our chests and yelled, "O-O-O-DIN!" (taken from what they did to greet each other in the film).

In the movie, this old crone, an old wrinkled wise woman gives Tony Curtis this fish made from a 'falling star', i.e. meteorite. It was magnetic and was considered to be major magic because it always returned to point north. With this fish always pointing north, the film had Viking ships cutting through pea soup fog and running back and forth between Norway and England like a cross-town bus. Stu went down in the pump room, built us an aluminum fish and die stamped "ODIN" on it.

He hung it from the MC box over the control room chart table. It dangled and swung back and forth. Every time some clown from the after battery would pass through the control room, he would give it a little 'start swinging' tap. This eventually drove the Chief of the Boat stark raving nuts! He would foam at the mouth...get red...veins would pop out of his neck. Words like, "God save us from these unruly children" and "In the Old Navy, the old man would rake your useless butts over the coals."

Why did ODIN stay where he was? Simple. The skipper liked it.

As time passed and we became more and more 'Viking', the exec put on his "Enough is enough" voice and announced over the 21MC that the crew of *Requin* had just gotten out of the Viking business. All stop. Don't answer anymore Viking bells...Over...El stop-o.

Ten minutes later, some idiot tapped into the 21MC and whispered,

"*ODIN LIVES... O-O-O-DIN...*"

The exec lit us up like a Christmas tree. From then on, we looked around for officers before giving each other the silent Odin salute.

When we came in and the exec opened his vertical uniform locker and removed his 'hit the beach' hat, it had grown a pair of cardboard horns. It had to be a miracle because the COB used everything but truth serum to get the rats to rat on whoever did it. I think the Chief finally recognized that the leadership of *Requin* may have pissed Odin off.

All the exec said was, "You sonuvabitches never comprehend when the game's over and it's time to pick up your toys and put them away!"

He was a deep thinker. We had no idea what in the hell the man was trying to communicate. We knew if he were really serious, he wouldn't be standing topside talking to the OD of the *USS Grampus* wearing a hat with cardboard horns attached to it.

Life was uneventful so we fought boredom any way we could. Most of the time submarine sailors won.

Forty years later, a group of late middle-aged bastards stood in the control room and watched Stu, the originator, replace 'ODIN'... And we yelled, "O-O-O-DIN..." and banged our chests.

We were young again and someone in the crew's mess yelled, "Jeezus, the idiots are at it again!"

I first joined Subsim in early 2001. Subsim was still on the EZ Board forum. I was looking for information regarding Jane's 688(I), and stumbled across the Fix My 688(I) petition. I rarely posted in those days, but I had read about the upcoming Silent Hunter II, and saw that the Wolfpack League was starting up and, as Neal knows, I became heavily involved in that project.

Tom "Takeda Shingen" Morris

Just Another Cargo Ship

Mariano Sciaroni

Seen from outside one wouldn't have distinguished a submarine, less a fearsome U-Boot. With a discolored sail from the aft deck to the tower, a false secondary chimney emitting thick, black smoke, a fresh and phony paint job, and some other well-made disguises, it looked like a poor man's fishing boat.

The men sprawled over the deck, certainly carefree, some of them were playing poker, others sunbathing on well-placed cots, and the daring ones water skiing from the stern of the sub.

Kaleu (for Kapitänleutnant) Ernst Bayern, Commanding Officer of the Type VIIc *U-132* sub, a tall Bavarian of fragile aspect, stood at the bridge scratching his beard very insistently and fanning his white hat.

As long as the U-Boot kept working smoothly, he didn't have interest for discipline, nor did he want to teach discipline to his men. At least, not unnecessary discipline.

Bayern had *U-132* in the described state for two reasons: to give the crew a time for relaxation after some hairy moments in the North Atlantic, and sure, to camouflage the ship's menacing silhouette.

He felt stealth could be accomplished in different ways, not only by having twenty meters of salty water over his head.

The Kaleu had always been a kind of rebel (at least for navy standards), but he hadn't been kicked out of the Kriegsmarine. Bayern had many contacts with prominent members of the party, Martin Bormann among them. Ironically, he despised the Nazis.

However, nobody could criticize his intelligence, sharpness of judgment, and personal courage. He had also demonstrated, more than once, his worth as a junior officer of the navy.

And above all, the commanding officer was a practical guy.

With those credentials, he was qualified for the underwater service (maybe his temper only fit in there), and promptly he joined in.

Now, Kaleu Bayern and the boat were in the warm waters off South America. Rio de Janeiro-based E-Dienst spies had informed him about a big merchant ship which, weighing anchor from the city and transporting corn, sailed for a final destination of Southampton.

The infinite wisdom of Doenitz had directed two submarines to intercept the ship, and any other which left Montevideo or Buenos Aires waters, removing them from patrols along the convoy's routes to the United Kingdom. But only *U-132* had arrived at the mandatory appointment with the milk cow (for fuel and food supply), a Type XIV sub in really bad shape. It had never known the luck of the other U-Boots. Possibly an accident or some long-range airplane had shaped its destiny.

Now they were near coordinates 1° N 20° W, navigating slightly above cruising speed. According to the precise information decoded by Enigma, they would intercept the cargo vessel in less than four hours. It wasn't necessary to consider extreme precautions this time. The British airplanes didn't come so far and no enemy warship activity was reported in the neighborhood. And after all, who would recognize them?

The Kaleu plotted an interception heading to the south. They were going to meet the ship slightly outside the assigned patrol zone, in the southern hemisphere.

Around 1700 hours a distant propeller sound was reported from the sonar station. Minutes later, visual confirmation of the expected target. With just a whistle from the Executive Officer, half the men took positions by the well-hidden deck weapons, while the others (following Kaleu's orders) kept up their mundane behavior, as if nothing was happening.

As the spies precisely told, they could see a big British-flagged freighter of about 9,000 tons. Considering the unusual camouflage of the U Boot, the British captain didn't realize he was going to be hunted until it was too late.

"This is the German submarine *U-132*. Stop your machinery and do not try to use the radio, you are far away from Tipperary, mister," Kaleu Bayern shouted, megaphone in hand, in his affected English. The British found it easy to obey. After all, they had no choice.

Obviously, the merchant captain was interrogated about the load and destination, but with not much desire since the information was well known beforehand. During the interrogation the Brit was imperturbable in his traditional English coolness.

After surveying the freighter, the crew was dispatched to Brazil in lifeboats. A torpedo was fired in order to sink the listless freighter. The torpedo hit in the middle. Everyone saw the explosion and then a column of water. But the freighter didn't sink.

A couple of minutes went by. Popcorn suddenly began to fill the ship. It left by the smokestack, it flooded the bridge, it arose by every open hatchway.

The fire and the oil from the ship's engine, of course, heated the corn grains into popcorn and sealed the hole caused by the torpedo. A movie snack turned the boat into a virtually unsinkable ship.

"We need a sugar freighter now," the Exec exclaimed.

The captain stared at him and then said with a big grin, "No, we will look for one carrying salt. And another one with beer."

The boarding party was sent again, but in this case "to catch" about 25 kilos of popcorn.

That night *U-132* crossed the equator, baptizing the young new wolves. Neptune's ceremony was observed by a crowded audience which, of course, ate until they dropped sick, full of popcorn. Kaleu Bayern and his crew enjoyed an unforgettable night. Maybe the last happy time of their short lives.

The war, and a moored mine, would find them whilst returning home.

⊕⊕⊕⊕⊕⊕⊕⊕

Does the story seem to have a fantastic twist? Well, it is based on real facts.

On October 8th, 1939, the PanzerSchiff *Graf Spee* tried to sink the *SS Newton Beach*, a freighter of 4651 tons filled with corn. But, as in the short story you have just read, the corn exploded and was turned into popcorn by the heat of the explosion, making it harder to affect the collapse of the ship.

And about the behavior of the U-boot crew, the camouflage, the baptism ceremony, and even the water skiing — they were related by the account of Oblt. Heinz Schäffer, commander of *U-977*, in his book *U-Boat 977* (1975, Tandem Publishing Limited ISBN 0553267302).

Kraka: The True Story

Jonathan Beck Jorgensen

Copenhagen International Boat Show 2006.

In October 2005 I got a call from Peter Madsen. He had an idea.

Is it possible to put Kraka into a submarine game?

Kraka was going to be featured on a stand at the 2006 International Boat Show in Copenhagen. For those of you not familiar with Peter Madsen, he is doing what most of us dream about. He builds his own submarines (with help, of course) and *Kraka* was his showpiece project.

His first homemade sub was *Freya*, a small 7.5-meter-long all-electric vessel. *Kraka* was a full-blown diesel/electric submarine with a hull shape inspired by the German WWII Type VII U-boat.

Peter often has great ideas, but some are borderline unrealistic. He is aware of this. Even though he thought inserting his sub into a game was hardly possible, he still remembered something about the Pacific Aces project I had spoken of.

Pacific Aces was a product of many months of labour by people from Subsim.com. The result was a full conversion of Silent Hunter II game from U-boats in the Atlantic to U.S. subs in the Pacific Theatre. The Pacific Aces team had to program all new 3D models of U.S. submarines and port

them into Silent Hunter II.

I told him that it was very possible to put *Kraka* into Silent Hunter II, like we did with the Pacific Aces project. But since Silent Hunter II was getting old, I thought it would be a better idea to see if it could be done with Silent Hunter III, still keeping the Silent Hunter II option as a backup.

I contacted some of the people who did such a great job with PA and found Aaken on an Italian sub forum. He was more that willing to do the 3D modelling. Okay, so we can get some 3D models that will work for both SH2 and SH3. Great.

For texturing I turned to Horsa, a long time Subsim contact who was happy to do the textures. But how the hell do I get a textured *Kraka* model into SH3? I remembered having read about a guy who had added his own models to the game. The name of that man was Sergbuto.

It soon became clear that we could put *Kraka* into SH3, but it could only be done as an AI unit. After some discussion we decided to go for SH3 because it looked so much better.

Months went by with modelling and texturing and the final result was very impressive. At the final stretch some people from Nethouse2000.dk stepped in and did a great job finalizing what we had created in a very professional way.

The presentation idea was to use a flat screen monitor and run a looping movie clip using Silent Hunter II footage. While we knew what the *Kraka* game model looked like when she was almost done, how would she look in the presentation?

The team had worked so hard on the virtual *Kraka,* and to top it off the real *Kraka* had just been given a total overhaul with new paint, new portholes and a new engine. I was very exited travelling to the show because I hadn't seen the final result. Everything had been so hectic all the way to the night before the show started.

So on one grey morning on the 25th of February I made my way to the show at Bella Center, Copenhagen, Denmark. Here she is! *Kraka* ready to meet the people. She looked so beautiful with her new coat of paint. Note the domes with portholes. You can also see the periscope and the snorkel....yes, this baby has a snorkel.

I was making my way around the hull looking for the flat screen that would hold our game presentation and all the months of hard work.

When I passed *Kraka's* stern I saw it; a flat screen monitor that showed a looping presentation of *Kraka* inside the Silent Hunter III game. Needless to say this presentation got a lot of attention. I spent a great deal of time answering questions about the presentation and *Kraka*.

Fortunately, this sub has a hatch at the bottom originally made for divers so they can enter the sub when it is submerged, but here it gave the people access to the inside of *Kraka*.

After the boat show it was only a matter of time before I was to embark on a voyage with *Kraka*. That is one of the perks of knowing Peter, the chief designer.

The 2006 *Kraka* tour

My older brother was going to live in Australia for a year doing research and was leaving during April, so I figured it to be a good experience for us to have time together before he left.

I talked with Peter and my brother and we agreed to meet on the morning on the 2nd of April. My brother was very excited because diving in a vessel was a very foreign feeling for him, having sailed a lot on ships. He had seen *Kraka* at the boat show so she wasn't a stranger to him, but the concept of slipping beneath the cold surface was.

First we had to get ready for getting underway. The sub had to be fuelled for the occasion, so we had to make a quick stop to get some diesel. After the refuelling we were ready to enter the boat.

My brother had time to get a good look at the map and the picture gives a pretty good impression of the cramped space. Space on subs is traditionally tight — on *Kraka* even more so.

After we had entered the sub, Peter instructed us in the safety features of the subs, just in case (an idea that later that year proved very valid, but that is an entirely different story).

In *Kraka*, you basically have four places to be. The forward torpedo "tube," the forward porthole, the tower (upper porthole) and behind the engine (the best place to see out the bottom hatch and portholes).

Being in the "tube" is very claustrophobic — I had one try and didn't like it so I crawled out and didn't go back. So we took station at the other three places. The tower is the place to navigate when you are running the boat on the diesels and the forward porthole is the best place to be when you sail on electric motor. The pilot controls the diveplanes from there as well.

At first, I was placed at the forward porthole and had a great view of the harbour sailing out. My brother had the warm space, sitting just next to the engine. Sailing out, my brother was ecstatic and I can't blame him. Even though I had dived with *Freya*, it was a totally different experience in *Kraka*. It was more real and more complete because it was that much closer to the real thing.

The principles behind both boats are the same; you sail and you dive, but just having a bigger area of operations and switching from electric to diesel motors made it very different.

It took about forty-five minutes to sail out of the inner harbour and out where we would have room to dive. When the time came to slip beneath the surface, my brother had the front row experience because he was looking out the tower, and man, was he excited! He later told me that doing that was somewhat barrier breaking for him, and I can imagine it being so with his viewpoint from the tower portholes.

Once below the surface we were virtually blind. We had to rely on basic navigation, and it was actually very hard. Because of the fact that *Kraka* doesn't have hydrophones or radar and because we were below periscope depth, we had no idea of what the traffic above was like. If a ship had sailed over us, it could have rammed us.

Being a diver, I have first-hand experience of what sounds are like under water. You have absolutely no idea of where the sound is coming from because of the high speeds that sound travels at under water. The sound from inside a steel hull is somewhat different from diving. I found the underwater sound close to what you hear in *Das Boot* and Silent Hunter III.

It was actually a very frightening experience because everything around us sounded like it was *very* close. On our morning of diving the traffic was light outside Copenhagen harbour, just a few ships and boats in the harbour, so I can only imagine what it must sound like when the traffic load is high.

Peter instructed us that he was going to dive the boat to the bottom. Knowing it was only ten meters deep at the most was a little reassuring, but the prospect of intentionally hitting the bottom was still a bit unnerving.

Seeing the sandy bottom through the portholes made me relax, even though we were travelling at what felt like considerable speed, three to four knots.

THUMP! We hit and bounced off.... THUMP...scraping across and coming to a stop. Now we had a break to chat and sip a coke on the bottom of the harbour.

After a while of chatting and discussing various topics, it was time to resurface. Slowly we climbed to the surface and I was anxious to see how close the contacts heard from below were.

A quick look around made it clear that they weren't close. Peter then started up the diesel and we began to make headway. At that time I was sitting at the back of the sub, right next to the diesels, and something started to smell. It turned out that Peter had forgotten to open the exhaust valve so the diesel fumes were venting into the boat. This was quickly solved and the boat had an authentic smell to it.

On the way in we decided it would be fun to sail right next to Danish Navy Command and Support Ship *Absalon*, which lay in the harbour. *Absalon* is, with its 137 meters (450 ft) length, the biggest and most advanced warship in Denmark. Approaching the Danish warship from periscope depth could result in some commotion because they had no way of knowing who we were or where we came from.

A submarine of only 12.5 meters isn't exactly an everyday occurrence in the harbour. From their point of view we could be an enemy midget sub.

Slowly we sneaked up from behind and sailed next to her, waiting in anticipation…but nothing happened!

Disappointed, we took a short trip around in the harbour. Along the pier everybody on shore took shots with cameras and camera cell phones. I guess a submarine of this sort is a rare sight in Danish waters. Now I know how monkeys in cages feel.

After the harbour trip we headed for the "pen." It had been an amazing experience for both my brother and me. With a sigh of relief we stepped onto steady ground. One experience richer, we headed for home in rainy weather in a car with out-of-order windshield wipers.

Looking back at it later, I realized how important a hydrophone and a periscope above the waterline is. Without these you are blind and deaf. I wait in anticipation for Peter's next project, the *UC3*, which will be a full-size submarine, capable of travelling to other countries. Next stop: pop the hatch in Scapa Flow!

You know you've been with Subsim a long time when…

….Your nightlight is very unique and very red.

Tesseraction Games and Enigma: Rising Tide

Kelly Asay

I had just climbed out of the pool after a long hot day of moving the last of the Tesseraction Games office contents into a solitary storage bay in a facility just off the Willamette River. Locking the door for the last time on the "SUB PEN," a nondescript warehouse building in an industrial park on Eugene, Oregon's west side, it was impossible not to reflect on the four years and six months I spent at the helm in our highly customized quarters.

Tesseraction Games was launched by a very diverse group of individuals with a glimmer of hope that we could rescue something for ourselves from the sudden closing of Dynamix. We pooled our severance checks, some of us put up cash we had saved up, and decided to give it a go. Members split off while others joined up in the commotion of an unplanned startup.

For several months we held our meetings at a pub on the river, meeting a couple of times a week, initially thinking we would make a racing game for the PlayStation 2. A little research and a cold honest look at our resources had us sitting back and looking for other project options. We needed to find a niche that held a sizeable market, but was neglected or underserved in some way. I had an idea.

It Started With Iron Wolves

If you played it, then I don't need to say anymore; if you never played it, you truly missed out. I was fortunate to have been a beta tester for the game and played it as my primary diversion until the server finally dropped offline a number of years ago. As far as submarine simulations go, it was a very poor example. As a graphics and audio showpiece, it couldn't even compete. Customer support, forget it. So what was the big deal? In a word: multiplayer. True, large-scale, diverse, and engaging massive multiplayer in a naval environment.

I proposed we take a look at the naval combat niche and consider making a game that fell somewhere between the simplicity of Iron Wolves and the ship simulation products like Silent Hunter and Aces of the Deep. Make a kind of "gateway" product that held some of the complexity of a simulation, but was accessible to the newest of players. After some due diligence it was apparent that no significant product had been released for the naval segment in a number of years. The ones that were released (and were done well) had been received very well and revealed a large group of neglected gamers. It didn't hurt that I was a longtime SSI fan and a couple of us on the team loved all things naval.

So we talked about it, got enthusiastic about it, and decided to set our sights not only on this market, but to try and create a truly unique title for it: Enigma: Rising Tide.

Ambitions

As an independent developer, you have a couple of paths you can take. Perform contract work for somebody else, create a pitch for a product and shop till you find a publisher. Or do what we did: raise some capital on our own, put in a couple years of nearly free work and create your own product. On this path there are also choices to make, particularly the choice of what level at which to compete. Since one's reach should (in nearly all circumstances) exceed one's grasp, we set our sights high. We would make an "A" level product utilizing as much high-end middleware as we could. It would be a massively multiplayer game and we would release it electronically via online partnerships.

It didn't quite work out that way.

Within thirty days of the start of Tesseraction Games, terrorists flew two jets into the World Trade Center and sent the U.S. economy reeling. When combined with the bursting of the dot.com bubble, a large financial dead zone appeared in the venture capital and investments market; precisely the waters into which we were headed.

I will summarize the results down to this: lack of sufficient funding would be our major issue. All our other issues, almost without exception, stemmed from poor funding. That said, we had managed to raise a little working capital, moved into an empty warehouse in the industrial area of west Eugene, and started work.

Sometimes it seems that not much went right. We had spectacular help from the game press community in getting the word out about us, the game, our plans, etc. Unfortunately we never seemed to be able to accurately convey just what Enigma was. Hardcore sim fans hated what they called "arcade-like" gameplay; non-naval enthusiasts thought it was too difficult. We did create a very loyal following out of those who either gave it a shot with an open mind, or understood it was meant to be a Naval Combat game, but this lack of clarity as to what Enigma was really slowed our acceptance into the naval game community.

As of this writing around 250,000 gamers have purchased Enigma. If you have followed the game or us at all, you also know that we never did make it to multiplayer.

A Long Way to Tipperary

In addition to capital in general, one of the things painfully absent from our budget was promotion money. To help build some anticipation and hopefully some revenue, we decided to release an early offline version; let people play with the same ships, ocean and AI that would be in multiplayer. They could download it or get a CD on-demand and we would raise a little badly needed capital to help us get to the multiplayer stage.

Turns out it wasn't great timing. The company we had made the releasing arrangements with went from a very small but growing concern (willing to make pretty much any kind of deal to help them grow) into a fast-growing profitable one. Our arrangement with them was based on sales revenue; they made their money from advertising. As a result of selling out their advertising capacity, we ended up with no advertising for the release. Consequently, only about 150 units sold, almost solely on our word of mouth effort.

Something did come out of this effort though. GMX Media contacted us about Enigma; they were looking for a single-player game. They offered money. Good timing, we thought, we really needed some. At this point we had a high stress level in the office; no one had received any money for a while and we were on the verge of losing people. We did some diligence on GMX and couldn't find anything untoward at the time. As a company we made the decision to alter course and create a single-player game we could release with GMX into Europe. GMX gave us a small advance. Combined with some additional capital from a couple of our members and a couple of our outside investors, it was just enough to hammer out the original single-player version of Enigma.

Along the way it became apparent that GMX didn't have much planned in the way of promotions for Enigma. We still didn't have a promotional budget, but we did have an idea; one that worked. We decided to release a series of demos of Enigma to help raise awareness and keep us in the press. Eventually we released seven different demos of the game; each with its own distinct graphic menu, its own ship or submarine, and two missions. Ultimately this resulted in a lot of teasing from the gaming community, but if you ask any celebrity, any press is good press.

The European Launch

When we created Enigma, we knew we would like to release in Europe. We also knew that there were certain issues that couldn't be utilized in the game without censure in some markets. An early design decision was to go with an alternate history timeline that would let us sidestep Nazi Germany. Dave Georgeson, my producer at Dynamix, was a history buff. He felt that sometimes the best way to teach the importance of seemingly insignificant events in history was to explore the possible outcome of changing one or more of these events. This was the approach we followed for Enigma. We picked a departure point, found surrounding small events, and changed them. We extrapolated what would likely have happened given fifty years of hindsight. In doing so our writers compiled almost 10,000 pages of historical documentation, 2000 pages of alternate history, and a well-written, historically accurate and historically feasible alternate history.

Sales wise our research had shown that the UK, Germany, and France were the big markets in Europe: together responsible for 80% of all product sold there. It seems we did forget to factor in one thing; in Enigma's timeline Britain *loses* World War One.

The first reviews for the release were done in British magazines, apparently by Brits. They hated Enigma. Actually that isn't totally accurate: they hated Tesseraction Games for making a game where Britain lost the war and Germany won. Our debut reviews were in the high 50s to low 60s compared to the high 70s to low 80s we would later get in the States. Both GMX and TAG panicked, GMX substantially more than us. They scaled back their launch plans and reduced their advertising efforts. They also made some bad choices in dealing with us.

Too Small To Fight Back

Because of the initial reviews, GMX gave us the impression that Germany, France, and the balance of the EU15 countries would be very slowly rolled out and in small numbers. They wanted to protect themselves against reviews such as those in the UK. This didn't turn out to be accurate.

Approximately five months into the European release, it was apparent we were going to have big trouble with GMX. We weren't getting complete royalty reports, when we got them at all, we weren't getting paid, and worst of all, GMX refused to fix any of it. During this time we had been working up Russian, Chinese, and North American releases with GMX. We notified GMX of breach in hopes that they would either comply or walk away. They did neither, but instead told us to sue them or bugger off. We terminated our pending deals for Russia, China, and the USA. GMX released in Russia and China anyway. To date we have never seen a cent of the more than 100,000 units that were sold into those unlicensed territories. We filed a breach of contract lawsuit against GMX in September of 2003.

We were in dire straits now. Staff attrition had begun, some friendly, some not. Stress was through the roof and we had been out of money for almost three months. As we all worked in the same room, everyone knew that we were in trouble. We needed something to happen fast. Fortunately E3 was upon us, that great, and now possibly extinct, circus of media excess and promotion, wonder, thunder and all the swag you were willing to stand in long lines for. It was here that we met up with Dreamcatcher Games. It was also here that we met Ken Kokin (*The Usual Suspects*) at Sasabune in Los Angeles (world's *finest* sushi). Ken came to Eugene and took part in Enigmacon.

Pointsoft, Dreamcatcher and Spike TV

Brian Gladman was the product manager at Dreamcatcher at that time and a good guy. He was enthusiastically interested in licensing an already completed game. So were his bosses. Dreamcatcher gave us a small advance on the release and we began what would become another in a series of uneasy relationships.

In the meantime we had managed to track down the company who was doing the publishing in Germany, France, and Spain for GMX, Stéphane Gonod's company Pointsoft. We got in touch with them directly and with the assistance of our growing team of attorneys, persuaded Pointsoft to work with us directly. It was good timing. It turned out that while the UK was decidedly *not* buying Enigma, Germany and France decidedly were. In fact, Germany was ultimately our best market, rivaling North America in units sold. Enigma had been in the Top 20 in Germany since its release. Pointsoft had been without product at this point for going on six weeks, and Enigma had dropped out of the charts during this product drought.

Pointsoft paid us a sum to get things rolling, so we briefly entered the manufacturing business, manufacturing 10,000 units of Enigma (German language) and shipping them overseas to them. During this time we coordinated additional manufacturing in Germany through Pointsoft. Pointsoft was paying our royalties and it was a refreshing change.

The release in North America was a pretty good one. Not a giant push, but well received and steady. Enigma would stay at full retail price for almost nine months. Interestingly enough there was almost zero advertising. The bulk of the promotions consisted of us doing interviews and press releases, visiting fan sites around the globe, and trying to fill a giant gap where our publisher tech support was supposed to be. Neal Stevens of Subsim Review, John Keefer of Gamespy, Richard Aioshi (then) of IGN and John Callaham (then) of Shacknews went over and above on their support for me and the game, and I owe them.

By this time many of the original team members had been displaced; some out of the industry, others just out of the area. A couple of guys suffered divorces, another a nervous breakdown (not me), more than a few contracted a permanent fear of startup businesses. A bright spot at TAG during this upheaval were a brace of brand newbies: Jason Lobo, Howard Day, and Bobby Wolfe. Jason had moved down to Eugene with his family from Wilsonville for the opportunity to work on Chapter Two of Enigma and was a research fanatic. Howard Day was a brash, self-taught artist with no experience in the industry. Our previous art team had received his portfolio CD and shelved it. I found the CD during a reshuffling of our floor plan, reviewed it and decided to take a chance on Howie. He has without doubt the loudest snore ever heard. During crunch time naps his snores would make the rolling door resonate. He is also a brilliant artist. Bobby Wolfe came along shortly after Howard. Bobby ultimately was responsible for the fantastic look of our Gold Edition.

E3 was once again rolling around and we heard about the "Into the Pixel" competition. Bobby and Howard created a series of three fantastic art pieces using the Enigma models and art assets, and we entered them into the competition. One of the pieces, "Merchant Raider," was one of the ten competition winners and we were invited to the Spike TV-sponsored E3 exhibit and press party. About a week later we received a phone call letting us know that unless we gave them a publisher contact, we would have to be removed from the show. We contacted Dreamcatcher and they agreed to be the contacts in exchange for getting to attend the party. It sounded okay to us at the time. Later it turned out that Dreamcatcher had submitted one of more than 320 entries in the competition…for their title Painkiller. We were the only independent studio that was represented in the exhibit alongside work from the likes of EA, Microsoft, Sony, and Nintendo.

My First Submarine

In 2004, Neal Stevens and Subsim Review invited me to attend their Sub Club gathering in Galveston, Texas. I was smart enough to go and got to meet some of the heart and soul of the Subsim player community. Spending two nights on the *USS Cavalla* was eye-opening. A couple of the merry Dutch pranksters managed to setoff *all* the alarms on the boat somewhere around 0200 hours. I am guessing submariners must have been very hard of hearing in their later years as it was loud enough that I couldn't actually remember where I was, who I was, or why I wanted desperately to *run*.

We got a look at Silent Hunter III and, though envious of their fine work, it was great to see that Enigma had revived publisher interest in the Naval simulation market.

The Bell Tolls

When royalty time came around for North America, we had apparently done far better than Dreamcatcher had anticipated. They weren't able to pay our royalties. We made payment arrangements with Dreamcatcher, but ended up having to involve the attorneys when they didn't make the payments on time. We had to make payment arrangements multiple times, putting a pretty severe strain on both our team and the relationship with Dreamcatcher. As if on queue, Pointsoft started having trouble paying their royalties as well. This caused us some serious issues. Howard and several programmers departed TAG. Bobby stepped up and handled the art beautifully, we got the GOLD edition done, we were making deals for its release and things were looking on the mend. We had managed to put together a multimillion-dollar agreement at E3 earlier that year for "Sink the Hood" and "Black Knights," we'd had a small but successful Enigmacon, and we were positioned to finally push to multiplayer, when Pointsoft filed bankruptcy owing us more than $250,000 in royalties. Okay, a little stressful, but we were positioned to survive it. It was then our lead engineer left. Ex-employees were heavily recruiting him and he had been under some pretty serious stress from lack of money and the daily grind of his responsibilities. Needless to say this killed the E3 deal.

We worked hard for the next eight months or so, bringing in new talent as we could, pitching to investors in hopes of raising capital, all while trying to make headway on Chapter Two. We were working on a deal with a new French publisher for Chapter Two when industry juggernaut Ubisoft, makers of Silent Hunter III, announced Silent Hunter IV. Not wanting to compete with a major studio in the same market space, our negotiations ended along with our hopes. We had won a default judgment in our lawsuit and it was then all over but the shouting. We would win the summary judgment. Our board of directors met in May and decided it was time to cut expenses and focus exclusively on the lawsuit. Our team was scattered to the four winds. I resigned as President of Tesseraction Games, but remained the Chairman of the Board.

In June of 2006 we won our $1,630,000 lawsuit against GMX Media and one of their directors and are now in the process of collecting it.

Never Give Up, Never Surrender

At GDC this year (2006) I had the pleasure of meeting and spending several hours talking with Rick Martinez. He was a founder and developer at SSI back in the day and still a subsim guy. I am vaguely aware that my fevered brain is busy in the background trying to work up a good strategy for talking Rick into creating the next milestone subsim with me.

I haven't touched on many other events of this last four-plus years: projects like Sealords; a great adventure begun with the Pacific Aces team at Subsim Review. Sealords also fell victim to TAGs financial problems. I haven't mentioned our two really great publishers: Graham Edelston (AURAN) in Australia and Volker Rieck (HALCYON) in Germany. These and the myriad of other events will just have to remain in the archive of cautionary tales to tell my children.

It does seem that, just sometimes, things get back around to where they started. A little hobby project of mine is turning into a real project. All this started at Iron Wolves and now I am creating Deep Six (www.deepsix-online.com). It is more Iron Wolves than Enigma; I call it a "casual" naval MMO. I am enjoying working on it immensely and hope to avoid many of the pitfalls I had to climb out of while making Enigma.

⊕⊕⊕⊕⊕⊕⊕⊕

As I head back out to the pool, I will leave you with a piece of one of my favorite poems:

> "...for the singing sheet is a siren sweet that tugs at the hearts of men, and down to the sea they must go once more, though they never come back, again."

Billy Bones by Skip Henderson

Back to the Future

The U.S. Navy Submarine Force and Its Search For Relevance in the Post-Cold War Era

Bill Nichols

Introduction

The submarine was originally conceived to be a naval weapon for destroying enemy ships. Submarines have proven themselves time and time again to be highly successful in this role. In the post-Cold War era, however, the United States enjoys supremacy on the high seas and the traditional mission of the navy's submarine force — to sink an enemy's warships and savage its merchant shipping — has evaporated. Today's submarine force is a community in search of a mission, one that will maintain its relevance in the twenty-first century.

The "Silent Service" faced a similar situation following World War II. Between December 1941 and August 1945 U.S. Navy submarines sank 1,314 Japanese warships and merchant vessels[2]. American submarines were so successful that they effectively worked themselves out of a job. In the last months of the war, sub skippers often returned from patrol with a full load of torpedoes, having found nothing larger than sampans and barges for targets. Once Japan surrendered, the Navy no longer had reason to maintain a large submarine fleet. Thus, in its 1947 plan the Navy allocated the submarine force only eighty boats[3], down from 263 at war's end[4]. The reemergence of the submarine in the decades following World War II

---

[2] Clay Blair, Jr., *Silent Victory – The U.S. Submarine War Against Japan*, p. 878 (J. P. Lippincott Company, New York and Philadelphia, 1975).

[3] Norman Friedman, *U.S. Submarines Since 1945 – An Illustrated Design History*, pp. 27-28 (U.S. Naval Institute Press, Annapolis, MD, 1994).

[4] Ermino Bagnasco, *Submarines of World War Two*, p. 213 (U.S. Naval Institute Press, Annapolis, MD, 1977).

resulted from the convergence of three powerful forces: technological advances, visionary leadership within the Navy, and the emergence of a new strategic threat. The history of how these factors helped transform the Silent Service during the Cold War gives insight into how today's submarine force can reinvent itself for the future.

The search for new missions after World War II

U.S. Navy submarines in World War II were designed to do one thing and to do it very well: sink surface ships. Although having performed admirably in many secondary roles (transporting raiding parties, conducting pre-invasion beach reconnaissance, and rescuing downed aviators) they were essentially "one-trick ponies." In the immediate post-war years it was obvious that a new mission must be found for the submarine force. It was unclear exactly what that mission should be.

Submariners have no lack of imagination and, "necessity being the mother of invention," many new operational concepts were tried. One of the first new concepts, the submarine radar picket (SSR), was an outgrowth of the Navy's World War II experience with using radar-equipped destroyers to give early warning of Japanese air raids and to control fighter aircraft. The radar picket mission required the destroyers to be positioned outside the protection of the taskforce, where they were most vulnerable to enemy attack. For example, of the thirteen destroyers and destroyer escorts lost during the invasion of Okinawa, eight were sunk while on radar picket duty[5]. Thus, the idea of using submarines to perform these hazardous missions was very appealing. Between 1945 and 1956, the Navy converted ten fleet boats to SSRs and acquired two more (*USS Sailfish* and *USS Salmon*) through new construction[6].

A major problem with the submarine radar picket concept was that the diesel-powered SSRs were not fast enough to keep up with the aircraft carriers they were supposed to be protecting[7]. The Navy's solution was to

---

[5] Theodore Roscoe, *United States Destroyer Operations in World War II*, pp. 470-485 (U.S. Naval Institute Press, Annapolis, MD, 1953).

[6] Friedman, p. 94.

[7] An amusing incident is described in *Power Shift – The Transition to Nuclear Power in the U.S. Submarine Force As Told by Those Who Did It*, by Dan Gillcrist (iUniverse, 2006). The picket submarine *USS Rasher* (SSR 269) was stationed at the rear of its taskforce and was struggling to keep up with the carrier. The task force commander decided to reverse course, putting Rasher directly ahead of the carrier group, which was bearing down on her at 20-plus knots. *Rasher*'s CO, instead of maneuvering around the taskforce, dived the boat and let the group

design a nuclear-powered picket submarine optimized for high surface speed. A total of four SSRNs were planned as escorts for the USS *Enterprise* nuclear carrier group. The first of these, USS *Triton* (SSRN 586), was commissioned in 1959. However, this plan was abandoned (along with the SSR concept) when carrier-launched radar early warning aircraft became practical[8].

The submarine troop carrier (SSP) concept was another "new" mission with origins in World War II. In a famous operation in August 1942, submarines USS *Argonaut* and USS *Nautilus* carried two companies of the Marine Corps 2nd Raider Battalion ("Carlson's Raiders") for an attack on Makin Island[9]. Although *Argonaut* and *Nautilus* were among the Navy's largest submarines, they could only carry 211 commandos on the Makin Island raid[10]. The Marine Corps desired a much more robust capability. In fall of 1946 they proposed a squadron of twelve SSPs with a combined load of 1,440 marines, four 75-mm howitzers, six 57-mm recoilless rifles, 378 tons of ammunition and supplies, and twelve amphibious tractors to carry the weapons and other equipment ashore. Only two SSPs (USS *Perch* and USS *Sealion*) were built before the concept of large, battalion-size raids was discarded in favor of smaller commando operations[11].

In addition to raiding operations, U.S. submarines also performed scores of reconnaissance missions in World War II[12]. This mission area gained increased importance with the emergence of the Cold War in the late 1940s. Although no submarines have been built specifically for intelligence collection, a number of boats have reportedly been modified for special operations[13] and electronic surveillance remains an important submarine mission today[14].

---

pass overhead. After they were clear, he surfaced and regained his position at the rear of the taskforce.

[8] Friedman, p. 96.

[9] Theodore Roscoe, *United States Submarine Operations in World War II*, pp. 156-158 (U.S. Naval Institute Press, Annapolis, MD, 1949).

[10] Blair, pp. 316-318.

[11] Friedman, pp. 87-88.

[12] Roscoe, *United States Submarine Operations in World War II*, pp. 508-522

[13] E.g., USS *Seawolf* (SSN 575), USS *Halibut* (SSN 587), USS *Parche* (SSN 683) and USS *Jimmy Carter* (SSN 23)

[14] Sherry Sontag and Christopher Drew, *Blind Man's Bluff – The Untold Story of American Submarine Espionage* (PublicAffairs, New York, 1998).

Anti-submarine warfare was a new kind of mission explored by the post-World War II submarine force. Although American submarines had sunk thirty-two Japanese boats during the war, these were all attacks against surfaced subs. After World War II, U.S. intelligence came to believe that the Soviet Union was building a large number of submarines using advanced technology derived from captured Type XXI U-boats.[15] The Navy concluded that, to prevail in a war with the U.S.S.R., Soviet subs must be sunk while leaving port — before they could reach open ocean. This required a platform that could loiter for long periods close to Soviet naval bases. Thus was born the concept of the hunter-killer submarine. Enormous effort was expended in the late 1940s and 1950s to develop new submarine tactics, sensors, and weapons (e.g., PUFFS passive triangulation sonar and Mk 37 wire-guided, acoustic-homing torpedo) for this mission[16].

The last area of experimentation to be discussed is the development of submarine-launched guided missiles. Nazi Germany's V-1 and V-2 missiles introduced a new aspect of military technology to the world, which both the United States and Soviet Union were quick to adopt. The creation of lightweight nuclear warheads that could be carried by the missiles of the day made these weapons even more appealing. Given its rivalry with the Air Force over the nuclear mission, it is no wonder that the Navy focused early on developing a sea-launched missile capability.

The short range of these early missiles meant that a ship would have to approach close to enemy shores to reach most targets. Surface ships would be vulnerable to attack by hostile aircraft and submarines, conceivably armed with nuclear bombs and torpedoes. Submarines would have a greater ability to penetrate enemy defenses undetected and reach their launching point.

A serious problem with this concept was that the available ballistic missiles were too large for submarines to carry and used dangerously volatile fuels and oxidizers. For these reasons, the first missiles carried by the new SSG submarines were Chance Vought Regulus I air-breathing cruise missiles, which were not too large (although only one or two could be carried) and used conventional aviation fuel. Regulus had a number of shortcomings as a submarine-launched missile, however. In particular, launching a Regulus required the submarine to forego its main advantage of

---

[15] In 1948 the navy's General Board concluded that the Soviet Union had the potential to build as many as 2,000 modern submarines by 1960. (Friedman, p. 75).

[16] Norman Polmar and K.J. Moore, *Cold War Submarines – The Design and Construction of U.S. and Soviet Submarines*, pp. 18-22 (Brassey's, Inc. Washington, D.C., 2004).

stealth. On arrival at its launch point, the submarine would surface, extract the missile from its watertight hangar, prepare the missile for launch, conduct the launch, and then radar-guide the missile for up to thirty minutes after launch. Also, because Regulus' maximum range exceeded the launching submarine's ability to guide it, a second submarine was needed to take over control in mid-flight and guide the missile to its target[17]. Despite these difficulties the Navy acquired a total of five Regulus missile submarines[18], which completed forty strategic deterrent patrols between October 1959 and July 1964, after which Polaris replaced them.

One lesson to be learned from the submarine force's experimentation with new missions in the 1940s and 1950s is the danger of trying to force-fit a military platform into a role for which it is unsuited. The submarine radar picket is a prime example. Neither the submariners who had to perform the mission nor the surface forces they were trying to protect were entirely happy with the concept. Furthermore, the idea of using nuclear power to give a picket sub high speed on the surface approaches absurdity. As soon as a better solution (i.e., airborne early warning) became available, it was quickly adopted and the submarine radar picket faded into obscurity.

Another lesson is that niche missions, such as transporting commandos and collecting intelligence, will never justify the cost and effort needed to maintain a large submarine force. Even at the height of the Cold War, only a handful of submarines were dedicated to these kinds of special operations. Instead, general-purpose attack submarines were assigned the majority of these missions.

Of the many new missions considered for the post-World War II submarine force, two came to dominate submarine design and force structure: anti-submarine warfare and strategic deterrence. These missions became important only because of the strategic threat posed by the U.S.S.R. during the Cold War. To gain mastery of these mission areas required a combination of technological innovation and visionary leadership.

Submarine innovation in the Cold War

Three areas of technological innovation were key to shaping the Cold War submarine force: nuclear propulsion, submarine-launched ballistic missiles (SLBM), and advances in submarine sensors and weapons. The

---

[17] Friedman, pp 178-179.

[18] *USS Tunny* (SSG 282), *USS Barbero* (SSG 317), *USS Grayback* (SSG 574), *USS Growler* (SSG 577), and *USS Halibut* (SSGN 587).

Navy's interest in nuclear propulsion dates back to 1946. In the fall of that year the Chief of Naval Operations, Admiral Nimitz, requested a report on the subject of nuclear propulsion for submarines. That report, completed in January 1947, stated:

> "Present anti-submarine techniques and new developments in submarine design have rendered our present fleet submarines obsolete.... The development of a true submarine capable of operating submerged for unlimited periods, appears to be probable within the next ten years, provided nuclear power is made available for submarine propulsion."[19]

In December 1947 Nimitz made a formal recommendation to the Secretary of the Navy to begin development of a nuclear submarine. In his letter, Nimitz noted that nuclear submarines would give the Navy a secure means of conducting offensive, i.e., strategic missile operations.[20] In August 1948 BuShips established a Nuclear Power Branch, headed by Captain Hyman G. Rickover. Captain Rickover soon put the Electric Boat Company in Groton, Connecticut on contract to design the world's first nuclear submarine, *USS Nautilus* (SSN 571). Construction began on June 14, 1952 with a formal keel-laying by President Truman. *USS Nautilus* was commissioned in September 1954 and began sea trials on January 17, 1955 with the historic message, *Underway on Nuclear Power*.[21]

The submerged speed and endurance made possible by nuclear power revolutionized submarine warfare. In exercise after exercise *Nautilus* demonstrated her ability to attack heavily defended surface forces with impunity. On one occasion she chased down a 20-knot carrier group by running at flank speed (21.5-knots) for more than ten hours to reach attack position. Sixteen hours later, she attacked a destroyer 240 nautical miles away.[22] *Nautilus* also opened the Arctic to undersea operations, becoming the first submarine to reach the North Pole in August 1958.[23]

Nuclear propulsion soon became the *de facto* standard for the Navy's submarines. The path from *Nautilus* to today's nuclear submarine fleet was not, however, so obvious. First the Navy had to find a compelling mission to justify the high costs of building nuclear submarines. Their capability

---

[19] Polmar and Moore, p. 54.

[20] Friedman, p. 102.

[21] Polmar and Moore, pp. 55-58.

[22] Friedman, p. 109.

[23] CDR William R. Anderson, USN, *Nautilus 90 North* (World Publishing Company, Cleveland and New York, 1959).

against surface ships was obvious, but no enemy fleets were to be found. Admiral Nimitz was certainly aware of this in 1947 when he recommended building nuclear submarines — not for the traditional anti-surface mission, but as secure platforms for launching strategic missiles.

The first nuclear submarines were in many respects experiments to probe the design space. The second nuclear submarine was *USS Seawolf* (SSN 575), which used liquid sodium as a reactor coolant (*USS Nautilus*' reactor was cooled with pressurized water). *Seawolf* was followed by series production of the *Skate*-class SSNs. The *Skates* were essentially smaller versions of *Nautilus* and suffered from reduced performance as a result (top speed of 18-knots vs. *Nautilus's* 23-knots).

Nautilus, *Seawolf*, and the *Skate* boats had conventional hull forms common to the Navy's post-war GUPPY diesel submarines. The Navy had also been experimenting with a new "teardrop" shaped hull optimized for maximum underwater speed. It was only natural for the next class of nuclear submarines to utilize this new hull shape. This became the *Skipjack*-class which, powered by Rickover's new S5W reactor, had a submerged speed of 29-knots.

The emergence of ASW as a submarine mission

In 1956 Admiral Arleigh Burke, the new CNO, commissioned a "blue ribbon" study (Project Nobska) on the future of anti-submarine warfare. The study participants concluded that the Soviet Union would have nuclear submarines by the mid-1960s, potentially armed with nuclear missiles. The Nobska committee made a number of recommendations, foremost among them being that the Navy should develop new classes of nuclear submarines for the anti-submarine mission.[24] Admiral Burke's Long Range Objectives Group (LRO) came to a similar conclusion, arguing that all future Navy submarines should be designed for anti-submarine warfare. Furthermore, the LRO reported, the Navy's quest for high underwater speed was a wrong priority — stealth and sonar performance would be critical to the ASW mission.[25]

---

[24] Two other recommendations from Nobska committee were adopted by the navy. Nobsca's concept for a high-speed, long-range anti-submarine weapon led to development of the Mk 48 torpedo. Also, Nobska member Edward Teller proposed developing a compact nuclear warhead, which became the inspiration for Polaris.

[25] Friedman, p. 133.

Sonar systems on U.S. diesel boats and early nuclear submarines of the 1950s could trace their design lineage to the German U-boats. Although effective against surface ships and snorkeling subs, they could not hear diesel boats operating on battery. It was uncertain how well they would perform against future Soviet nuclear subs. Furthermore, the British had recently discovered that relatively easy silencing techniques could significantly reduce the detectability of snorkeling diesel boats.

Submarine weapons were another problem area. The Mk 37 torpedo was designed to attack noisy, slow-speed, snorkeling diesel subs. Its low speed and limited maneuverability made it useless against surface escorts and fast nuclear submarines. The World War II vintage Mk 14 torpedo was still being carried for use against surface ships, but no weapon was available that could attack a nuclear sub.

The *Thresher* program was the Navy's answer to the ASW problem. *Thresher*-class subs[26] combined *Skipjack's* S5W reactor plant and streamlined hull form with an entirely new sonar system consisting of a large, low frequency passive array wrapped around the submarine's bow. The passive array was expected to yield detection ranges out to the first sonar convergence zone (~ 30 nautical miles). *Thresher* also carried a new weapon system, Subroc, designed to destroy enemy submarines at long range. Subroc was a torpedo tube-launched missile that flew to its target point, where it re-entered the water and detonated its nuclear warhead.

Thresher was also the first nuclear submarine to extensively incorporate silencing into its engineering design. Silencing both increases stealth and enhances performance of the sub's own passive sonar by reducing the self-noise factor of the sonar equation.

Just three months before *USS Thresher* was commissioned more fuel was added to the ASW fire. In a May 1961 threat assessment, the intelligence community predicted that the U.S.S.R would have 578 modern submarines by 1971, including fifty SSNs and thirty-six SSBNs.[27] The U.S. began a massive building program of its own. Between 1961 and 1974 the Navy commissioned fourteen *Thresher/Permit*-class SSNs and thirty-seven SSNs of the follow-on *Sturgeon*-class. ASW became entrenched as the primary mission of the attack submarine force and retained that status until the end of the Cold War.

---

[26] Later known as *Permit*-class after the loss of *USS Thresher* on April 10, 1963.

[27] Friedman, p. 122.

The emergence of a strategic strike as a submarine mission

In the late 1940s a struggle ensued between the Navy and the Air Force over which service would have primary responsibility for strategic nuclear strike. The Navy's concept was to fly bombers from aircraft carriers. Because early nuclear bombs were large and heavy, a large, heavy aircraft was needed to carry them. The Navy's aircraft carriers, left over from World War II, were too small for the new bombers. The Navy wanted an expensive, new supercarrier for the nuclear mission.

The Air Force, on the other hand, believed that the strategic nuclear mission was its birthright. It planned to acquire 100 Convair B-36 "Peacemaker" intercontinental bombers. By the end of 1948, the Air Force had taken delivery of twenty-one B-36A bombers.

The post-war defense budget could not afford two major programs. In April 1949 President Truman cancelled the Navy's 65,000-ton supercarrier, *USS United States*, just days after its keel had been laid. The result was a long, nasty, public battle between the Navy, Air Force, and Congress. In the end the Navy lost not only its supercarrier, but also its prestige.

Although the supercarrier was gone, the Navy still aspired to have an important role in the strategic strike mission. When Admiral Burke became CNO in 1955, he immediately ordered the Navy to develop a sea-based ballistic missile capability and placed Rear Admiral William Rayborn in charge of the newly established Special Projects Office.

The initial design concept was to put up to six Army Jupiter missiles in the sail of a submarine. The Jupiter intermediate-range ballistic missile (IRBM)[28] was a sixty-foot tall, fifty-ton missile fueled by kerosene and liquid oxygen. The main reason the missile was so large was because its W-49 warhead weighed more than 1,500 pounds.

Putting Jupiter in a submarine was clearly going to be a problem. First, Jupiter's large size meant that it could only be housed in the submarine's sail, thus only a few missiles could be carried. Also, the missile could only be launched from the surface, which would expose the submarine to attack. Finally, the liquid fuel was dangerous to store and handle.

Two technological breakthroughs breathed new life into the program. The first was a new, lightweight warhead from Livermore National Laboratory (originally conceived by Edward Teller during Project Nobska).

---

[28] Not to be confused with the Army's Jupiter-C, which was a modified Redstone missile.

The lighter warhead suddenly made it possible to use a smaller, solid-fuel rocket instead of the large, liquid-fuel Jupiter. The missile's smaller size meant that more could be carried. Lockheed was awarded a contract to build the missile, named Polaris.

The second breakthrough was to use compressed air to eject the missile from the submarine. Once clear of the sub, the missile's motor would ignite and it would be on its way. This "cold launch" technique made it possible to fire missiles from underwater.

Admiral Rayborn quickly restructured his program to take advantage of the smaller missile and its submerged-launch capability. The revised SSBN design, adopted in June 1957, would carry sixteen missiles inside the pressure hull. In a crash program, two *Skipjack*-class SSNs under construction were cut in half and a 140-foot missile compartment added aft of the sail. The first Polaris submarine, *USS George Washington* (SSBN 598), was commissioned on December 30, 1959 and began her first strategic deterrent patrol in November 1960.

Forty-one Polaris SSBNs were built over a seven-year period. They were upgraded periodically during the Cold War to carry new missiles (i.e., Poseidon and Trident) capable of carrying a larger number of warheads to greater distances. The "Forty-One for Freedom" were eventually replaced by the larger *Ohio*-class SSBNs, which remain in service today.

Submarine missions in the post-Cold War era

With the fall of the Soviet Union in 1991 the Silent Service's primary Cold War missions, ASW and strategic nuclear strike, declined in importance. Just as in the years following World War II, submariners again struggle to find new missions to justify their existence.[29] Some of these are carryovers from the Cold War. Others are the result of new capabilities being enabled by technological advances. Lurking in the background are potential new strategic threats, any of which might lead to new missions for the submarine force.

Desert Storm introduced a new warfighting concept, strategic strikes by precision conventional weapons, which the Navy in general and the submarine force in particular have been quick to adopt. The Tomahawk cruise missile entered the Navy in the 1980s in three versions: nuclear land

---

[29] The navy is struggling to keep fifty SSNs in service, compared with a peak of 100 during the Cold War.

attack, conventional land attack, and anti-ship. Tomahawk was viewed as an inexpensive way of increasing the combat effectiveness and flexibility of naval forces. A submarine-launched variant was developed and many *Los Angeles*-class SSNs were upgraded with twelve vertical launchers for Tomahawk in the forward ballast tank.

Since the end of the Cold War, Navy submarines have launched Tomahawk land-attack missiles on numerous occasions against targets in Iraq, the Balkans, Afghanistan, and Sudan. Until recently, the submarine's primary shortcoming regarding cruise missile attack has been its relatively small magazine size (twelve vertical launch tubes plus room for up to twenty six weapons in the torpedo room). This is currently being rectified by the conversion of four *Ohio*-class boats to SSGNs. Each SSGN will carry 154 cruise missiles, as well as up to sixty-six special operations troops and their equipment.

Another mission that has recently become important is submarine support to aircraft carrier striking groups. During the Cold War, the Navy created a mission for its *Los Angeles*-class submarines called Direct Support. SSNs in the Direct Support mission were assigned to carrier battlegroups and were expected to protect the force from submarine attacks. The problem with this mission was that the battlegroup commander was often unable to communicate with his submarines. Thus, if the battlegroup changed course (for example, to conduct flight operations) the submarine could easily find itself out of its assigned patrol sector. In the worst-case scenario, it might even be attacked by friendly forces because it was not where it was expected to be and was thus mistaken for a hostile submarine.

Today's submarines have much better communications capability in the form of secure satellite communications, high data rate antennae, tactical datalinks, and Internet access. The Silent Service is silent no more.

Emerging technologies may open new mission opportunities for submarines. For example, conversion of Trident missiles to a high-accuracy conventional weapon has been proposed for quick reaction global strike. In this concept, each *Ohio*-class SSBN would have the nuclear warheads on two Trident II missiles replaced with two possible conventional warheads: a standard "slug" useful in penetrating buried targets and a "flechette" designed to attack larger targets on the surface. The missile's flight time would be less than thirty minutes and could hit targets up to 6,000 miles away with an accuracy of ten yards.[30]

---

[30] "DoD Defends New Sub-Launched Missiles," *Inside Defense* (March 10, 2006).

Unmanned vehicles are also being considered for submarines. A number of unmanned underwater vehicles (UUV) are currently in test and submarines have even controlled unmanned aerial vehicles (UAV) during exercises. The Defense Advanced Research Projects Agency (DARPA) even has a concept for an UAV that can be launched from and recovered by submarines.[31]

The future shape of the Navy's submarine force will ultimately be determined by the strategic threat to the United States. Today's submarine missions in the Global War on Terror are different from what might be needed in a new Cold War with China as an adversary. Ultimately, submarine missions will be constrained by the availability of technology. Conversely, new technologies may open new mission opportunities for the submarine force.

The importance of visionary leadership cannot be over emphasized. The Navy's development of nuclear propulsion would have been much different without Nimitz's vision and Rickover's technical direction. Would the Navy have ever succeeded in building SSBNs if Admiral Burke had put someone other than Rayburn in charge of Special Projects? Where are today's Nimitzs, Rickovers, Burkes, and Rayburns?

Finally, the question must be answered, "Does America even need nuclear submarines?" Obviously, the missions of the Cold War are of little relevance in today's geopolitical environment. However, nuclear submarines remain the only platform that can operate stealthily, for a prolonged time, within sight of an enemy's shores. This is a military advantage that should not be surrendered lightly.

I received a copy of Sub Command before Xmas 2001 and found the Subsim sometime after while doing a Google search on submarine sims. I keep coming back due to the subject matter and the good company. When I was young I thought cruising through the ocean at 400 plus feet at ultra-quiet in trail of a cold war enemy SSBN was stressful. Doesn't even compare with keeping up with my three sons (ages six, seven, and eleven).

Neptunus Rex

---

[31] "Cormorant Unmanned Air Vehicle (UAV),"

The Lucky Lighter

Jason Lobo

December 20, 1939

Dearest Anja,

It seems like a hundred years have gone by since I've held you in my arms. Since this is my first real "war patrol," I don't know what to expect. It's been rather boring up to this point. Joining the Navy was not as exciting as I thought it would be. We get up, go on watch, eat a meal, work on equipment, and try to keep each other entertained in what little free time we have. Then it starts all over again. Even though I've been at sea for almost a month, don't worry about me. I have good comrades and a good Captain. We all look out for each other.

Matrose Gunther Kohler set his pen down and rubbed his eyes. Although at the present moment he doubted his "sense of duty" to the Fatherland, he still believed he had made the correct decision by joining the Navy, just as his uncle had done in the First War. Gunther had enlisted in mid-September of 1939 with high hopes of glory and the U-bootwaffe was making good on the promise of glory. On September 17, HMS *Courageous* was blown from the water by *U-29*. Then, on October 14, Gunther Prien, the commander of the famous *U-47*, sank HMS *Royal Oak*. The U-boats had the same role as in the First World War: to strangle the British Isles. Just like the kracken, the U-boats would squeeze the life out of the Isles until they cracked. The tentacles of war gripped the world again.

A loud voice came over the U-boat's speakers startling Gunther from his sleepy, sickly state. "Third watch report to the bridge in ten minutes." As Gunther began to climb out of his bunk, the boat suddenly pitched, slamming Gunther's head on one of the torpedo racks.

"Hey, Grunni, are you okay?" a shipmate yelled out from another bunk.

"Ja, I think I'll live. Besides, I don't want to miss leisure time on the bridge," Gunther said as he began to put on his foul-weather gear.

Gunther, or "Grunni" as he was called by his shipmates, was one of the newest and youngest crewmembers. He put his hand on his forehead. There was a little blood, but he figured he would be okay. You could always tell the new cadets, fresh from naval boot camp, because they were still too young to grow a beard, and Gunther had a baby's face.

His raging head cold wasn't helping his morale. Gunther blew his nose in his already fouled handkerchief and threw it back on his bunk. He put his notebook with his letters to his new, young wife under his sweat soaked pillow. Gunther had married only four months earlier, just before entering the Navy. Everyone at home had advised against it, but Gunther couldn't live without his childhood sweetheart. Since he was one of the newest crewmembers on the *U-55*, his sleeping arrangements were the worst on the boat, the forward torpedo room. It no longer bothered Gunther that he was sleeping between two G7a torpedoes. He smiled to himself, as he recalled in his mind, his mother's expression when he told her he slept on 280 kg of explosive. He was still waiting for a convoy so they could get rid of the torpedoes, then he could stretch his arms.

Military service was a family tradition. His Uncle Otto rarely, if ever, spoke of his war experiences. Uncle Otto had been horribly scarred by mustard gas and was missing his left hand. Gunther knew he didn't want anything to do with the Army. Living in a cold, muddy trench wasn't his idea of a good time. Gunther never knew his Uncle Rolf. He was just a portrait on the wall in the living room at his parents' house. Uncle Rolf was the only family member who didn't come home. Rolf was burned to death when his Albatross D II fighter was hit in the fuel tank and burned uncontrollably in a spiraling dance of death somewhere over the Somme. Flying was out for Gunther.

In Gunther's mind, the sea was the only heroic place to serve. Uncle Werner told such heroic sea tales of his U-boat days in the Mediterranean during the Great War, young Gunther was prepared to enlist in the Navy at the age of nine. His sense of family duty and tradition were the only things keeping him going during his first real patrol at sea.

His father had reluctantly approved of Gunther joining the Navy. He thought he should help more on the family farm since he was such a strong boy. In addition, he had a new wife. Gunther's father was too young at the time to enlist in the Great War so he felt empty for not serving his country as his older brothers had. There was no countering Uncle Werner's influence on Gunther. Uncle Werner was a decorated U-boat commander, made famous for several brave surface engagements.

Gunther grabbed his binoculars and worked his way back to the conning tower.

"Gunther Kohler reporting for watch, sir," he said to the 1WO, the Erste Wach Offizier.

"That's a nice cut you have there, Grunni, do you want to see Willi first to patch that up?" the 1WO said in a concerned tone.

"Nah, I'll be fine. It's stopped bleeding," Gunther said as if nothing had happened.

"Suit yourself. Have a nice watch."

Icy, North Atlantic water crashed down the conning tower hatch as the second watch came off watch. Each man slid down the ladder, landing with a wet thump on the floor. Then the third watch climbed over them, up the ladder to the exposed bridge.

The cold, gray Atlantic slapped the *U-55* with each icy wave and swell as if it were an insolent child who had just sassed off at an angry parent. But *U-55* was a stubborn boat and kept pushing her way through each wave. She was a Type VII B, a vast improvement over the U-boats Gunther's uncle went to war in. She could dive deeper than the best German submarine of the Great War. Her purpose was the same, to destroy enemy shipping. The last man climbed up the narrow ladder to the bridge and then closed the heavy outer hatch with a loud thud. Gunther hooked his lifeline and took the lens covers off the binoculars. He scanned the horizon once and put the covers back on the binoculars. He thought to himself how pointless this was. A wave crashed on the bridge, soaking him. He could hardly breath with this damn cold. The sun was beginning to set, but it didn't really matter because the overcast sky shrouded out any remnants of sunlight.

A voice out of the spray and darkness. "Hey Grunni, how's your head? I heard you hit it pretty hard."

"Aww, it still hurts a little, but I'm too cold to worry about it," Gunther said, adjusting his hat and wiping his face.

Another voice broke out. "I'd do anything for a cigarette right now."

A wave crashed against the conning tower, drenching the men with forty-degree water.

"Hey…Romeo…" a deep, raspy voice called out. "I'll pay your bordello bill if you can light a cigarette in this weather." There was no mistaking that voice; it was Mueller. Gunther tried to avoid Mueller. He was the kind of guy you could get in a lot of trouble with, that is, if you hung out with him long enough. Mueller was on a first-name basis with the local port police and the base doctor. There was no vice that Mueller didn't like to indulge in.

"You're on! I was really looking for a reason to go out with Gabrielle. I hear she's really something, just out of my price range." The voice exclaimed in the dark. Laughter broke out on the bridge from the cold sailors. The man dug into his pocket for his cigarettes and somehow managed to get one in his mouth.

"Do any of you rats have a lighter?" the shivering sailor called out.

"I do." Gunther took his sea gloves off and fished in his pocket for the lighter, his Uncle Werner's lighter. His fingers were already stiff from the cold and he could barely feel the lighter. He pulled his hand out of his pocket. The wet, metal lighter slipped out of his numb fingers and bounced on the deck aft-wards. "Scheiss!" Gunther exclaimed as he unclipped his safety line and lunged for the loose lighter.

The U-boat heaved and slammed into another large ocean swell, again covering the exposed bridge. The watch rose back up after crouching to dodge the wave.

"Hey, Grunni, hurry up, the cigarette is going to be too wet to light up," Mueller yelled. He turned back from his lookout position to see what was taking him so long. Gunther was gone.

"Man overboard!" Mueller screamed out. Everyone turned to look where Gunther had been standing, but it was too late. Gunther Kohler was gone and there would be no finding him in this storm at night.

⊕⊕⊕⊕⊕⊕⊕⊕

Adrenaline surged through Grunni as he flailed wildly in the water. He waved and yelled to get his comrades' attention, but the U-boat melted into the stormy darkness. Drowning simply wasn't going to happen to him. Despite the constant teasing, he always wore his life preserver when he went on the bridge or on deck. He pointed out to anyone that made fun of him that if he ever fell in, it would make rescuing him easier.

There he was, alone, bobbing up and down in the North Atlantic, waiting for death. He found himself surprisingly calm. He knew about hypothermia and its lethal effects, so he waited to drift into unconsciousness and oblivion. He rolled over on his back and stared at the shrouded heavens. He thought about his young wife for a minute as blurred memories raced through his mind. He saw a Christmas tree, snow, candles, and a little boy standing staring at it. He saw his parents' house, the farm, the barn, the dairy cows, and Misha, the family dog, in the barn barking. He saw a young boy, running in the house with his muddy boots. The boy ran up to the wall with the family photographs on it. He could smell a spice cake his grandma was baking. The images stopped racing by. Uncle Rolf standing by his Albatros DII. Rolf's red scarf fluttered in the wind. Rolf took off his flying gloves and reached out. Come Gunther, come fly with me.

"Rolf..." Gunther called out as he lifted his arm out of the water. "Uncle Rolf..."

Something hard hit Gunther on the head. He was in the water and he was cold. Reality had returned. He spun around to see a very hard shadow behind him, a small lifeboat. Adrenaline again surged through his numb arms and legs. *My comrades have not forgotten me!* His heart exploded with a newfound life and blood began to flow again. He swam around to the side of the boat. His trembling hands seized the rails. He managed to get his torso over the side, and using his last bit of strength, heaved his leg in. He collapsed in the boat with a wet thud, like a fish that had just been landed. Saved!

He took several deep breaths then sat up. He couldn't see anything, so he groped around the bottom of the boat. Water sloshed back and forth. The boat was deserted. He found some sort of a rubber tarp. He pulled it over himself, covering his entire body. What strength he had, he expended it getting in the boat and working his way under this heavy tarp. At this point, there was no mental or physical energy left. He was out.

⊕⊕⊕⊕⊕⊕⊕⊕

Machinist First Class Rowan Tanner of the merchant ship *Rhodesian Star* closed the heavy door behind him with a loud clang. He had been working on a broken pump for the last four hours and was covered with lubricating grease. He produced a dirty rag and wiped the sweat from his brow, leaving a large grease smudge on his already dirty forehead. He mumbled to himself a time-honored curse about lousy damned pumps and walked into the ship's mess. He staggered over to the pantry, opened it, grabbed a steel cup and poured himself some coffee. He hated this ship's coffee, but at the moment there was nothing else to drink. He stirred a spoonful of sugar into the cup and took a sip.

"How's the pump?" the ship's cook, Alfred, asked in his distinct South African accent, setting a beef sandwich down in front of him.

"Have you ever tried to change ball bearings in a rocking ship? It's bloody impossible. Trying to keep the tools and parts from rolling away," Rowan said acidly.

"That's a nice bruise on your head, how'd you do that?" Alfred asked, joining him at the table.

"Ah, I tipped over the box with the bearings in it. Bearings everywhere. I was scrambling to gather them when I lost my balance and hit my head on the bulkhead. Hurts."

"Bulkheads are never flat and smooth, nor very soft, you know. You should find a better spot to dash your brains."

Rowan mumbled with a mouthful of sandwich. "I'm going to lie down and get some sleep."

"What about the pump?" Alfred asked.

"It'll keep for a few hours."

"Do you want anything else to eat?"

"Have you made any of your Christmas cakes yet?" Rowan asked, perking up. Alfred was well known for his baking abilities. Even though they were on a large, smelly tanker, his cooking could make the Queen smile.

"It's next on my list of things to do. I've got some leftover pudding if you'd like," Alfred said, getting up from the table.

"Nah, I guess I'm really not that hungry. Thanks for the sandwich." Rowan got up and went to his cabin.

The *Rhodesian Star* was a small South African flagged tanker owned by the British Petroleum Company It plodded alone, northward into the war zone. It was the tanker's second trip to Venezuela to pick up the vital lifeblood of Britain's military and economy — oil. The first trip had been uneventful, but now the German U-boats patrolled the Western Approaches like a cat waiting for a juicy mouse to venture out of its hole. In the last few weeks, submarines operating in the region had savaged two convoys. It had been several days since the *Rhodesian Star* had picked up a distress signal of a torpedoed ship, but it didn't matter for they couldn't render any aid. It would be a tough decision to even stop to pick up any survivors. Stopping could mean destruction, but there was that "code" among sailors to help fellow sailors in distress. After the first distress signal it became very clear that this run wasn't going to be easy, so the captain had a meeting to decide what to do if they encountered any problems. After a heated debate, it was decided by one vote to stop to pick up survivors. If they were to stop, it would only be for five minutes.

Rowan Tanner was no stranger to the hardships of the sea or the terrors of war. Rowan came of age during the First World War when, in 1916, the coal ship he was sailing on, the *Christopher Wilson*, crossed paths with the notorious Max Valentiner. Valentiner was a notorious U-boat captain. He had a reputation for killing survivors. Rowan learned this first-hand when his ship had been sunk and the lifeboats used as target practice for the gun crew. Somehow he had survived the experience to sail another day.

Rowan took his greasy shirt off, revealing an old, faded tattoo that he had picked up in some forgotten port years ago. He put on a striped T-shirt and then a heavy wool shirt and climbed into his bed. Rowan had his own personal cabin. He had earned it. It was a special perk from the captain, who knew how to keep an old sea dog happy. Rowan had been a merchant sailor his entire life. He had seen the world many times over and learned many useful skills in his time as a sailor. He had served on the *Rhodesian Star* for the last ten years and had become a key member of its crew. In addition to being the top mechanic on the *Rhodesian Star*, he was also the unofficial doctor. Rowan closed his eyes and worked through in his mind what he had fixed on the broken pump and what still needed to be worked on. Within minutes he was fast asleep.

The years couldn't wash away nightmares that still lurked in his subconscious. The First World War had permanently scarred his soul and this new conflict had resurrected the past demons. The tremendous explosion ... jumping into the cold water. He could still taste and smell the burning fuel. He swam...rifle shots...the lifeboat shot to pieces...the U-boat submerging ... the periscope ... the stillness.... He awoke in a cold sweat, his heart racing. He became aware he was safe in his bunk and relaxed. Then it occurred to him something was really wrong, it was quiet, the ship's engine had stopped. *Why in the hell would we be stopping in a war zone?*

He quickly put his coat on and pulled his boots on. He quietly closed his cabin door behind him and headed up to the bridge. The sunshine felt good on his face and the cold air washed his lungs of the stale ship air. He opened the door to the bridge and walked in.

"Morning, captain, why have we stopped?" Rowan said, in an irritated and concerned tone.

"Morning Tanner, it's a good thing you're here. We've stumbled across a lifeboat from the *Walter MacPherson*, a ship that was torpedoed a few days ago. There's an unconscious man curled up in it. You want to check him out?" the old captain asked.

"Sure. Ian, go get a blanket and pillow ready in the mess, we'll put him there if he's alive," Rowan said as he and the captain walked out of the bridge. They worked their way down the ladder to the deck, then on to the starboard side, where the lifeboat had been hooked and tied to the side of the boat. When the captain and Tanner arrived, the crew had just pulled an unconscious man from the boat. Tanner walked up and gently removed the unknown man's foul weather hat, revealing a young man. Tanner checked his pulse.

"He's still alive. Get him below decks and find him some dry clothes." Rowan looked at the captain and said, "I wonder who our mystery man is?"

"That's rather odd of someone to survive that long in a lifeboat. You'd think he should have died by now. Let's get on our way before we get a torpedo, too," the captain said.

The castaway was stripped of his wet clothing, dressed in some warm clothes, and laid out on the table. They then piled two heavy blankets on him.

"I wonder when he's going to wake up, he's been through quite a storm." Rowan said, staring at the unconscious man.

"How long do you think he was in the life boat?" Alfred queried.

"Who knows? He's bloody lucky to be alive, if it's not too late for him already," Rowan said, sitting down and drinking a cup of coffee. "Bring me his pants. Let's see if he has any papers or ID on him."

Alfred walked over to Rowan and handed him a lighter. "He was clutching that when we pulled him out of the boat." Rowan studied the lighter. He flipped the cap open and tested the flint. "Hmm, still works. Look at the writing on it, German," Rowan said, handing the lighter back to Alfred. Inscribed on it was "Wir jagen wie ein Loewe." On the other side was an Iron Cross with "U-92" etched under it.

"Do you think this chap is a Jerry?"

"If he is, we should throw him back. German pirate bastards," Rowan grumbled bitterly.

"Ah, Tanner, maybe he's not German. He probably picked the lighter up in some foreign port of call," Alfred added, trying to settle Rowan down. Alfred had heard of Rowan's war experiences and his flaming hatred of anything German.

"You're probably right, Alfred. Call me if he wakes up. I'm going to finish working on that bloody pump," Rowan said and walked out of the ship's mess. Rowan looked at the lighter and put it in his pocket.

⊕⊕⊕⊕⊕⊕⊕⊕

He smelled gingerbread.

There was a rhythmic thumping noise… ship's engines. Gunther's eyes slowly opened. He was staring at a Christmas tree. He was confused. *Where am I? Am I dead? Where's Uncle Rolf? Where's Mama? Can I have some cake? I didn't know heaven had diesel engines?*

He sat up and it slowly came back to him. He wasn't lost in the ocean anymore. He pushed the blanket off and looked down at himself. He was wearing different clothes. He called out: "Hallo?"

Alfred suddenly looked around the corner. He saw the man sitting up. He took his oven mitts off and rushed into the mess.

"My God, you're alive!" Alfred exclaimed, rubbing his forehead.

Gunther looked at him, still a little dazed and not understanding. "Wo bin Ich?" he asked.

Alfred's jaw dropped. *He's German. Rowan's not going to like this.* Alfred left the mess and came back in with a large slice of cake and coffee laced with brandy and gave it to the man.

With a mouthful of cake, Gunther managed to get out a "Danke." Alfred rushed to the engine room. He opened the door and meandered his way though the deafening noise to where Rowan was working. Alfred signaled to Rowan to come.

"Our survivor is awake," Alfred yelled out above the noise of the engines.

"That's great, did he say anything?" Rowan said with an eager face.

"Rowan, he's *German*," Alfred said in a concerned tone. "He's still really confused, doesn't know where he is."

They both rushed to the mess. Rowan burst in, the door slammed open with a loud clunk. Word spread quickly on the ship. Within minutes, the captain entered the mess. The man was babbling something in German.

The captain spoke first. "Go get Lewis, he speaks German, doesn't he?" Rowan glared at the German. "Rowan, take it easy. He's just a boy." Rowan left the room and came back with a small, black leather bag, his medical kit. Lewis walked into the room.

"What's he saying?" the captain asked.

"Something about his family…more of his mom's cake," Lewis attempted to translate.

"Tell him to shut up so I can take his temperature," Rowan snapped, digging in his bag. Rowan shoved a thermometer in Gunther's mouth and waited.

"My God, he has a temperature of 105 degrees. This man is really sick," Rowan said. "Ask him his name."

"Mein Name ist Gunther, bist du Onkel Rolf?" Gunther said in a confused tone.

"What did he say, Lewis?" Rowan said.

"He said, his name is Gunther and he wants to know if you're his Uncle Rolf."

"It's obvious that he's totally delirious. The fever is affecting his brain," Rowan said, looking at Gunther.

Gunther looked at Rowan. "Onkel Rolf, ist es Weinachten? Darf Ich mehr Tort haben, bitte."

All eyes were back on Lewis. "He asks, if it's Christmas and if he can have some more cake. And he still thinks you're his Uncle Rolf," he said with a trace of a smile.

Rowan turned and looked at the Christmas tree that was in the mess. Alfred had put the tree up a week ago to get the tanker's crew into the Christmas spirit. He looked at Gunther again, his features softening. This was a simple, lost teenage boy. He was a boy who had been lost at sea, just like he had been. He wanted his family. He just wanted to be home. Rowan stared at him longer.

"Lewis…" Rowan said, breaking the silence in the mess, "tell him it is Christmas and I'll get him some more cake." Lewis translated and Rowan came back with a large slice of cake. Gunther perked up and wolfed the cake down. He thanked his long lost uncle for the cake and went back to sleep. The crew in the mess were silent.

Alfred asked if anyone wanted some cake, since they were there and the cake was still warm. Rowan took a piece and silently went back to his cabin. He poked at the cake, but he couldn't eat it. He put the cake on the table, turned the light out and went to bed. Had he "saved" his hated enemy, or was it Gunther who had saved him?

⊕⊕⊕⊕⊕⊕⊕⊕

There was a knock at the door. Rowan awoke. "What is it?"

"Rowan, it's that German boy. He keeps asking for you." It was Alfred.

"Asking for me?"

"Yeah, you *are* 'Uncle Rolf', you know," Alfred said with a slight chuckle.

"Well, I guess it's better than being mistaken for his mother."

When Rowan entered the mess, he was shocked. Gunther was deathly pale.

"Rolf, ist es Weinachten jetzt?" Gunther mumbled.

"Weinachten, what's Weinachten?" Rowan asked, checking Gunther's temperature.

"I think it means 'Christmas'," Alfred said.

"Go wake Lewis up," Rowan said, putting a cold cloth on Gunther's head. Rowan looked at Gunther. "Ja, Weinachten…Weinachten."

Ten minutes later, Alfred appeared with an irritated Lewis in tow.

"Rowan, I hope you're not going to do this to me every time he says something."

"Whatever, just translate what he's saying. He keeps repeating *fliegen*, what the hell does *fliegen* mean?"

Lewis looked at Rowan with a confused expression. "He keeps saying something about, he's ready to *fly* with you."

"Fly? Why would he want to fly with me? Tell him we'll go flying tomorrow. God, that fever must really be causing some serious problems in his brain."

Lewis translated. Gunther smiled and his head slumped back.

⊕⊕⊕⊕⊕⊕⊕⊕

He smelled the sweet smell of gingerbread.

He still saw those lights before him. It was the happiest time of the year. It was Christmas. Mamma was cooking. Uncle Werner was here, and Uncle Otto. It was warm and safe. And there would be presents.

Later…later, there would be presents.

Now, he would just very much like to sleep a bit.

Slowly, slowly, the Christmas lights faded like a ship slipping into deep night fog.

⊕⊕⊕⊕⊕⊕⊕⊕

"My God…" Rowan surged forward and checked the young German's pulse. There was none. "He's dead," Rowan said turning back to the other two men.

"What do you mean, he's dead? He was just bloody talking to us. He probably just fainted," Alfred said in a disbelieving voice.

"Here, you check his pulse, maybe I'm wrong." Alfred pressed two fingers on Gunther's throat. Seconds passed.

"Wait, where is it? I can't feel it," Alfred said, frustrated.

"Dammit! He was just a lad, maybe seventeen, I'd say. I'm going to get some air." Rowan pulled the blanket over Gunther's lifeless body. He stormed out of the mess, up the stairs, and threw the door open to the deck outside. He walked towards the stern, one of the few locations where smoking was permitted on a tanker carrying flammable liquids.

It was still dark outside and a bitter wind was blowing. Rowan was in the act of buttoning up his coat when there was a tremendous explosion. The blast slammed him against the side of the ship.

Rowan lay on the deck stunned from the hammer-blow. The flames quickly brought him to his senses. *Oh God, not again...*

Rowan was familiar with torpedo explosions. The torpedo had ripped into the number two bunker, instantly igniting the aviation fuel. The ship erupted in a massive ball of flame skywards, like a prehistoric volcano, raining molten chunks of metal and debris everywhere. The second torpedo slammed into the number five bunker, which was carrying lubricating oil. Oil gushed out of the gaping hole into the sea. The oil bled out and ignited from the heat of the first fire. The entire port side of the *Rhodesian Star* looked like an apocalyptic scene from Dante's *Inferno*. There were men on fire, running and leaping into the burning water. There was no escape.

Rowan limped around to the starboard side of the aft forecastle to shield himself from the heat of the massive fire. There was no time to hesitate. *I must get off this ship now!*

The *Rhodesian Star's* back was broken and she was starting to slip beneath the waves. The night air was filled with hissing steam and metallic groaning as the heated decks buckled from the stress. He wasn't wearing his life preserver and he would sink like a rock if he jumped overboard now. Rowan quickly stripped down, out of his coat and heavy sea boots. He then scrambled over the railing and dangled precariously over the edge of the ship. As far as he could tell there was no oil slick where he wanted to jump. So with all of his focused energy, he pushed away from the burning ship.

He fell for what seemed like a thousand years until he splashed into the icy water. He kept going deeper and deeper until the forward momentum finally stopped. He kicked his legs and pumped his arms until he had some upward motion. He fought the urge to take a breath as he kicked rapidly. His head broke the surface and he flailed for a bit as he gulped air. He could feel the heat of the burning ship on his face. He started to swim away from the ship, he didn't want to get dragged under by the suction or get caught up in the oil slick.

He swam as hard as he could for a minute and stopped. He figured he was safe, so he turned around to see what was going on. The bow section of the ship had already slipped under the waves. The stern desperately clung to the surface, but the trapped air that was keeping it afloat would escape soon enough.

Rowan looked around for anybody else who might have escaped. He saw something floating in the water and swam towards it. It was half of a charred lifeboat. He pulled himself onto the remnants of the overturned boat.

Rowan watched his ship burn. The flames still raged, lighting up the night sky for miles. What was remaining of the *Rhodesian Star* was a twisted, distorted wreck from the heat of the flame. The stern section finally lost its buoyancy and slid under the oily waves. Rowan noticed some movement in the corner of his eye. He turned to see what it was and was struck with terror when its shape came into the light. It was a German U-boat. The scattered fires of burning oil illuminated the forward part of the submarine. He could see the net cutter, the deck-gun, the conning tower and its crew. The U-boat was running on its electric motors so it made little noise other than the waves splashing at its side.

Rowan cowered in terror on the remnants of the lifeboat, watching the U-boat. He realized it was heading towards him. Rowan wanted to swim away, but where was he going to swim? He didn't have the strength to swim far. He didn't want to drown, so he clung to the boat. The U-boat slowly approached him. He looked up at the men in the conning tower.

"Hallo?" a voice yelled out through a megaphone from the conning tower. Rowan saw men climbing down from the conning tower onto the deck. *This is it*, he thought, *I really am going to be shot*.

A rope hit him in the head. He grabbed it and held on. He was pulled over to the side of the U-boat. A pair of hands reached out and jerked him off the remnants of the lifeboat. He lay curled up on the deck of the U-boat, staring at his rescuers. Two strong sailors, dressed in the standard rubberized coat and pants, wrapped a blanket around him and escorted him through the access hatch behind the conning tower. Rowan was led into the officer's mess, where a bearded man wearing a white cap met him.

"You are aboard the *U-55*, I'm Kapitanleutnant Moehlmann, the commanding officer. Here, let me offer you some schnapps to warm you up," the officer said in a thick German accent, taking off his gloves and opening a bottle of very expensive Swiss schnapps. Rowan reached out, took the glass and drank it like a true sailor.

"Would you like some more?"

"You speak English?" Rowan asked, drinking his second glass.

"Yes, of course. It was part of my education."

"I thought you submariners killed your survivors," Rowan said in a venomous tone.

"No, I'm sorry to disappoint you, we're not bloodthirsty pirates."

"Like hell you aren't. I was in the First War. I know all about you murderous bastards," Rowan said, ready to physically lash out.

"Well, since you don't appreciate our hospitality, let's cut through the pleasantries. Tell me, old timer, what was the name of your ship and its destination?"

"Piss off, you bastard!" Rowan said, wiping his nose.

"Mueller, search him." Now a pistol was pointed at Rowan. The sailor began to pat Rowan down like a common criminal. Mueller felt something in Rowan's pocket.

"Empty your pockets, please," the captain ordered. Rowan reached into his pocket and pulled out a small, shiny object. Mueller stared in amazement. He snatched the object from Rowan.

"It's Gunther's lucky lighter," Mueller said in a hush.

All eyes were on Rowan. "Where did you get that lighter from?"

"We picked up this man in a life raft. A man named Gunther, or something like that. I tried to save him, but he died from the shock and exposure. He was totally delirious when we picked him up. He went down with the ship, when you torpedoed us."

"We lost Gunther overboard about five days ago," Moehlmann said. You saved him?" He looked at the lighter. "Go put this with his personal items." The captain handed the lighter to Mueller and he looked back at Rowan. "You personally saved Gunther?"

"If you want to call it that. I was the ship's doctor. He was very sick when we found him. He died peacefully though. He kept talking about his family."

Captain Moehlmann looked thoughtfully at Rowan. "We cannot take prisoners, and I cannot let you die at the hands of the sea since you saved one of our crewmembers. If this is acceptable to you, here is what I propose. We will sail as close as we can to land and drop you off. Since you like schnapps so much, I'll give you a bottle from my personal collection. I'm sorry we can't do more for you; it is war after all. But I want to repay you for saving one of us. We'll wait until sundown, then we will drop you off tonight."

The hours beneath the waves passed slowly. Rowan and Moehlmann exchanged sea stories. Rowan recounted his last encounter with a U-boat and Moehlmann understood why Rowan feared and hated Germans.

The time had arrived. The sun had set over an hour ago and Moehlmann deemed it safe enough to surface. For that day, Christmas Day, the war had been put on hold. Rowan was grateful to be alive and the crew of the *U-55* was thankful to know that Gunther wasn't lost at sea. They were also thankful to be reunited with a part of Gunther, his lucky cigarette lighter.

"Well, my friend, it is time for you to go," Moehlmann said. "We will surface in five minutes and I'll have the crew take you to a nice, quiet beach. After that you're on your own, but at least you won't be swimming. You should have enough food to keep you going for a day or two."

"I want to thank you for rescuing me and for your hospitality," Rowan said, buttoning up a coat the Germans had given him.

The boat shuddered and moaned as it broke the surface. The sea was calm on the clear, moonless night. The crew opened the hatch with a hiss and began to prepare a small rubber raft. It was gently lowered into the water and lashed to the deck railing of the U-boat. Rowan emerged from the U-boat and stretched his legs. He shook the commander's hand and climbed down into the raft. The raft was untied and slipped off in the darkness towards the beach.

⊕⊕⊕⊕⊕⊕⊕⊕

It was a somber day. Gunther's mother had received the news of her lost son the week before. She opened the package that arrived that morning which contained her son's notebook, uniform, various personal items, and his lucky lighter. She took out the official photo of Gunther's naval graduation and hung it on the wall next to the photo of Uncle Rolf. Gunther finally got to meet his Uncle Rolf.

Experiences of a Civilian Submariner
Donald Ross, Ph.D.

Prelude

Although I didn't know it at the time, April 29, 1952 was to be a decisive day in my life. On that day, on Eleuthera Island in the Bahamas, scientists from Bell Labs demonstrated their new Lofar low-frequency passive detection system to a group of high-ranking Navy officials. A U.S. diesel submarine provided services, snorkeling in rectangular box patterns at increasing distances from the island. Two decisions resulted from the successful trial of the new sonar.

On that day the Navy decided to proceed with the development and deployment of what was to become the SOSUS network. The second result took longer to materialize. The target sub had been requested to operate two diesels at the same speed. Several hours into the day, the sub received an amazing radio message: *the port diesel was running ten rpm too fast*. A tachometer check showed this to be true. Naturally, the skipper reported this incident to his superiors.

In time it trickled up the chain of command. Whatever it was, the submarine Navy wanted it, too. So, in 1953, a task was added to BTL's (Bell Telephone Laboratories) contract to investigate the feasibility of applying the new Lofar low-frequency signal-processing technology to submarine platforms. Knowing that this was coming and expecting flow related noises to be important, Bell recruited me for this project in view of my background in fluid mechanics and flow noise.

Before I joined Bell Labs in October 1953, I had never been on a submarine or had any contact with submariners. For the preceding eight years, I had worked on the design of quiet propellers for torpedoes at the Penn State Ordnance Research Lab. At Penn State, I had also been deeply involved in the hydrodynamic design of the large Garfield Thomas Water Tunnel, which was built there in the late 1940s. My only connection with submarines had been in 1947, when at a Naval Hydrodynamics Symposium in Washington, DC I was chastised for suggesting that shaping submarines more like torpedoes, with centerline propellers, would increase both their propulsion efficiencies and their cavitation-free speeds. A senior naval officer patiently explained to me that submarines would always have two screws, for maneuverability as well as for safety.

Indoctrination

The BTL submarine project began the day I arrived at Bell. The project was staffed by three engineers and a technician. One of the engineers, Herman Straub, was a Naval Reserve submariner who had served on subs during the war and had even been a skipper at the end of the conflict. The second engineer, Larry Churchill, was skilled in electronics and, like me, had been hired specifically for this project. I was the scientist of the group. As we were all of equal rank, we decided to run the project as a triumvirate without a formal head. All decisions were made by consensus and, although we had many heated discussions, we never had a fight. When we dealt with the Navy, we were told that they were more comfortable with a single chief, so we would pretend that Herman was our Head.

Submarine assistance to the BTL project was assigned by CNO to Submarine Development Group Two in New London. Since Larry and I had never been on a submarine, we decided to do a one-day familiarization ride. It turned out to be on a snowy Monday in January 1954. When we arrived in New London by train that Sunday evening, the snow was so deep that we had to trudge up Main Street to our hotel. However, by Monday morning the snow had stopped, and we were able to get a taxi to the Sub Base. I don't remember the name of the sub, but it was to be on a daily op in Long Island Sound and included a qualification exam for one of the officers.

Lt Cdr Bill Banks, the Skipper of the *Cavalla*, was the examining officer. At one point, he removed the fuse from the "Christmas Tree" in the control room. Without its vital information, the officer lost control and the sub touched bottom. I learned that hitting bottom in shallow water wasn't too serious. On that first trip, I also learned that periscopes leaked a lot, which was no problem as long as the leak was less than could be handled by the bilge pump. The ship was conducting training drills, which I thought I could safely ignore. However, I was in the forward torpedo room when the drill called for that compartment to be assumed flooded. All hands moved aft. They closed the hatch and the pressure started to rise. I started to shout and pound on the hatch, not knowing what to expect. Soon a voice over the loudspeaker reassured me that all would be well.

On the *Cavalla* (SSK 244)

Following that first experience in January, we made frequent trips to New London to plan our first field trial with the personnel of the Development Group. At that time, the *K-1* and three SSK conversions were assigned to the Group. One of these, the *Cavalla* (SSK 244), was tasked to work with us. At the pre-trial meetings we explained that our first goal was to determine just how much of the length of the sub could be used for a receiving array and that this required measuring the noise field of each piece of machinery. To do this, ideally all items would at first be turned off and then each operated one at a time. The *Cavalla's* Skipper, LCDR Bill Banks, said that he thought that he could accomplish our objective by hovering on a layer of cold water. By the time of our trial, he had perfected this technique.

We were assigned to the *Cavalla* for two weeks in May 1954. The first week the sub was at the pier in upkeep. The Bell Labs team used this week to mount about three dozen hydrophones throughout the superstructure from bow to stern. A total of about a mile of DSS-3 cable was strung from the hydrophones through a special torpedo-loading-hatch cover to preamps located in oak boxes arranged in racks in the forward torpedo room. When we went to sea the following week, Bill Banks found a useful layer, balanced the tanks and then hovered for significant periods with the sub in dead quiet mode. We were able to record the noise field of each machinery item. Thanks to the excellent participation of *Cavalla*, the test was 100% successful. The experience of working together was so good that the sub wanted to schedule another week with us within a few months. But we explained that we would need more time to analyze the data and could not profitably go to sea that soon.

From our first *Cavalla* tests we learned that only the region forward of the sail was suitable for a low-frequency hydrophone array, limiting us to a length of about 100 feet. With this length limitation, we would have to choose a sonar band centered at about 500 Hz, in order to have sufficient directionality as well as array gain. This apparently ruled out Lofar, which operated at much lower frequencies. (Years later towed arrays solved this problem.) We were about to give up all hope of applying Lofar technology to submarines when we read in an ONR-sponsored Jezebel report about a novel application of Lofar called "demodulation Lofar." In this technique, an octave wide input band is demodulated and the resultant signal fed to a standard Lofar. Since the mid frequencies contain most of the same tonal information in their modulation spectra as do the low frequencies, this system promised delivery of Lofar-like signals while using much shorter arrays than those used by SOSUS. We called the new demodulation system DEMON Lofar.

It was February 1955 before we were ready to try out our new technology on *Cavalla*. This time we again mounted hydrophones and loaded equipment in New London during an upkeep. The boat then transited to Bermuda, where we joined her. Initial tests with DEMON were encouraging when we found that, contrary to regular Lofar, own ship's machinery did not produce interfering tones. The display was blank except when there was a target. After several days of solo ops off Bermuda, we had the services of a target submarine which snorkeled for about ten hours per day about 100 to 120 miles from *Cavalla*. Since our array in *Cavalla's* bow only formed one beam, perpendicular to the sub, we operated in slow circles in order to sweep the array beam. After finding a potential target, *Cavalla* would steam straight, keeping the target on its beam, while we analyzed the signal using both DEMON and a MOD-3 regular Lofar.

We repeated this procedure for five days, using each night to move to a new operating area closer to New London, until the final day's ops were carried out in shallow waters south of Nantucket. *Cavalla* was able to detect the target sub consistently, probably the first time that a submarine was able to track another at such long ranges. Also, quite unplanned, we detected and classified a carrier task force at about 150 nm out. One wonders whether these detection records set by *Cavalla* in 1955 might still stand.

After the 1955 *Cavalla* trial, the decision was made to proceed with the engineering development of the DEMON Lofar system as the classification component of a new generation passive-active sonar system being developed by the Navy's Underwater Sound Lab in New London.

On the *Nautilus* (SSN 571)

The tests on *Cavalla* had demonstrated the feasibility of adapting Lofar technology to a conventional submarine platform, providing the sub could operate quietly. However, by 1956 it began to look as though the Navy would be switching to nuclear power for submarines, and it was not at all certain that our system would work on a nuclear. The *Nautilus* had a reputation of being very noisy, being much more detectable than a snorkel sub. To investigate its self-noise field, the Bell Labs crew installed about a dozen hydrophones in its superstructure, and in March 1956 Herman Straub and I rode her during a ten-day submerged transit from New London to Florida.

The time on *Nautilus* was a memorable experience. Once again we sailed from New London on a snowy, foggy winter day. It was Monday, March 19, and we were the first boat out. Just outside the river we received a radio message that the weather was too severe and we should return to port. However, our radar was free of targets, and *Nautilus* skipper CDR Dennis Wilkinson radioed back that the visibility was sufficient to proceed, even though he couldn't see his own bow.

This trip was intended to test a new air purification system, and during it, we established a new record for time submerged without taking in any fresh air. As I was the only civilian aboard, Herman being a Navy reserve officer, I briefly held the civilian world record for submergence time. By the time we took in air, the CO and CO_2 levels had become dangerously high, especially in our part of the boat, which may account for some early minor memory loss that I experienced. During this trip I became fully knowledgeable about all systems, making frequent trips through the reactor compartment to the engineering spaces. On one occasion I sat on the top hat for several minutes in order to be able to tell my grandchildren that I had once sat on a nuclear reactor. My radiation badge showed no ill effects.

BTL had been assigned a very small space in the forward torpedo room, only large enough for an Ampex twelve-channel tape recorder and one analysis Lofar. Our sleeping spaces were on top of two torpedoes. We recorded all of the hydrophone signals for later analysis. Since the tape had to be changed every seventy-five minutes, we never got to sleep for more than that at any one time for about eight days. Along with changing tapes, we kept a detailed machinery log and also investigated the sources of any strange noises.

In our time on *Nautilus* we became thoroughly familiar with all the details of its plant, and in the process came up with several ideas on how the sub could be operated in a manner less detectable by both SOSUS and other submarines. We reported these ideas to the ship's officers at meal times. Therein lies a funny story. *Nautilus* had the largest wardroom of any then-existing sub. On the first day, Herman and I had places at the far end of the mess table, next to the junior supply officer. The next day, we found that our napkin rings had been moved one place further up the table. This trend continued for several days until we were seated to the left and right of the skipper, having replaced the Exec and the Chief Engineer. Clearly the captain was very interested in what we had to say. The *Nautilus* became a quieter sub from then on; and some fifty years later, I heard from Admiral Wilkinson that our input had been a factor in the Navy's decision to proceed with a nuclear sub force.

Following the *Nautilus* trip we spent months analyzing our recordings and I then wrote the first comprehensive report on nuclear submarine noise. There were some interesting follow-ups. Some months later I was called in by Eldon Bissett of the fledgling acoustics branch of Naval Intelligence to interpret Lofargrams that had been made of sounds from a new Soviet submarine. The Russians had claimed they had a nuclear sub. Not knowing that the *Nautilus* plans had been given to the Soviets by a spy, our Navy had denied their claim. However, after a day of studying the Lofargrams, I concluded that the spectra were indeed from a twin-screw, geared, steam-turbine drive and were therefore consistent with nuclear propulsion. This was indeed the first NOVEMBER.

There was a dispute between the SOSUS group at Bell Labs and the radiated noise measurement people at DTMB. Tones from *Nautilus* had literally burned the paper at SOSUS stations while at distances in excess of a thousand miles, implying radiated noise levels at least 10db greater than those of any diesel. Yet DTMB's measurements showed the strongest tones from *Nautilus* to be similar in strength to those from diesels. I personally knew all the people and believed both groups. I was finally able to explain the discrepancy by considering the different operating depths of diesel snorkel subs and nuclear subs. SOSUS has been designed to detect diesels, for which the effective source depths were about fifty feet, generally well within the surface layer. All of that system's transmission loss measurements had been made for sources at this depth. However, the *Nautilus* operated at depths well below the surface layer, in the deep sound channel, for which distant sound propagation was very much better. The same strength source in the deep channel produced a much stronger received signal than from near the surface. This effect of operating depth on detection ranges became a saving grace for SOSUS when Soviet nuclear submarines became quieter.

On the *Bang* (SS 385)

It was January 1957 before the BTL crew was ready for another sea trial of the DEMON Lofar classification sonar. This time it was the *Bang* (SS 385) that served as the test platform. Again, as it was winter, the operation was conducted starting in Bermuda, and again a snorkeling target was employed. I have but little memory of the *Bang* trials. By this time, the work of the initial project team was pretty much over. Shortly afterwards I was assigned to the SOSUS performance team, and Herman Straub prepared for retirement. Actually, while on pre-retirement leave using up his accumulated vacation and sick days, Herman died in a Swiss hospital from a skiing accident.

Larry Churchill and I had joined Bell Labs in our thirties, instead of right out of college, as was the norm. By 1957 we began to realize that because of our late arrivals there were no real career paths ahead of us. That year, I was appointed to a subcommittee on submarine silencing of the Committee on Undersea Warfare at the National Academy of Sciences. There I met Richard Bolt, one of the founders of Bolt, Beranek and Newman (BBN) in Cambridge, Mass. One day he asked me if I might be interested in interviewing with his firm, which I did. In July, I notified BTL that I planned to join BBN, but offered to stay for six months to produce a Handbook for the SOSUS project. It was the end of January 1958 when I moved with my family to Boston.

Navy Lab Head

For almost a decade, while at BBN, I contributed to submarine silencing, but did not ride any subs. (For details of this period see the article "Submarine Silencer"). In July of 1967 I became the fourth Head of the Ship Acoustics and Vibrations Laboratory (SAD) of the Naval Ship Research and Development Center (NSRDC) (formerly DTMB) at Carderock, MD. Responsibility for all of the Navy's noise trials in the Atlantic was assigned to my group. This was a period of heavy construction of new nuclear submarines — all of which underwent noise trials — which we conducted in two sheltered bodies of deep water in the Bahamas using the converted YW-type water barge MONOB-1. As noise measurement and analysis had been one of my major interests, I became quite involved in the trials program, frequently working with the group on MONOB and occasionally joining our personnel on the submarine. All told, during the four plus years that I headed SAD, I participated in about a dozen nuclear submarine noise trials.

During this period, my department also played a major role in the submarine Acoustics Intelligence (ACINT) program. During special ops conducted by our attack submarines in close proximity to the Russians, our subs used their own sonars to record the sounds radiated by Soviet subs. SAD was responsible for the calibration of the sonars, enabling the recorded signals to be expressed quantitatively whereby radiated noise levels of Soviet subs could then be estimated. On one memorable occasion in 1968, the USS *Dace* (SSN-607) returned from a special operation with the first photographs and acoustic recordings of two new Victor and Charlie class Soviet subs, but without having had its own sonars calibrated. Instead of returning to the New London Sub Base to the usual fanfare, the *Dace* docked a few miles down the river at the Navy's New London Underwater Sound Laboratory. Only the Skipper's wife knew that the sub had returned. I headed the multi-lab team that then boarded *Dace* and took her to sea to conduct the sonar calibrations, after which the sub arrived officially at the Base. The skipper of the *Dace* at that time was Cdr. Kinnaird McKee, who later became the Navy's youngest admiral and the successor to Adm. Rickover as Head of the Navy's Nuclear Propulsion Program. While with him on *Dace*, I was strongly impressed with the high state of morale as well as the capabilities of his officers and crew.

Project HERMAN

In late April 1970, President Nixon cancelled all special missions by attack submarines because of collisions with Soviet subs. On Saturday May 2, Adm. Dennis Wilkinson, as COMSUBLANT, called a group of us from several organizations to his office in Norfolk and explained the problem. We were to find and fix the cause of the collisions. Wilkinson assigned three 637-class submarines to us for two weeks. In my position as head of the lead lab, I was given operational control. Since the Admiral was to be in Washington on a selection committee, he supported my authority by giving me his home phone number, saying: "If any one questions your authority, have him call me. I'll set him straight." No one called.

That was on Saturday. On Sunday, I called a meeting of key personnel to organize the trial team and to plan the test agenda for what soon would become known as Project HERMAN, in memory of former Bell Labs team member Herman Straub. That same day, the *Whale* (SSN 638), the *Sunfish* (SSN 649), and a third sub, left New London for Charleston, which was to be the project's home base. There, the *Whale* was outfitted with special equipment to allow her to mimic the acoustic signature of a Soviet sub, and the *Sunfish* was prepared to act as the trailer. The third sub was to remain in port to be a backup should either of the other two have a problem. After two days, the *Whale* and *Sunfish* were ready to sail to a remote location in the Bahamas for five days of simulated trailing encounters.

Even before the beginning of the tests, we had a pretty good idea of what was causing the problem. The BQS-6 spherical sonar array was equipped with an auto-tracking system that gave the vertical as well as the horizontal direction of the strongest signal emanating from the contact.

While conducting trailing operations, a U.S. sub would quietly creep into the baffle area behind and presumably below the Soviet. Invariably the sonar was pointing up, implying that we were deeper than the Soviet sub. The problem was related to surface bounce. Actually, the strongest signal was not coming directly from the contact, but was coming from the surface. The trailing sub really didn't know where the Soviet was, and sometimes a collision would result. What we did in our tests was to confirm the problem and then to develop a procedure to overcome it.

On May 16, exactly two weeks after the original assignment, I was flown to Norfolk in the Admiral's own plane to present the test results and to help write a new op order correcting the problem. That afternoon, Admiral Wilkinson flew to Washington and briefed President Nixon at the White House, and the President approved the resumption of special operations. Mission accomplished.

Those two weeks of Project HERMAN were by far the most exciting of my entire career. In November 1971, at the instigation of Admiral Wilkinson, I received the highest naval citation that can be given to a civilian, specifically for leading this project.

How It Felt to Work on Submarines

While with Bell Labs, I had become a civilian submariner, fully accepted as such by the submariners with whom we worked. Because of the nature of our project, I had had to learn every system about as thoroughly as did the crew. I knew how to respond to emergencies anywhere on the sub. After the *Cavalla* trials, I was awarded a plaque with bronze dolphins, which I still treasure.

I have often been asked how it felt to be submerged for long periods. Wasn't it claustrophobic? The answer was of course "No!" Actually on numerous occasions when tests were over and the sub had surfaced, I stayed below to record data or to pack up equipment. I liked being on a submarine. The freedom from the telephone was great. One could concentrate on what he was doing without interruptions. I also appreciated the all-male society and the cooperative spirit that prevailed. I couldn't help but notice the sincere interest that the senior officers expressed in the problems as well as successes of each member of their crew.

When we weren't busy, we could join the crew to watch a movie, often shown backwards; or we could park in the wardroom where we would consume a bowl of ice cream and socialize. Each sub had its own game, usually played after meals, determined by the Captain. On *Cavalla* it was cribbage, while on *Nautilus* it was poker. I was quite proud that by the time *Nautilus* had arrived in Key West my accumulated losses at poker were nil.

When we were using Bermuda as our base, we got to spend some time ashore with the ship's officers. Often we would find a bar, where liar's dice would become the standard way of selecting the payee for each round of drinks. Our relations with the officers and crew were invariably good, being based on mutual respect. I was also very appreciative of the way the families supported each other when the men were at sea. After each of the Bermuda trials, the Bell Labs team would entertain the ship's officers and their wives at a restaurant in New London.

Another issue was seasickness. Submarines roll and toss quite badly in heavy seas, even when at periscope depth. Only once was I slightly queasy. I learned to shift my weight with the roll of the ship, and this carried over to dry land when we arrived in port. As a result, I was "land-sick" for a day each time that we came back from a week at sea.

Another question concerned fear of a fatal submarine accident. My answer was that the many times when we were operating out of New London, I would drive there from Whippany in Herman's Volkswagen Beetle. Like CDR Wilkinson, Herman knew only one speed: "full speed ahead." After driving with Herman, being on a submarine felt relatively safe.

....The radar detector in your car is rigged up to look like a Biscay Cross.

Chief Mac and the Contact

Dave Stoops

Now this is no bull....

It had been a long, slow patrol. We were on the last legs of the second patrol of a summer Westpac. Although I was a nuke electrician, I had been assigned to the forward IC division as a result of shortages of qualified forward electricians and had been charged with getting two new non-quals trained to stand the auxiliary electrician forward watch. I enjoyed the AE Forward watch — I had the run of the whole ship, and duties were minimal. I had to take atmosphere readings once a watch and blow sanitaries when we came up to periscope depth. It was a real treat to talk the cook out of a half loaf of fresh hot bread as it came out of the oven on the midwatch. (This was the only time the bread was edible — by breakfast time it had dried out to the point that as far as I was concerned, it was only good for toasting or patching holes in the deck). When we picked up a contact, I would also become part of the plotting party.

The Ballast Control Panel watch was Chief Mac (name changed to protect the guilty) and Mac was generally sharp as a tack. He was a career torpedoman, had high standards, and had a private pilot's license to boot. One day while he was on watch, he asked the OOD if it would be all right if he qualified OOD (Officer of the Deck). The OOD told him that he would ask the old man and get back to him. Sure enough, on the next watch, the OOD told Mac that the skipper had thought about it. He said that while Mac could never really stand the watch as an enlisted man, the fact that he qualified would look good on his already impressive record and therefore, it was fine if he wanted to qualify.

Mac went after the opportunity with zest and enthusiasm. Whenever something was happening, Mac would request to relieve the OOD under instruction and would assume the watch while the OOD would sit at the Ballast Control Panel. Mac generally performed without error.

One day, after cruising submerged to a new patrol location, we were to come to periscope depth, shoot the garbage, get the sched, blow sanitaries, and everything else that needs to get done. Mac needed the procedure for his qualification and requested permission to relieve the OOD under instruction. Permission was granted, and soon Mac was barking out commands to the helmsman and planesman as to course and depth.

At the proper time, Mac yelled out, *Up Periscope!* and the quartermaster promptly threw the valves to bring the scope up. As it rose, Mac grabbed the handles smartly, wrapped his arms around the scope in true John Wayne fashion, and began his 360-rotation to look for other ships in the area. He didn't get very far before he yelled, "Gotta contact! Mark this bearing!"

Of course, at this time, the whole control room came alive. We finally had a contact after weeks of inactivity. I had been sitting at the fire control panel, but got up and walked over to the plotting table, opened up the contact log book, made sure all the equipment was up and running and the input to the bearing indicator was selected to the scope. Mac continued to report his find.

"Surface ship. Riga class warship! Estimated range....*5000 yards*! Estimated speed, 10 knots."

The OOD was just as eager to see the target as the rest of us, but Mac was in the spotlight and wasn't about to give the scope up so soon. He reported an angle on the bow, heading, etc., and when he had milked the situation for all he could, he stepped back and proudly offered the scope to the OOD. The OOD bent over and squinted into the scope. He was quiet. He flipped the scope to high power. He flipped the optics back to low power. He swung a little to the left, then to the right. Finally, the silence was broken:

"Dammit, Mac! That's a *#%*$*# *rock formation!*"

To this day, our ship's contact log will tell you that V-208 was a rock formation.

All ships can dive, only submarines can surface.

Ron Martini

Ron Martini is a retired submarine veteran and one of the pioneers in the submarine website genre. His "Ron's Submarine website" and BBS were among the first on the web and certainly became the best-known sub-related website of the nineties. I first became aware of Martini's website from the "Special thanks" section of the Aces of the Deep manual. I had the pleasure of working with Ron during the initial *Cavalla* restoration effort. Ron held an auction on his BBS and raised $2200 — which was about $2000 more than we had up to that time. Subsim pays homage to a submarine website trailblazer.

--Neal Stevens

Ron Martini was born six months before Pearl Harbor and raised in beautiful Sheridan, WY at the foot of the Big Horn Mountains. Ron put in a half year of junior college, then talked his cousin into joining the Navy.

Ron attended boot camp and EM "A" School in San Diego then journeyed cross-country to Sub School in the winter of 1961. He caught up to his cousin again in San Diego, where he spent eleven months on the *USS Catfish* SS-339. It was the boat that was later sold to Argentina and was lost by them in the Falklands War.

In early 1962 Ron received orders to Nuclear School at Vallejo. The same guys he went to "A" school with all gathered there once again. Ron spent six months there and then six months in Idaho at the A1W prototype.

Ron requested orders for an SSN out of Hawaii and the Navy obliged him with orders to the *USS Patrick Henry* SSBN-599 Blue crew in Groton to start 1963. The young Ron put in six patrols on her and returned her to the yards at EB for core change and missile upgrade in 1965. Back to school, Ron was sent to EM "B" school from there. Just prior to checking in at

Great Lakes, Ron married his wife (also from Sheridan) and they set up residence in Zion, IL.

He received orders from there to new construction, Portsmouth and the *USS Grayling* at the end of 1965. Ron was the only one there and after five weeks, they called him in and said he was way too early! He was handed orders to the USS *Patrick Henry* Gold crew, which really pleased him. Ron did seven patrols on her and then "walked out the gate after eight" in 1968.

He worked for Safeway Stores (food chain) from 1968 until 1997, retiring back to his hometown Sheridan, Wyoming.

Ron took up computing around 1982-3. He played with Sinclairs, Commodores, and settled on a Tandy 1000SX in 1986. He had the biggest hard drive in Sheridan at 80 MB. He began attending meetings of a computer user group there and in '88 broke off from them (they were Apple people) and formed SMUG: Sheridan Microcomputer Users Group. They built a shareware library of 5 ¼" disks and checked them out like library books. At one time there were some 300 members.

In 1994-5, a local computer storeowner called Ron and said he was bringing the Internet to Sheridan and he would give Ron space for a webpage and a bulletin board free if it would bring people into his site.

The first page was located at www.rontini/wavecom.net. In those days, there were only six other homepages about submarines on the Net. One was authored by an instructor at Sub School that folded after pressure from the Navy because he had included some shots of DBF systems. Another homepage was out of Connecticut, which lasted five or six years. Another by a gent nicknamed Hot Rod Rodriquez in Kings Bay, who was a Lt. Cmdr.

Yet another webpage was on the *Sturgeon* boats. The author was a non-submariner with the passion and he told Ron once he had ONI come to his home and question him. It appears he guessed pretty well the speed and depth on the *Sturgeon* boats and they didn't like it. Then there was the granddaddy of them all, Don Merrigan's Silent Service page. He had info on all the boats and was the seller of Jim Christley's Submarine Force Data Book that is still the bible of all boats up through to the *Virginia*. Don was also Nat. Jr. Vice Cmdr. of USSVI and *American Submariner* editor for a time. Don's page folded tents in 2005. Ron followed Don Merrigan's lead and started his Submariner's BBS in 1996. That makes Ron's website the oldest page of the fleet with over 1000 links to all things submarine.

Ron's Submarine World Network has drawn attention from Navy Times, Houston Chronicle, CNO's offices, CSL offices, NavSea offices, and the New London Day, among others. The Navy Times did two articles drawn from the BBS. The first was a humor bit about how to "live like a submariner at home" (actively maintained by Ron) and the other concerned a problem that developed with giving Midshipmen silver dolphins. Ron proudly points out that he played a major role in getting that practice stopped.

In late 1996, Gil Raynor approached Ron about starting an online submarine memorabilia shop. Merchandise grew from the first mouse pad and 3 1/2" disk labels to over 400 items at the shop at: submarineshop.com

Ron says, "My impetus for doing the page goes back to the wonderful time I had throughout my eight years on the boats. Never regretted a minute of it and that includes the first two minutes of missile alarms before the CO said it was a drill. That was two minutes of terror, especially during the mid-60s and the Cuban Missile crisis, Kennedy's assassination, and the Israel-Egypt war that took us into the Med for our only time."

Ron retired in 1997 and the long winters in northern Wyoming are wonderful for keeping up with his interest in submarines and their history. Thanks to Ron's efforts, a USSVI Submarine Library was established at the Arkansas Inland Maritime Museum in North Little Rock in 2006.

His secret? According to Ron, "A labor of love contains no work."

You know you've been with Subsim a long time when...

You've posted more topics in the Radio Room forums than XabbaRus!

Cavalla Makes its Mark in Naval History

Capt. Ernest J. "Zeke" Zellmer

After the devastating attack on Pearl Harbor, the Imperial Japanese Navy sank the pride of the British Pacific Fleet, the battleships *Prince of Wales* and *Repulse*.

In a span of mere months, the Japanese stormed Hong Kong, British Burma, Malaya, Borneo, Singapore, and the American-held Philippines. Admiral Yamamoto and the Japanese navy appeared unstoppable.

The Battle of Midway turned the war in the Pacific around, followed by the U.S. invasion of the Solomon Islands at Guadalcanal and the Allied thrust into New Guinea. The Japanese position was weakened by serious shortages of oil in the home islands, where the fleet was stationed. U.S. submarines had destroyed much of the Japanese tanker fleet. However, unlimited supplies of oil were available in Borneo.

The Japanese strategic position remained strong because of bases in the Bonin, Mariana, Paulau, and Caroline island chains and in the Bismarck Archipelago. These islands provided "unsinkable aircraft carriers" extending in an arc south and eastward from Japan to New Guinea. Intelligence on the opposing forces was critical to both sides. Each had similar tools: air search, submarine reconnaissance and communications intercept.

Air search from the "unsinkable carriers" proved to be most useful to the Japanese. Beginning in mid-May, Japanese submarines were strategically stationed to warn of U.S. force movements.

By the time of the Battle of the Philippine Sea on June 19, the Japanese had lost seventeen of those submarines. In a remarkable two-week period, the *USS England*, a destroyer escort, sank six Japanese submarines. (The *USS Stewart* in Sea Wolf Park, Galveston, is of the same destroyer-escort class.)

U.S. submarines were not sitting ducks for the Japanese destroyers. During May and early June, they sank eight Japanese destroyers and damaged at least two others. *USS Harder*, in three days, sank three destroyers and damaged two. The Japanese Fleet grew short of escorts.

The Japanese reorganized their fleet under Admiral Ozawa. It was designated the Mobile Fleet and would be based near the vital source of oil.

On May 11, the elements of the fleet left the Inland Sea and steamed south toward Tawi Tawi, just east of Borneo. On May 16, the *USS Bonefish* reported six carriers, four or five battleships, three cruisers and many destroyers at Tawi Tawi.

Admiral Spruance, commander of Operation Forager, and Admiral Marc Mitscher, Commander of Task Force 58, learned where the Mobile Fleet would be based. On May 27, MacArthur invaded Biak Island at the western end of New Guinea. That same day, Japanese planes scouted Tulagi, where the U.S. Southern Force was staging for its part in the Marianna invasion. At Kwajelein and Majuro, the Japanese planes saw the main invasion force.

The stage was set. On June 11, the curtain was raised as elements of Task Force 58 attacked Saipan, Tinian, and Guam. Vice Admiral Kakuta's Base Air Force was greatly hurt, but reports to his superiors were optimistic and misleading. His lack of candor contributed to Admiral Ozawa's eventual defeat.

On June 12, a naval force that had been sent south toward Biak was directed to head for the Philippine Sea to rendezvous with the Mobile Fleet. Ozawa departed Tawi Tawi heading north. His departure was reported by U.S. submarine *Harder* and *USS Redfin*. On June 15, Ozawa exited San Bernardino Strait (in the central Philippines) and was reported by U.S. submarine *Flying Fish*.

The Biak relief force heading for the rendezvous was detected 200 miles east of the Philippines by *USS Seahorse*. Admiral Spruance knew the Japanese were coming to fight. Locating the opposing fleet frequently was vital. In the open sea, a task force can move 500 nautical miles a day in almost any direction.

From this time on, the Japanese would know, almost continuously, the location of Task Force 58. It would be within search range of the "unsinkable carriers." Ozawa was still beyond air search capability of Task Force 58. ComSubPac (Commander Submarines, Pacific Fleet at Pearl Harbor) ordered his submarines to search for the Mobile Fleet.

Seawolf Park's own memorial submarine, the *USS Cavalla*, and other subs were directed to scout across the estimated track of Ozawa. Just before midnight June 16, *Cavalla* spotted a convoy of two tankers and three destroyers. She sped ahead of the convoy and dove. *Cavalla* was detected and forced deep as the convoy passed by. A destroyer searched for *Cavalla* for about a half-hour. When clear, *Cavalla* surfaced, made her report and, with no contact on the convoy, decided to resume her search.

ComSubPac felt the convoy was a replenishment group heading to a rendezvous with the Mobile Force. *Cavalla* was directed to pursue.

After an unfruitful pursuit, *Cavalla* received new search orders. At 7:57 p.m. June 17, *Cavalla* made radar contact at 30,000 yards. She found the range closing rapidly. At 15,000 yards, a large carrier could be seen visually, and *Cavalla* submerged to hunt beneath the waves. Radar had seven large blips while sonar heard at least fifteen different ships. *Cavalla's* captain, Commander Herman Kossler, made the difficult decision to abandon the attack and carry out ComSubPac's general orders: *Cavalla* would track, watch and report as soon as she could surface.

By 9:30 p.m., the main body had passed, but two escorts remained and delayed *Cavalla's* surfacing. At 10:45 p.m., *Cavalla* radioed its report and started her second chase. The task force was moving at *Cavalla's* maximum surface speed; it was not to be overtaken. That left the *Cavalla*, on its first patrol, with a frustrated crew. Two opportunities of a lifetime, and we had yet to sink a ship. ComSubPac gave his submarines new orders: Shoot first, report later.

After relocating Task Force 58 on the afternoon of June 18, Ozawa decided to attack the following morning. He would strike Mitscher while the Mobile Fleet remained beyond Mitscher's range. Kakuta's planes searched through the night and made intermittent contacts with Task Force 58. Several of the Japanese snoopers were shot down.

Very early on June 19, an American long-range seaplane from Saipan made radar contact on a large number of ships in two groups. Then the fog of war descended — the plane's radio report never got to Spruance or Mitscher until midmorning when the plane returned to base. Well before dawn on June 19, both forces launched search aircraft. About 7:30 a.m., the Japanese sighted Task Force 58, and Ozawa prepared to attack. The American forces were nearly 400 miles distant. Ozawa changed course to maintain that distance; his planes could now hit Mitscher's forces, but Mitscher could not reach the Japanese fleet.

About 8:30 a.m., the Japanese launched the first of four strike waves. Just before 10 a.m., the first wave was detected 140 miles from Task Force 58. Mitscher's fighters flew out to meet them. The air battle raged; only a few of the Japanese aircraft broke through the fighter screens to face the guns of the fleet. Attacks by Ozawa were over by midafternoon. Mitscher's planes also had to fight a string of small attacks by Admiral Kakuta. The American defense decimated the Japanese planes, but Ozawa's ships eluded Task Force 58. Ozawa lost 250 planes and Admiral Kakuta lost another fifty. Mitscher lost twenty-nine planes, but many of the aircrews were rescued. Only the *South Dakota* was hit by a Japanese bomb; it continued operations. Ozawa did not escape without serious ship losses.

As the strikes were being launched, the submarine *USS Albacore* made contact with the Mobile fleet. It was moving rapidly as it launched planes of the second wave. Just as *Albacore* was ready to fire from 2,000 yards, the torpedo data computer failed. Capt. Blanchard had to react quickly. A wide spread of six torpedoes was launched with last-second bearings cranked in by hand. One torpedo hit the *Taiho*, the flagship of Admiral Ozawa. *Taiho* continued on with the fleet. (*Taiho* was Japan's newest and largest carrier.) *Albacore's* crew felt a huge disappointment; a single hit was unlikely to sink a carrier. But fortune favored the Americans. *Taiho's* damage-control teams should have been able to contain the flooding, repair broken gasoline lines and put out the fires. They mishandled the crisis. An officer ordered fans to carry the gasoline fumes out of a hold. The vapors spread throughout the ship and, inevitably, a spark caused a massive explosion that sank the carrier.

Later that morning, it was my submarine's turn to get lucky. At 10:12 a.m., an enemy plane closing on *Cavalla* caused her to submerge. Four small planes were seen through the periscope and sonar heard propeller noises on their bearing. *Cavalla* headed that way to investigate. Ship masts appeared, and *Cavalla* went to battle stations. Four ships were visible: a carrier with two cruisers on its port bow and a destroyer on her starboard beam. The destroyer was in position to cause trouble. Positive identification of the ships was important. We did not know the whereabouts of Task Force 58 and could not risk hitting one of our own carriers. The carrier, the *Shokaku*, was large and similar to U.S. carriers, especially when viewed head on through a periscope that dipped beneath the waves occasionally. The last look before reaching the firing position gave proof positive - the Japanese naval ensign was visible.

Six torpedoes were launched. Three tore into the *Shokaku*. The close, unfriendly destroyer dropped four depth charges as *Cavalla* went deep. *Cavalla* ran as quietly as she could. Because the main induction line (the 36-inch diameter pipe through which the engines got air when *Cavalla* was on the surface) was flooded during the initial depth-charge attack, *Cavalla* was very heavy. The submarine increased speed (and noise) to maintain depth control. Before being able to check her descent, *Cavalla* was nearly 100 feet below her test depth and we hoped (prayed!) the safety factor would keep the hull from imploding. Three destroyers continued to hunt for *Cavalla*, dropping 106 depth charges. Over time, *Cavalla* drew away from them.

At 2:18 p.m., four large explosions were heard in the direction of the attack. Sonar reported loud, rumbling noises for many seconds thereafter. A jubilant crew believed the explosions spelled the end of the carrier. *Cavalla* surfaced and triumphantly reported to headquarters. The carrier that had attacked Pearl Harbor lay under the sea.

Task Force 58 spent June 19 in a series of defensive actions that decimated Ozawa and Kakuta's air power. Admiral Mitscher then proceeded westward at best speed to engage the mobile fleet. Ozawa, still optimistic, retired toward a refueling point to reorganize his forces. Before dawn on June 20, both fleets sent out search planes, but neither was successful. At noon, Mitscher launched another search, this time equipped for extended range. Two searchers made contact reports about 4 p.m. Mitscher decided to attack even though the range was extreme and it would not be possible to recover returning aircraft before dark. The American planes reached Ozawa's fleet with about a half-hour of daylight left. While Ozawa had few fighters left, his ships' anti-aircraft fire was intense.

When the attack was over, all of Ozawa's three divisions and the replenishment group had been hit. He lost a carrier, the *Hiruna*, two tankers and more planes, and had two carriers, a battleship and two cruisers badly damaged. At the start of the battle, the Japanese had 550 planes and nine carriers. The U.S. had 950 aircraft and fifteen carriers. The Japanese were outmaneuvered, outgunned and outfought.

At the end of the battle, the Japanese had lost more than 400 planes, three carriers, and two tankers and had major damage to several ships and many planes. Japan would lose the Marianas Islands, which the U.S. would use to base B-29s to strike mainland Japan. U.S. naval air and surface forces had won a remarkable victory in the "Turkey Shoot of the Marianas."

Originally published in the Galveston Daily News. Heber Taylor, Editor: "June 19, 1999 on the 55th anniversary of the decisive day of the First Battle of the Philippine Sea, a pivotal battle that crushed Japanese naval power during World War II. The USS *Cavalla* is on display in Galveston's Seawolf Park. This account of the battle was written by Zeke Zellmer, who was on board the *Cavalla* as its communications officer."

The Flanders U-boat Flotilla 1915 - 1918

D. Swetnam, MA

The Germans never intended to carry out commerce warfare with submarines during World War I. The concept of operating them out of an advanced base in Flanders was probably never even contemplated. It was the failure of the Germans' initial strategy which led to the occupation of Flanders and the use of the ports. Allied negligence in the evacuation of the ports made the setting up of an advanced base there that much easier. The establishment of the flotilla in March 1915 led to a long and destructive campaign and much effort to contain them. The Flanders U-boat flotilla (1915-1918) has not been dealt with discretely by a British researcher. Mostly the activities of the flotilla became overshadowed by the High Seas Fleet flotillas. Only in the Naval Staff Monographs (Historical) are the flotilla's activities separated from those of their High Seas Fleet comrades. As I researched deeper into the subject, I realized that the Flanders boats deserved to have their story told separately. Once they were established in their safe haven of Bruges, the U-boats were able to proceed stealthily into the Channel and North Sea and eventually the Bay of Biscay to attack Allied trade and fishing fleets.

The Germans, faced by hostile powers France and Russia, bound by treaty, feared a war on two fronts. In 1906 General Schliffen formulated a plan to defeat France quickly as they had done in 1870, so the German army could then be transferred to the Eastern Front to deal with the Russians. In order to enter France quickly, they had to go through Belgium. This brought Britain into the war under the terms of a treaty of 1839. The combined British and French armies slowed the German advance, and as the Germans neared Paris they had over-extended their supply lines and slowed the advance even more. The Allies now regrouped and mounted an attack on the Marne which halted the German advance. After withdrawing to the Aisne to regroup, the German northern flank headed towards the coast.

French and British troops were then rushed into Belgium through Dunkirk, Ostend and Zeebrugge in early October 1914,[32] but they were allowed no time to organize before the Germans assaulted Antwerp and drove them back out through the ports at which they had landed week before. During the evacuation they committed a major strategic blunder. They left the port facilities intact, and most importantly, they failed to destroy the locks of the canals leading to Bruges.[33] The Admiralty had wanted to destroy them, but it was vetoed by the military authorities, who wanted them to be kept intact as they expected an early return. This had happened after a previous evacuation in September, but this time the military situation was far worse, and this should have been recognized. [34]

Once on the coast the Germans came within range of naval gunfire and so Rear Admiral Hood assembled a scratch force consisting of the obsolete cruisers *Brilliant* and *Sirius*, destroyers, the trials gunboats *Excellent* and *Bustard*, sloops *Rinaldo* and *Wildfire* and the requisitioned Brazilian river monitors *Humber*, *Mersey* and *Severn* to bombard them.[35] This was so effective that at first the Germans considered the Flanders ports as being untenable. The British army command quickly realized their error and on 23 October, 1914 the Dover command was given the authority to destroy Ostend. An advance by the French army delayed the operation and on the 24th the bombardment squadron was diverted to support them. Encouraging progress made by the French delayed the operation until the Germans mounted a massive counter-attack which drove the French and Belgian forces back. Bad weather denied the troops any naval support and disrupted the bombardment of Ostend. The Germans bought up heavy guns which were positioned in the sand dunes, and they made approaching close inshore dangerous.[36] These guns were later augmented by guns stripped from obsolescent warships, a process which was to last into 1916. They were manned by a German Marine division who were garrisoned on the Belgian coast from the latter part of October 1914.

U24, *U27* and *U30* left the Ems on 22 October 1914, *U19* on the 24th and *U28* on the 26th for operations off the Belgian coast. The Harwich Force supported by elements of the Grand Fleet were also at sea. On the

---

[32] Corbett J 1938 pp 186,187

[33] ibid. pp 214, 215

[34] ibid. pp 222, 223.

[35] Corbett 1938 p222

[36] Corbett 1938 pp228/9

night of the 24th *Badger* (Cdr Fremantle) rammed *U19* (Kolbe) off the Dutch coast, forcing it to put in to Zeebrugge for repairs.[37] If the Germans needed a demonstration of how useful Zeebrugge could be to U-Boats, this provided it.

On 15 November a second Marine division was sent to Flanders and the whole force was renamed the Marinekorps Flandern.[38] Admiral Ludwig von Schröder was appointed Supreme Commander Flanders on 3 September, 1914,[39] directly responsible to the Kaiser.[40] This was to give the Flanders units a high degree of autonomy.

Fearing a landing, U-boats were sent to Zeebrugge from the Ist and IIIrd flotillas of the High Seas Fleet. *U12* (Forstmann) arrived first on 9 November and sank *Niger* (Lt Cdr Muir) off Deal Pier on the 11th.[41] *U11* (Suchodoletz), *U5* (Lemmer) and *U24* (Schneider) followed. *U11* was lost off Ostend on 9 December and *U5* on the 18th, both victims of mines. *U24* arrived on 23 December and sank *Formidable* (Capt Lorly†) off Portland on 1 January 1915.[42] *U12* and *U24* returned to Wilhelmshaven on 20 January 1915.

A second wave was dispatched in January to act against the cross-Channel troop transports. *U7* (König†) set off first, but was sunk in error by *U22* (Hoppe) on the 21st. *U29* (Plange) followed, but was damaged in a collision at Zeebrugge and when Plange fell ill, the boat was returned to Wilhelmshaven. *U14* (Dröscher) arrived at Zeebrugge on 1 February and was damaged in an air attack on the 2nd and sent to Bruges to be repaired. Further air attacks followed and as the boat was proceeding along the canal to Ostend, it was strafed and nine crewmembers were injured. *U10* (Stuhr) sailed for Flanders in January, but turned back.

In November 1914 the CinC France, Sir John French, inaugurated a plan to outflank the German army with an amphibious assault on the Belgian coast. The Admiralty decided to include an assault on Ostend and Zeebrugge. The army concurred as they considered the U-boat base to be a serious threat to their cross-Channel communications. However, in January

---

[37] Gibson & Prendergast 1931 p15.

[38] Ryheul J. *The Flandern U Boat bases and U Bootflotille Flandern.*

[39] Scheer 1920 p 345.

[40] TNA :PRO ADM 137/3876.

[41] Ibid. p23

[42] Ryheul J. *U Bootflotille Flandern*, Grant 1969 p183, 2002 p 20

1915 Kitchener stated that the army had insufficient guns and ammunition for such a venture and he possibly was worried about the state of amphibious warfare which had not advanced since Nelsonic times. Finally obstructionism by General Joffre caused the cancellation of the joint operation on 28 January. It now became a purely naval operation. Admirals Jellicoe and Wilson were willing to undertake this, but the threat of losing old battleships caused Fisher to reject the proposal outright. Consequently the Germans were left to develop their new base virtually unopposed.

During this period the Germans had developed small coastal attack and minelaying submarines. These were ordered in October and November 1914 from Germaniawerft, Kiel (*UB1 – 6*), A G Weser, Bremen (*UB7–17, UC11-15*) and Vulcan, Hamburg (*UC1-10*). The continuing bombardments by the Dover Patrol made the German Admiralty reluctant to use the Flanders ports as a submarine base and it informed von Schröder of this decision on 11 February, 1915. However the mounting of 50mm, 88mm and 105mm guns on the Belgian coast was providing a measure of protection and there were plans in hand to mount guns as large as 380mm. The German Admiralty reversed its decision on 4 March, 1915.[43] German engineers were sent to Flanders to start preparing the three ports as U-boat bases. Korvettenkapitan Karl Bartenbach was made commander designate of the U-boat flotilla.

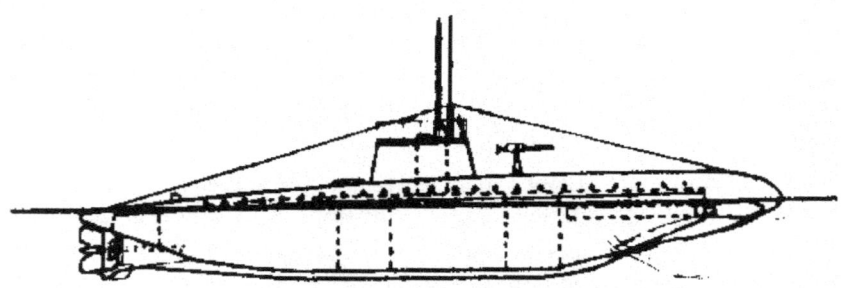

UBI class built by Germaniawerft and Weser

Source: *Gröner German Warships 1815-1945; Used with permission*

The *UB1* class coastal attack submarines, carrying just two torpedo tubes, were sent in sections by rail to the Hoboken yard at Antwerp for reassembly.[44] Here they were attacked by RNAS aircraft three times

---

[43] Ryheul J. *The Flandern U Boat bases of Bruges Zeebrugge and Oostende and U Bootflotille Flandern*

[44] Ryheul, J *Flandern U Boat Bases* and *U Bootflotille Flandern*.

between 12 March and 1 April, 1915. A French intelligence report claimed two submarines destroyed, one damaged, with forty-two Germans killed and sixty-two wounded.[45] However, as all the designated boats arrived at Zeebrugge the first part of this claim can be discounted and the German records contain no mention of these raids. *UB10* (Olt Steinbrinck) was the first boat to arrive on the 27 March, followed at intervals by *UB4* (Olt Gross), *UB10* (Olt Steinbrinck), *UB5* (Olt Smiths), *UB12* (Olt Nieland), *UB6* (Olt Haecker), *UB13* (Olt Becker) and finally on 10 May by *UB2* (Olt Fürbringer) and *UB17* (Olt Wenninger). *UB2* had not been dismantled and reassembled as by now this was considered too time-consuming.[46]

The first patrol was made by *UB4*, which sank the Belgian Relief steamer *Harpalyce* on 10 April, despite these vessels being granted a safe conduct by the German Admiralty. *UB10* sailed on the 13th and torpedoed the Dutch steamer *Katwijk* on the 14th. *UB5* was on patrol at the same time and sank the British steamer *Ptarmigan* on the 15th. On her second patrol *UB4* sank the Greek steamer *Ellispontos*, all these sinkings taking place off the Noord Hinder Light Vessel [47] and all without warning, which was contrary to International Law and the rules of engagement of the current Restricted U-boat Campaign.

UB6's first patrol had an unexpected consequence. On 1 May, *UB6* sank the destroyer *Recruit* in a torpedo attack off the Galloper Light Vessel and withdrew. The Admiralty made deployments to intercept *UB6*. Four trawlers on patrol off the Hinder were alerted and four destroyers were sent from Harwich to support them. The trawlers were attacked by the torpedo boats *A2* and *A6* from Zeebrugge. *Columbia* was sunk by torpedo; *Barbados* and *Chirsit* were damaged by gunfire. *A2* and *A6* were intercepted off the North Hinder by the Harwich destroyers *Laforey* (Cdr Edwards), *Leonidas* (Lt Cdr Grubb), *Lawford* (Lt Cdr Scott) and *Lark* (Lt Cdr Hughes-White). In the ensuing action the torpedo boats were disabled and sunk. *UB6* returned to Zeebrugge without further incident on the 2nd.[48]

The first minelaying boat, *UC11* (Olt G. Schmidt) joined the flotilla on 26 May and laid the first minefield off the Goodwin LV on 31 May. It claimed its first victim the following day when the destroyer *Mohawk* was damaged. This was to be the start of conveyor belt minelaying operations

---

[45] TNA :PRO Air 1/2099

[46] Fürbringer 1999 pp11-14

[47] Ryheu J, *Flandern U Boat Bases* & *U Bootflotille Flandern*, Lowery & Dufeil M/S

[48] Gibson & Prendergast 1931 p 39, Corbett J 1920 p401

between Folkestone and Lowestoft, which caused the loss of two hundred and twenty vessels. *UC1* (Olt von Werner), *UC2* (Olt Mey), *UC3* (Olt Weisbach), *UC5* (Olt Pustkuchen), *UC6* (Olt von Schmettow) and *UC7* (Olt Wäger) joined the flotilla between 25 June and 12 August.[49] Time spent at sea in the *UC1* class was very exhausting as the crew could get very little sleep.[50] The patrols generally lasted three days, which was probably as much as a crew could endure and still remain efficient.

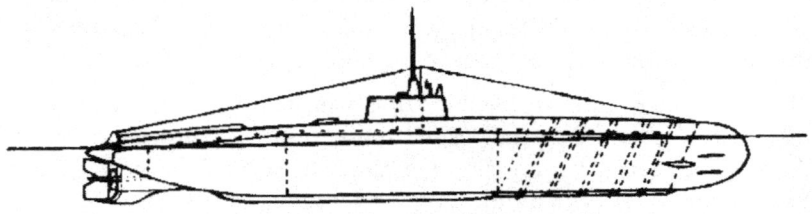

UC1-15 Built by Vulcan (*UC1-10*) and Weser (*UC11-15*)

Source: *Gröner German Warships 1815-1945; Used with permission*

The Admiralty could not account for the appearance of these small minefields that appeared around the coast until *UC2* became the victim of one of her own mines and the wreck was examined by divers.[51] (*UC9* (Olt Schurmant) and *UC68* (Olt Detetaut) of the flotilla were also lost in this way as well as other boats from the High Seas Fleet and Mediterranean flotillas.[52]

In early 1915 the Admiralty had strung a net between Dover and Cap Gris Nez. This net led to the sinking of *U8*. Other boats had become entangled and escaped with difficulty. The unexplained loss of the new *U31* and *U37* in the Dover straits at this time led the German High Command to ban the use of the Straits of Dover, but Bartenbach was never happy about this restriction. When *UB6* found a passage through the nets, he sent *UB2*, *UB5*, *UB6* and *UB10* to operate off Le Havre, Boulogne, Folkestone and Dover. None of the boats achieved any successes and *UB2* broke down on the return passage and literally drifted through the gap in the nets with the tide. Despite this, Bartenbach put the restriction back in place.

---

[49] Gibson & Prendergast 1931 p38, Ryheul, *Flandern U Boat Bases* and *U Bootflotille Flandern*.

[50] TNA:ADM 137/3876

[51] Grant 2002 p25

[52] Grant 2003 pp45-47.

UC11 laid a minefield off the Sunk LV on 9 June. *Lady Salisbury* (1446) sank on one of them soon after. *Brazen* and *Vulture* attacked a periscope near the spot without success. Three destroyers were sent from Harwich to assist in the hunt and *Fervent* was missed by a torpedo. This could not have been *UC11* as it was not fitted with torpedo tubes. The number of destroyers was increased to twenty-three. *Erna Bolt* (1731) sank in the Sunk field that evening. This increased anti-submarine activity did not prevent *UB2* (Olt Fübringer) from sinking the fishing smacks *Edward* (52), *Qui Vive* (50), *Britannia* (43), *Welfare*(45), *Laurestina* (48) and *Intrepid* (59) off the Galloper L.V. on the night of the 9th/10th.[53]

On the 10th the Nore Command put in force a large anti-submarine hunt, consisting of five destroyers, six torpedo boats and Auxiliary Patrol trawlers. Early that morning *TB12* ran into the Sunk minefield and her bow was blown off. *TB10* came to her assistance and took the wreck in tow, but struck another mine and was blown in two, the parts sinking in a few moments. An anti-submarine hunt by *Vulture* and other destroyers failed to find anything and attempts to salvage the wreck of *TB12* were unsuccessful. These events illustrated the Royal Navy's inability to deal with the U-boats and Fürbringer's *UB2* was the first of the Flanders boat to attack and sink fishing craft off the east coast.

Now confined in the North Sea, the flotilla turned on the fishing fleets. The *UB1* class, lacking a deck gun, was ill equipped to attack these vessels; instead the vessels were destroyed by explosive charges. The Scottish fisheries were also being attacked by the High Seas Fleet flotillas. These attacks had to be stopped as anxieties were being expressed over the supply of fish.[54] The Royal Navy employed decoy trawlers which towed a submerged submarine instead of a trawl net. This scheme accounted for two U-Boats off Scotland, but when the scheme was tried further south *C33* (Lt Carter) and *C29* (Lt Schofield) were lost with all hands when they strayed into minefields.[55]

Then four smacks were taken up and armed with concealed three-pounder guns.[56] Their first encounter came on 11 August when the decoy smack *G & E* was attacked and hit *UB6* (Olt Haecker) with a single shell. *UB6* returned to Zeebrugge on 13 August too late to warn Olt. Gross of

---

[53] Monograph vol. v p254

[54] Corbett 1940 p129 & Gibson & Prendergast 1931 pp44.

[55] Evans 1986 pp 42 – 43.

[56] Dittmar 1972 pp126 – 129.

UB4.[57] *UB4* had possibly already sunk three fishing smacks when on 15 August, 1915 the armed smack *Inverlyon* (Skipper T Philips) was approached off Smith's Knoll. The U-boat was allowed to come within thirty yards' range before the gun was exposed and fire commenced. Hits were scored and the U-boat quickly sank with all hands. *UB2* (Olt Fürbringer) nearly became a victim on 23 August when he attacked the decoy *Pet*. *UB2* and *UB16* (Olt H. Valentiner) left Zeebrugge on 6 September[58] to act against these decoy smacks. *UB16* had a brief engagement with *Inverlyon* and then missed the vessel in a torpedo attack. After this, sinkings of fishing vessels ceased until *UB16* sank three smacks off Lowestoft on 18 January 1916.[59]

These activities were carried on against the background of the restricted submarine campaign which was mainly carried out by the High Seas Fleet boats in the western approaches. Ironically, the ebb and flow of the U-boat war was determined, albeit indirectly, by a nation that was not even a combatant. It was fear of America entering the war on the Allied side that ended the restricted campaign of 1915 on 20 September. However, in November the Flanders flotilla was given permission to resume attacks on ships entering ports between Le Havre and Dunkirk. *UB10* passed through the Straits in December and sank *Belfor* and *Huntly* off Boulogne on 20 December[60] and the Flanders Flotilla continued the trade war alone.

In late 1915 Admiral Bacon started to plan an attack on Ostend using monitors and trawlers to carry troops into Ostend harbour. However, whilst planning was in an advanced stage, the Germans began building a new battery at Knocke which would have enfiladed the landing sites at Ostend and so the attack was cancelled.[61] One flaw in the plan was that Zeebrugge would be left unmolested so that there was nothing to stop the Germans from attacking the transports of the second and third waves in mid-Channel as they advanced from Le Havre and Dover.

The larger *UBII* class boats were ordered in the spring of 1915 and completed November 1915 – August 1916.[62] They suffered from a low

---

[57] Corbett 1940 p129.

[58] In his autobiography Fürbringer has his dates confused. He claims that this operation occurred on *UB2*'s previous patrol which took place between 28 and 30 August 1915 (Lörcher in correspondence 2001)

[59] BVLS, Lörscher, Lowrey & Dufeil.

[60] Gibson & Prendergast 1931 p59 & 61.

[61] Bacon 1919 pp209 – 222

[62] Gibson & Prendergast 1931 p64.

surface speed of nine knots, with a cruising speed of five to six knots[63] Despite this, they became very popular with the Flanders officers.[64] The UBIIs had a better diving time and were fitted with a hatch between the conning tower and inner hull, making then safer in the event of any damage to the conning tower.

The first of these boats, *UB18* (Steinbrinck), *UB19* (Becker) and *UB29* (Pustkuchen) joined the flotilla in February and March 1916. Now the zone of operations was extended as far as Southampton and Le Havre.[65] The *UBI* class, which had never passed a line Dungeness–Boulogne, were relegated to outpost and training patrols.

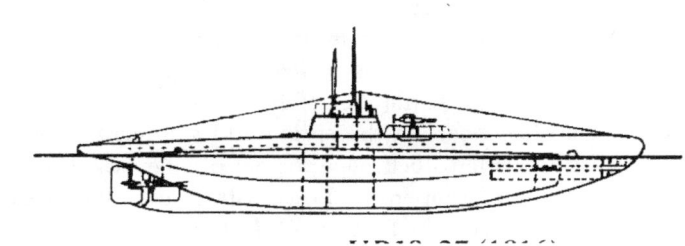

UB18-27 Built by Blohm & Voss 1915-16

Source: *Gröner German Warships 1815-1945; Used with permission*

The second restricted campaign started on 13 March, 1916 after a long period of wrangling between the German politicians, generals and admirals.[66] However, the Americans again protested after the sinking of a number of neutral vessels and *UB29* (Olt Pustkuchen) attacked the cross-Channel steamer *Sussex*[67] off Dieppe on 24 March. The bow was blown off the crowded vessel and many of the civilian passengers were killed, including some Americans.[68] President Wilson declared: "Unless the Imperial Government should now immediately declare and effect an

---

[63] ibid. p110.

[64] Gibson & Prendergast 1931 p252.

[65] TNA :PRO ADM 137/3874 UB35.

[66] Marder 1965 p346.

[67] Some sources claim that *Sussex* was sunk, but she was towed into Dieppe. There were contradicting claims that the vessel was British, or French. The vessel was in fact owned by the Chemin de Fer de l'Etat, but managed by the London, Brighton & South Coast Railway. (Lloyd's Register 1914)

[68] Jameson 1965 pp188-9.

abandonment of its present methods of submarine warfare against passenger and freight-carrying vessels the Government of the United States can have no choice but to sever diplomatic relations with the German Empire altogether." [69] The Germans initially tried to claim that a mine had been responsible. However, fragments of a torpedo in the damaged part of the ship conclusively proved that a U-boat had been responsible. Pustkuchen claimed that the vessel was either a minelayer or a troop transport, but the attack was carried out in excellent visibility, so he must have been able to see that the people on deck wore civilian clothes.

Under advice of his foreign minister, the Kaiser ordered that from 29 April all U-boats follow International Law and exercise the 'Stop and Search' rule. In exasperation Scheer immediately recalled all the High Seas Fleet boats.[70] These boats were used instead in support of operations that led to the Battle of Jutland. A number of Flanders boats were detached to aid the High Seas Fleet boats in the spring and summer of 1916. On 23 April, 1916 six UB boats left Zeebrugge for patrol billets off Lowestoft and Harwich to cover a bombardment of Lowestoft by the High Seas Fleet in support of the Easter rising in Dublin. Only five arrived in their billets, *UB13* (Metz) presumably having become a victim of recent British minelaying operations. However, *UB18* (Steinbrinck) sank *E22* (Dimsdale) on the 25th. In the morning of the 24th *UC1* (Ramien), *UC5* (Monrbutter), *UC6* (von Schmettow), *UC7* (Haag) and *UC10* (Nitzsche) were also sent, but only *UC6* and *UC7* were able to get through new nets.

This was Admiral Bacon's latest scheme for containing the U-boats. The net laying started on 8 April and was completed by 26 May by which time two lines of moored deep contact mines forty miles long and fifteen miles wide had been positioned thirteen miles off the Belgian coast.[71] It was not until 26 April that German minesweepers were able to clear a path. *UC5* sailed and passed under the nets and her commanding officer, Mohrbutter, who was later to be described by his interrogating officer as a "very conceited young officer without much brains," [72] ran *UC5* hard aground on Shipwash Sand on 27 April and it was captured intact by HMS *Firedrake*.[73]

---

[69] Sides A.1998 p2 para 3: retrieved 8/3/04 from http:/www.uboat.net/forum UB123

[70] Jameson 1965 pp188-9.

[71] Grant 1969 pp61-2

[72] TNA:ADM 137/3876 UC5

[73] TNA:ADM 137/3876 UC5

In May Admiral Scheer planned to bombard Sunderland and to cover this operation all the available minelayers (*UC1*, *UC3*, *UC6* and *UC10*) were ordered to lay mines off Orfordness. *UC3* never returned from this operation, possibly a victim of British mines laid off Zeebrugge on the 26th. *UC6* (Ehrentraut), *UC7* (Haag) and *UC10* (Saltzwedel) sailed on the 2 July on another minelaying operation from which *UC7* never returned. Possibly she became a victim of new British mines laid off the Thornton Ridge on July 3 and 4.[74]

The Flotilla played no part in the Battle of Jutland. Despite the disputed outcome of the battle the status quo was unaltered. So Scheer decided to try again on 18 August. Now the flotilla was involved. It formed two patrol lines thus: Flanders Line 1 off Swarte Bank consisted of *UB39* (Fürbringer), *UB23* (Voigt), *UB18* (Steinbrinck) and *UB29* (Pustkuchen), and Flanders Line II off Terschelling Bank consisted of *UB37* (H Valintiner), *UB19* (Becker), *UB16* (Hundius) and *UB6* (von Heydebreck). During the High Seas Fleet sortie, the Harwich Force passed between the lines several times, but the U-boats made no contacts.[75] After the boats had been released from working with the High Seas Fleet, Bartenbach sent *UB18*, *UB23*, *UB29* and *UB39* back to commerce raiding. All four were deployed to the western Channel and the Bay of Biscay, where they sank fifty-five vessels under prize rules between 3 and 14 September.[76]

The first of the new *UCII* class, *UC16* (von Werner) and *UC26* (von Schmettow) arrived in September 1916. They were not as popular, as their diving qualities were bad, but they carried mine warfare into the Bay of Biscay. The original allocation of *UCI* class boats had been reduced from ten to just two and so redeployments took place to increase the number. *UC11* was recalled from the Kiel training flotilla, where it had been since October 1915.[77] *UC8* was also sent from Kiel, but stranded off the Dutch coast on 4 November and was interned. At Pola *UC14* was dismantled and put onto a train for Flanders where the boat served until its loss in October 1917.[78] *UB12* was dry-docked at Ostend in December and the fore end was

---

[74] Grant 2002 pp33 –36.

[75] Marder 1966 Chart 16

[76] Gibson & Prendergast 1931 p110, Appendix 5 pp13 – 15 .

[77] Retrieved 27/12/03 from http://uboat.net/wwi/boats/index.html?boat=UC+11

[78] Retrieved 27/12/03 from http://uboat.net/wwi/boats/index.html?boat=UC+14

cut off and a new minelaying fore part, which had been constructed by A G Weser at Bremen, was grafted on.[79]

A new net and mine net barrage was strung from the Goodwins to the Outer Ruytungen by September 1916 and was extended to Snouw in December after the destroyer raid of 26 October.[80] These nets were reinforced by 2,010 deep mines laid in three rows to the south, laid between 17 December 1916 and 8 February 1917. There were many problems with the barrage and consequently it failed to stop the passage of the U-boats.[81] It was finally removed in November 1917.[82] To replace it Admiral Bacon proposed to lay a mine barrage across the Dover Straits, but there were not enough mines available. Therefore a new Belgian coast barrage was laid in April 1917.

On 25 July, 1917, 120 deep mines were laid along a line eighteen miles off the coast and drifters laid fifteen miles of mine nets and this was extended on the 27th. It did not act as a deterrent, as *UC61* (Gerth) left Zeebrugge and passed this and the Goodwins-Snouw barrage at night before stranding on Cap Gris Nez. *UC65* (Olt Steinbrinck) passed through on the 26th before sinking the minelayer *Ariadne*.

Simultaneously Bacon was planning an amphibious assault over the Belgian beaches. After detailed surveys had been made, six landing places were selected, three main and three alternatives. The troops, 557 officers and 13,000 other ranks, all their guns and equipment and nine tanks, were to be carried on three large pontoons, each propelled by a pair of monitors lashed together. The trigger for the landing was to be a breakthrough by the army in the third battle of Ypres. Once this happened the embarkation procedure was to be put into operation and the landing was to take place at dawn two days later. However the breakthrough never took place and the forces for the operation stood down on 15 October.[83]

What would have happened if the breakthrough had occurred at a time when the tides were not favourable? The tidal range of the Belgian coast varies from seventeen feet nine inches at high water springs and fourteen feet nine inches at high water neaps.[84] Bacon's pontoons could land at

---

[79] TNA :ADM 137 3874 UB110

[80] Gibson & Prendergast 1931 p117

[81] NSM(H) vol. xix Home Waters Part ix pp 170, 171.

[82] Gibson & Prendergast 1931 p222, Humphries 1998 p127

[83] Bacon R .nd pp223-258

[84] Reed's Nautical Almanac. Bacon gives the high water neaps figure as eleven feet.

depths between fourteen and sixteen feet.[85] Therefore the landing could only take place ten days either side of the high spring tide and these conditions would have to be at dawn. If this were missed there would be a delay of up to two weeks before the right conditions occurred again. With an operation planned for the autumn months, the weather could be unpredictable and force a postponement of an operation even though all the other conditions were ideal.

It would be interesting to speculate as to whether or not the landing could have succeeded. Certainly there was the element of surprise. Bacon's plans went to great lengths to ensure that the enemy did not get advance warning of it or sight it after it had been set in motion. Then, to an army heavily engaged in repulsing a frontal assault from the Ypres salient, a strong force landing behind their lines would have caused them to switch troops to counter this new assault and lead to confusion. Perhaps the resultant weakness on the front line could have led to that elusive breakthrough.

In the late spring and summer of 1917, a reluctant Admiralty re-introduced the convoy system. Once this had happened the time of easy victories for the U-boats was over. Now if they wanted to sink merchant ships they would have to engage an escort. *UB18, UB36, UB74* and *UB110* were sunk in engagements with convoy escorts. The fate of *UC75* and *UB36* demonstrated what could happen if a U-boat inadvertently or deliberately entered the columns of the convoy

On 13 April, 1917 Felixstowe flying boat SP8661, piloted by Squadron Leader T Hallam, took off for the first of the 'Spider's Web' patrols.[86] Eventually there were a series of up to four octagonal patrol lines, centered on the Noord Hinder Light Vessel when weather permitted, covering over 4,000 square miles of sea. During the first three weeks of these patrols eight U-boats were sighted and three unsuccessful attacks were made, the third probably on a British submarine. However, there is a strong probability that *UB32* (von Ditfurth) was sunk by an attack of flying boat SP8695 on 22 September, 1917.[87] Other similar air patrols were later started in the Channel Approaches and Western Approaches.

In the autumn of 1917 the *UBIII* class came into service. They were much larger and faster than the *UBII* class, with a greater range. Five

---

[85] Bacon R nd p 235

[86] Hallan D 1979 p106

[87] Grant 2002 p63

torpedo tubes were fitted as compared to two in the *UBII* class and ten reload torpedoes compared to four were carried. Both classes carried 88mm deck guns, but some *UBIII* boats had it replaced by a 105mm gun in 1918. They were later found to be prone to diving accidents[88] and this may account for the unexplained loss of a number of them. The *UCIII* class arrived too late to see any operational service.

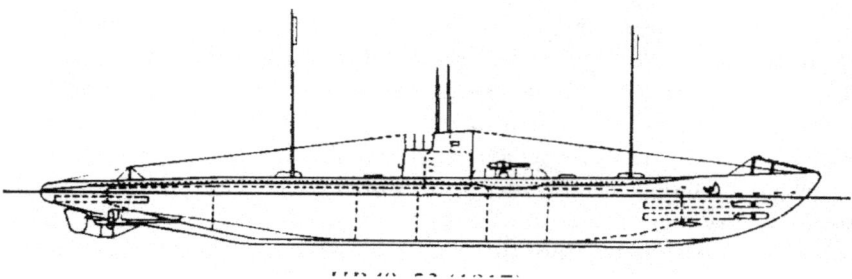

UB48-53 First group *UBIII* class, built 1916

Source: *Gröner German Warships 1815-1945; Used with permission*

Admiral Bacon's plans for a deep minefield across the Dover Straits were finally realized by the end of 1917. However, to make the minefield really effective, the Straits had to be illuminated at night as the U-boats were simply passing over the mines on the surface. Keyes advocated this, but Bacon ignored it. Consequently Keyes replaced Bacon at Dover on 1 January, 1918 and he had the Straits illuminated by flares. He also revived a modified version of Tyrwhitt's plan for a simultaneous attack on Zeebrugge and Ostend with blockships. It had been strongly criticized by Bacon and rejected by Jellicoe. The rapid clearing of a passage around the blockships was recognized. Bacon continued to believe that the locks could be destroyed by gunfire.[89] Three blockships were allocated to block Zeebrugge and two for Ostend. *Thetis*, the lead ship of the Zeebrugge trio was to ram the lock gates and scuttle herself in the entrance. However, when the raid took place on 23 April, 1918, the plan went wrong and *Thetis* was so badly damaged by gunfire on her approach that she sank in the harbour. The captain of the following blockship, *Intrepid*, blindly followed his orders and scuttled his ship in the fairway. Another chance of damaging the locks had been lost.[90] Within two days of the raid *UB12* was able to pass out of the

---

[88] *Wreck Detectives* Channel 4, 8 August 2004 Produced by Paul Griffin

[89] Marder 1970 p49

[90] Scheer assumed that the target had been the lock gates. (Scheer 1920 p339)

canal. This was the smallest operational boat in Flanders, but soon dredging had cleared channels around the blockships deep enough for all submarines to pass by. By June all craft could pass the blockships day or night at all states of the tide. For the first three weeks, only small craft could enter the channel by day. *UC17* was able to enter the canal on 4 May and *UB74* sailed from Zeebrugge a week later. [91]

By the summer of 1918 morale was cracking. "A great many submarine officers had nervous breakdowns ... trouble with crews as well,"[92] "cases of insanity, though they actually occurred amongst the submarine personnel, were not numerous."[93] Stephan of *UC17* and Dobberstein of *UB40* had suffered nervous breakdowns after channel patrols.[94] In early 1918 Fürbringer was hospitalized for 'nervous exhaustion.' Prisoner of war interrogations revealed that a number of Flanders commanding officers had been sent on sick leave during 1918. Rhein of *UB30*, Howaldt of *UB40*, Loch of *UC70*, Lange of *UC11*, Steinbrinck of *UB57* all relinquished their commands due to illness. Stosberg (*UB78*) died of an unspecified illness in May 1918.[95] This reveals an incredible amount of illness and it leads to speculation that the strains of command were becoming unbearable.

Interrogations also revealed that unfit commanders were being sent back to sea. Warzecha of *UC1* was incapacitated by heart trouble, but returned at the end of 1917 and relieved Steindorf in command of *UC71*, Olt Stier was reported to have been have been given the command of *UB16* in January 1918 without any previous submarine experience. "Current rumour ... has it that he is a consumptive with a very short life before him which may be an explanation of his extraordinary recklessness."[96] Stier transferred to *UB30* in April 1918 and it was sunk off Whitby with all hands on 13 August, 1918.[97]

When the Allied armies breached the Hindenburg Line, the Flanders bases were now threatened. Admiral Scheer went to the Kaiser and received the royal assent for withdrawing from Flanders on 29 September. On 1

---

[91] Grant 2002 p87

[92] TNA :PRO ADM 137/3899 p33

[93] Gibson & Prendergast 1931 p283

[94] Grant 2002 p124.

[95] TNA :PRO ADM 137/3876 UC11.

[96] TNA :PRO ADM 137/3899 p23.

[97] Grant 2002 p128

October Radio Bruges instructed boats on patrol to return to German bases.[98] The evacuation proceeded without any interference from the allies. Ten U-boats, eleven destroyers and seventeen torpedo boats sailed up the North Sea.[99] The Flanders bases were finally evacuated on 5 October. The vessels which could not return to Germany were scuttled: *UB59* was scuttled at Zeebrugge, *UB40* at Ostend and *UC4* and *UB10* were taken out to sea and scuttled. The destroyers *G41, V47, S61, V67, V69, V74* and *V77* were also scuttled.

The Allies committed several acts of negligence. First, by not demolishing the port facilities and locks of Zeebrugge and Ostend they handed the Germans a perfect advanced base. The lessons of history were there. In June 1794, part of the British army was retreating from the French and a detachment was cut off in Ostend. They were evacuated, but the locks were not destroyed.[100] In 1798, it was discovered that invasion barges were traveling to Dunkirk via the Bruges–Ostend canal. An attack was mounted on Ostend which destroyed the locks and rendered the canal inoperable.[101] Having realized their blunder, a number of amphibious operations were planned, but cancelled.

The longer the Allies delayed, the better entrenched the Germans became. An early assault on the Belgian coast could have turned the flank of the western front and so changed the whole history of the war. The Allies allowed the Flanders flotilla a foothold by their negligence and the Germans took full advantage of the situation. Once established in the Flanders ports, the flotilla raided shipping relentlessly in the southern North Sea, the Channel, and the Bay of Biscay. Neutral flags were no protection. Nothing was safe — fishing smacks, pilot cutters, Belgian Relief ships and hospital ships were attacked and sunk. The containment of the flotilla was badly mishandled, much faith being placed in net barrages, which were ineffective, clumsy and labour-intensive to keep in a good state of repair. Mines finally defeated the flotilla, but until the design of the German mines was copied, British mines were unreliable. Finally it took the removal of an admiral to enable the proper measures to be taken to close the Dover Straits to U-boats.

---

[98] Grant 2002 p93

[99] Scheer 1920 pp343, 345.

[100] TNA: PRO WO 1/170

[101] *Gentleman's Magazine* 1798 pp 432, 433.

Ninety-four U-boats served in the Flanders flotilla at various times. Between them they sank or damaged over 2,500 allied vessels. In the process seventy-four of them were lost, mostly with all their crews. The exact fate of *UB12, UB17, UB20, UB54, UB104, UB107, UB108, UB113, UC1, UC9, UC14, UC16, UC21, UC36, UC50, UC62* and *UC77* are unknown and some of the other accepted fates may also be incorrect. Five were interned in Holland and Spain. The flotilla tied up a huge amount of allied resources that could have been put to better use in other theatres and caused a tremendous amount of destruction and loss of life.

It was the army that eventually dislodged the Flanders flotilla. Breaking through the Hindenburg line and the rapid advance through Belgium made the base untenable so the Germans withdrew, destroying anything that they could not take with them.

You know you've been with Subsim a long time when...

....You'd pay good money to watch a submarine race.

Silent Hunter II Play Test Report

During the long and difficult development of Silent Hunter II and Destroyer Command, I was offered a chance to play the alpha and offer some comments to the dev team. What follows is the first of a dozen play test reports submitted by me to the Ultimation dev team and SSI.

Neal Stevens

This report contains observations from twenty-two hours of play testing. All nine historical single missions were completed. They will be covered individually and in more detail in the next report. This report will deal with the general feel of Silent Hunter II and player observations.

Test System: Gateway Performance 500, Intel Pentium 500 with 160 MB of RAM, nVidia TNT2 graphics card with 32MB RAM, SoundBlaster 64PCI, 6XDVD/CD-ROM, Win 98.

What I saw that I liked.

3D is a winner. Some refinement needed but overall great. The exterior views will make SH2 much better than the original.

Unrestricted view of the ocean, sky, U-boat from the bridge.

Wonderful sounds of engines, motors, screws, and sonar. Really, really good stuff. Especially the way the sound differs from screen to screen. The engine's sound is different on the bridge than in the sound room, underwater exterior, etc.

Interface is great. Need F hotkeys as before, but I'm guessing this is coming. Love the sliding panels.

Great sounds when torpedoes hit.

If the random mission generator works (mine didn't) it will be a big hit.

Great wave action, another selling point. Beats Aces hands down.

It has an imaginative set of missions. More on them individually when I do thorough mission play tests next week.

First-rate artwork. Oh boy, remember the days of 256 colors? Wow, lovely stuff.

Game manual is adequate and the writing style conducive to use and enjoyment.

- Page 8 – Single Missions menu: The manual says the Single Mission menu is displayed in the U-boat pens — mine was not, rather I had a war poster window? There was no Mission setup in the Office window either. It went directly to the briefing folder.
- Page 16 - Periscope section needs to mention how to raise the periscope and what the candystripe float indicates.
- Page 16 - periscope screen is default view, not control room.
- Page 17 – need more info on the left sliding panel, the ship viewing buttons.
- Page 29 – explain how to activate the diesels. Is it automatic depending on your status of dived vs. surfaced?
- Page 32 – The explanation of the Target Position; it states: "This gauge shows both relative target bearing (also known as 'angle off the bow'). I'm not sure this is correct. The angle off the U-boat's bow is the "bearing" to the target. No calculations are needed to obtain this part of the shooting solution. The *angle on the bow* is a term that refers to the angle "on the target's bow," the angle between the target's heading and the bearing of your sub, which is used by the skipper to determine the target's course. [See angle.gif] You better check into this — it is the kind of error you don't want "hardcore" sim guys to discover.
- Page 37 – In the second paragraph, get rid of the word "guesstimate," replace with "estimate." And describe how to use the markings to estimate range.
- Page 42 – torpedo depth settings. It would be nice if a player could manually set the torpedo depth while still using the automatic TDC. SH1 allowed this.

- Page 43 – The Uzo Station; it says the Uzo is primarily used for night... someone better put a small light on the dial markings, they sure are hard to read at night.

- Page 45 – Left sliding panel, doesn't mention "To U-boat," "Next ship," etc. features.

- Page 52 – In paragraph four, you should mention that if the player comes to a dead stop, he can listen 360 degrees.

- Page 56 – The damage control station, about the colors; so what does "yellow" indicate? My blueprint would frequently have yellow sections that the tool tip described as "okay" but could still be added to the repair list. Is this routine maintenance or something?

Gameplay

Periscope, UZO, and bridge right/left view arrows do *not* modulate the speed at which the viewer turns by how far you drag the mouse left or right. The hydrophones do this quite well.

Enemy AI seems very weak. I was fired on by escorts many times but they never chased me down to depth charge me. I will note this in the next report after more testing.

Ship ID log in the right sliding panel; it would help a lot if the rotate buttons would turn the small ships continuously when the mouse button is held down. Lots of clicking is required to make one sweep.

Way too many contact reports! One every couple hours is enough. At least don't let them all go into the log.

Need more oral crew reports for things like, "We're being shelled!", "We've been hit", "The flooding is getting critical", "Smoke on the horizon!", "Alarm!, Schnell! Schnell!", "Torpedo sighted!", "Depth charges in the water, sir!" (both — shouted the first time and whispered at the onset of each new round), and so on. Especially when the enemy fires guns. I could hear water splashing, but none of my crew was cursing or warning me we were sinking, and there were no impact sounds on my boat.

What? No control room screen? Blasphemy! You did such a good job with the captain's quarters (which, by the way, I recognized as the *U-505* in Chicago) it's a shame not to have a control room station. Alas!

I never found crush depth. Shouldn't I hear creaking sounds? A rivet popping (this boat has a welded hull, okay, so I'm listening for Hollywood rivets) or line bursting?

I found the red target arrow way *too* big. I promise everyone will love you if you change it to a small, thin red X or +, or at least a ^. That's plenty for an eagle-eyed skipper. The current oversized marker belongs on Broadway, not in a subsim.

Several times my sub fired a torpedo when I surfaced or dived to PD. No idea why — a bug, perhaps? In the reality and physics testing session (next week, after the mission design testing) I will try to recreate this and determine how/when it occurs.

Several times, at 5000 meters range the TDC was urging me to shoot with 100% solutions.

Vessels do not burst into flame or explode (I am guessing that hasn't been implemented yet), however, they do sink very well, slowly and realistically. On this note, it should be added that some should sink very quickly and some should break in half, as in SH1.

I am very concerned about the night vision the enemy surface ships' lookouts have. Are they on a straight carrot diet? I found enemy surface ships would open fire on me while running on the surface at 5000 meters. At night. Before radar (March 1940). Before I had attacked. It doesn't work this way. The U-boat is a small, low-profile vessel and should not be easily detectable without radar. After the first ship is torpedoed it is possible for an escort to "catch sight" of a U-boat if he is close.

I had my pressure hull gauge go straight to 100% while in transit a few times. Something or someone needs to explain why to the player.

Give the right sliding panel pushpin a hot key to lock it in.

The lock target button doesn't work very reliably. Many times I could see a target in the UZO and could not get the "L" key to bring up the gargantuan red arrow.

The humungous red arrow is visible through the periscope while submerged with the scope down.

Mission 6, in the map describes a surface group as a "Big Mofo group." What the heck do you mean by this? I'm guessing it is programmer humor and will be refined to "Large Muthas" later.

The aircraft description 'Ukfunbomber' sounds like another example of too much Everquest during breaks.

When I get 100% hull damage, I lose forward propulsion. Shouldn't the U-boat sink instead?

Along with the sliding right panel, why not add a sliding torpedo timer clock, in the space below the main right panel? Or at least a hit key could bring it up and stick it there. This would be very nice.

Deck gun doesn't seem to work right. The elevation doesn't change my chances of hitting the target much. Seems like I always hit it. More testing on this coming.

The UZO needs a small lamp on the degree dial at night. It would be cool if this worked through a hot key — the player pushes the Q key, the small light illuminates the dial, release the Q key, it goes out.

In Historic Single mission #8, the enemy began firing on me at night at 9000 yards. I know this was close to the first use of radar by the Allies, but they would never fire on a radar contact that was out of visual range. They would streak over and if they caught the U-boat on the surface, fire and ram. If the U-boat managed to dive, the escort would begin sonar ranging and dropping DCs. This needs attention to be respectable.

Where are the flares depicted in the intro movie? Searchlights? Hey, if it's too hard to make the streaking and firing of flares, after the first merchant if hit, maybe code in a few starshells.

How do you charge batteries? Once I went to electrics by diving (and usually test firing a torpedo inadvertently) I could not get back on diesels, and I couldn't get the red and green buttons to do anything.

Why can you fire aft torpedoes at a target in front of you?

Did I mention the red target arrow is *too* big? And omnipresent?

The machine guns are super, with reloads and mouse control. Everyone will love this feature. Why do machine gun rounds make the same sound hitting a ship as the 88mm deck gun?

Historic mission #9: the word buoys is spelled "bouys" and the word "opening" is spelled "openeing" and I'm doubtful Kapitn-Lieutnant is spelled correctly too. Is this part of the Enigma code, spelling words in new and interesting ways?

Historic mission # 10 is MIA, and historic mission #11 has all the same spelling errors as #9. In fact, it seems to be the same mission. Now that ain't right, as we say in Texas.

Seems like every night there is a full moon. Where are the lunar phases?

What? No record player? Haven't you guys seen *Das Boot*? You better correct this before release...unless the first expansion disk is going to be the "SH2 Tipperary mission disk."

What? You cannot send messages by radio? How will the Allies and their Huff-Duff stop us now?

Glitches - Aspects the dev team must be aware of, but I'm going to bring them up anyway.

While typing in the Captain's log, I was surfacing, firing torpedoes, ordering many new speed changes, and whatnot. The crew thinks I've gone mad. And they laugh at my spelling. Seems like the key commands are still enabled in the Captain's Log window, they shouldn't be.

The ocean fills are pretty funky. When the scope is down, I can see the bow of my U-boat, even at night. Many times I can see the sun and moon below the horizon through the scope (screenshots available upon request). In the exterior 3D, the below ocean portion is white. The enemy ships disappear frequently, and often the whole screen is light blue with only the ship's smoke visible.

The limited visibility option doesn't do much to limit visibility. Can still watch torps streak away. More on that in the next section.

AI is weak!

Red target arrow is scaring me when I'm not expecting to see it. It's a real big Mofo.

The Lat-Longitude info for contacts on the map is mixed up. It seems the contact message in the window bar is reversed from the map pop-up tip. See sample2.gif. The map pop-up tips read range in nautical miles while the click and draglines read in meters and kilometers. There may be a reason for this.

Captain's log text file saved on the hard drive is abbreviated — seems like before any player inserted notes.

Additional comments

You should consider a Commander Realism option - for serious players and enhanced gameplay. All subsims border on being one-dimensional if the gameplay is not held to a fairly realistic standard. I'm not talking about how pretty it is or if you have balsa wood in the holds of 12%

of your ships. I am talking about how much of a God's eye view and the level of info the player is allowed. That is what realism settings are for, right? Yes, I know. But there needs to be an additional realism level, one that would elevate SH2 beyond its predecessors. There needs to be a Commander's level realism option, for serious players (and those who fancy themselves as such, yours truly included).

To ensure respect of many players who feel as I do that these sims are too arcade-like, there should be three types of visibility settings:

1. Unlimited visibility – player can see any and everything for those girls who just wanna have fun.

2. Casual gamer limited visibility – You could say this is the option that prevents you from seeing surface ships while underwater.

3. Serious player limited visibility – what we have been asking for since 1994. In the Commander realism level, this would be the only visibility allowed. No torpedoes on the map. Very infrequent contact updates (so what if we miss the convoy and have to chase for days — we want that!). Broad sonar lines indicating escort and merchant bearings (looks like you have that part in — good). Contacts should not always be identified as enemy, friendly, neutral. Make a percentage of them undetermined (gray) and let the player decide if they are flying the Union Jack or Swedish flag. Paste a flag in there that can only be seen at 1500 meters or less. That has to be easier to create than a zillion ship types, which you have already done quite nicely. I want to feel the necessity and risk of getting in close enough to see who they are and then deciding to shoot or not. Imagine closing to 1200 yards and seeing the German flag. That would be so great — it adds to the gameplay without adding complexity to the program.

For the serious Commander realism level, the map view can never zoom in to show the ship-shaped icons. Only the square icons can appear on the map. It would be okay to have a square icon for each ship. It's okay for their position to be accurate on the map as long as the player is within visual range or sound range, and their accuracy should be reflected as such. It appears SH2 uses the sound station to update the approximate bearing of the contacts — that is really a big improvement, thanks. This is the way to do it.

Visual sightings are called out by the crew well after the time that the target appears. Make the map view so that the player must take regular periscope observations to get really good map updates. Do not let the player leave the scope up and go to the map and watch the course of ships. He must look in the scope, see the targets, and then lower the scope, return to the map (if he needs it) and see the last update. If he doesn't swing the scope and see the destroyer coming up his stern, he doesn't get a map update. When he hears the high-speed screws getting louder, he will say, "Hmm.... What could that be? I better have a look." I guarantee this would add to the repeat play and intricacy to the sim, a level that doesn't exist in SH1 or Aces of the Deep. This in itself would constitute a really big bonus for many players who do not want the "compress time-walk up to the convoys-point and shoot" type of play.

Absolutely *no* target selections from the map view. The player must see and select his target from the scope or UZO. I feel the automatic or manual settings on the torpedo data computer should be left as they are. This gives the serious player the option to use manual or auto in the Commander's realism level. No time compression more than 4x while in visual range. No crew confirmation "Ship is destroyed." Compel the player to see his victim sink or risk letting a cripple get away.

Last, make the AI routine on enemy ships "see" the periscope if the player leaves it up for more than ~8 seconds at less than 1500 meters. You certainly have AI routines that make the enemy "see" a surfaced U-boat at certain ranges; they should see the scope at close range on the Commander's realism setting, too. This should work on a sliding scale, starting with ~20 secs at 2500 meters down to ~5 secs at less than 1000 meters (or somewhere in this area, depending on the weather). Once the destroyer "sees" the scope, he should aggressively punish the player by DC. Make sure this is explained in the manual.

Keep up the good work.

Neal Stevens, Subsim - Oct. 15, 2000

I can't even think of any other websites that I visit regularly after all this time! I ran across Subsim back when I heard SHII was being made and it rekindled my interest in SH1. I've been around off and on ever since. This is one of the few communities that have really stood the test of time.

Richard "Kresge" Christensen

Blood & Honor

Gerard Cuomo

25 June 1943

1730 GMT

It is summer once again, June 1943. Most of our mighty 6th Army has just surrendered to the Kommunisten, and the great Afrika Korps is no more. The Tommies bomb our cities every night. A few months ago the Americans joined the fray and began bombing our factories by day. From here in St. Nazaire, we see their giant air armadas form and head for the Fatherland. We hope and pray that fate will smile on our families and keep them safe, but each of us knows they are suffering a terrible toll with no respite in sight. The "Happy Time" has come and gone. We are now losing U-boats at an ungodly rate. I sometimes close my eyes to visit with my fallen comrades, but they are so many now that it is like looking at a crowded square in Berlin — nothing but a sea of faces. The numbers are mounting so fast, most of us leave port with the understanding that we won't return. When we do make it back, each of us celebrates it as a personal miracle. Yet even with this knowledge, we sail. We sail, and we fight. No one shirks his duty. We sail with pride. We sail for each other. We sail for our families. We sail for Deutschland.

As I take my position on the bridge, I see the chief has already prepared the boat for our evening departure. Each man is at his post, the lines have been cast off, and the diesels are humming. She is the *U-103*, a Type-IXB, and the greatest boat I have had the honor of commanding. This shall be our fourth war patrol together. Aside from a few minor transfers, this crew has been relatively unchanged as well. With this boat and these men, I have sent over 30,000 tons of enemy shipping to the bottom. And here we stand again. Orders in hand, all eyes on me awaiting a nod to get underway. I take a deep breath and close my eyes. The air is sweet with the smell of saltwater and diesel. To us, it is the most familiar odor we know. On the distant pier, a band is playing *Das Deutschlandlied*. The gulls overhead are screeching at one another. The crowd of family and friends are shouting their farewells. And with a final prayer for our souls, I give the order.

"Ahead Slow, Chief. Take us out."

Adolf Carlewitz is the chief of the boat and a great sailor. He served under Scheer at Jutland and is a man whom both the crew and I respect very much. His knowledge of our boat is unsurpassed, as is his steadiness under fire.

"Ahead slow, jawohl, Herr Kaleun," he replies and passes the order below to the conn.

Moments later the tempo of the engines increases, and I am greeted with a healthy waft of diesel exhaust. Slowly, we begin to creep from our pen, increasing to hold turns for about five knots. It is calm tonight. The harbor waters are as smooth as a mirror. I take a moment to enjoy the sights, for I know they will be few and far between in the weeks to come. The farther away we slip from the docks, the more my thoughts turn away from home and family. When we clear the final channel marker, my mind is clear, like that of the hunter. Mind, body, and spirit all in balance to achieve the kill. From now on, I only think of the men, the boat, and our mission. The time for pleasantries is over.

25 June 1943

1955 GMT

With the sun almost set, and land completely gone from sight, I set the maneuvering watch and head below. Again I am greeted with those all too familiar sights and smells. When first underway, they are more prevalent. As the patrol goes on, they grow undetectable as we become acclimated to this demanding environment. The air is a vile concoction of diesel, body odor, cigarettes, and smoked meats. Everything I touch is damp from leaks and condensation and bathed in dim red light. Below, men are moving about with a purpose. Equipment is being checked and stowed. Gauges and valves are being inspected. Above the roar of the engines there is a cacophony of voices. None but the closest are distinguishable. It is strange, but among this chaos, I feel at home. I make my way over to the navigator's station to check his plot.

"Mr. Hartenstein, show me what you have."

Udo Hartenstein is a twenty-one-year-old Lieutenant from Munich. Good kid. He knows his job and has my utmost confidence. Our orders are to patrol in and around grid square EK71. This will have us hunting off the

African coast near Dakar. By the look of things, the course he has plotted will keep us about 300 km off the Spanish shore and well clear of most the Gibraltar traffic. While a bit on the conservative side, I am pleased with his chosen route. I wish to minimize the possibility of our position being compromised prior to reaching our patrol area. This trip will be different. No longer are we safe to pursue every contact, or engage every target of opportunity. With a single radio transmission, anti-submarine patrols will swarm the area. Now they come armed with radar, neutralizing our old friend darkness. Running surfaced at night to recharge is no longer safe. Yes, things are different now.

"Very good, Udo, well done."

"Fritz, maintain course, make turns for ten knots."

Fritz Friederichs is my Executive Officer. We've been together for two years now. He is a very competent officer and a good friend. We work together as a cohesive unit. We know each other's thoughts, and can usually anticipate each other's actions. I'm sure the crew notices this teamwork and it inspires confidence and teamwork amongst them as well.

"You have the conn, I'm going to bed."

"Jawohl, on course and ten knots, Gute Nacht, Herr Kaleun," his voice booms down from the bridge through the polished brass voicetube.

One last scan around the conn. Everything seems in order. With luck, we will pass through the coastal shipping lanes tonight undetected. By morning, we should be in open sea. A final pat to Udo's back, and I head for my rack.

26 June 1943

0657 GMT

"Herr Kapitän, we are detecting radar signals."

The voice is that of Chief Wolf Degen standing over me. I immediately jump to my feet and follow his lead to the control room. Fritz is taking a report from the radioman and relaying the bearing information to the topside lookouts.

"Mr. Hartenstein, what is the depth under our keel?"

Having already known what information I would require, Udo responds without delay.

"Depth is greater than 200 meters, sir."

I am about to accept the situation report from Fritz when from the bridge--

"Alarm! Unidentified aircraft bearing one-eight-zero!"

"All ahead flank, emergency dive!"

The command relay begins before I even finish giving the order. First Fritz, then Chief Degen. I quickly sound the claxon, and the entire boat comes alive again with voices and movement. Unlike last night, the movements are faster. The voices are more strained and deliberate. Valves are being spun. Levers are being pulled. Men, one after another in rapid succession come flying through the aft hatch moving forward. The objective is to get as much weight to the bow as possible to expedite our descent. With Olympic quickness, the topside watch comes sliding down the conning tower ladder, the last man announcing the hatches are secured. In less than a minute, the boat begins to submerge, painfully slow at first. We are all fixated on the depth gauge as if the entire crew is attempting to will the boat below the surface. The downward pitch finally begins to increase, as does our rate of descent.

"Make your depth seventy-five meters," I command.

"Seventy-five meters, jawohl."

Now all is eerily quiet except for the hum of the batteries. Even though we know there is nothing to see but pipes and wires, we are all looking up as if trying to pierce through the structure and catch a glimpse of our fate above.

"Did you get a good look at the aircraft?" I ask the watch officer.

"Yes, sir. Single ship, two-engined seaplane, possibly a Catalina."

The PBY Catalina is a British long-range ASW patrol aircraft. It can carry both contact munitions and depth charges. More importantly, it carries a radio. It must now be assumed that our location has been called in and more ASW assets will surely be getting underway to hunt us. They seem to be getting very lucky lately. It is almost as if they know exactly where to look for us no matter which routes we choose to take.

"Do you think they saw us?"

"I believe so, sir. She was nose-on directly to our stern."

Well, that settles it. Our only hope now is that she is unarmed.

Fritz calls, "Level at seventy-five meters, Kaleun."

"Very well, Fritz, ahead slow."

"Jawohl, ahead slow."

Now we sit and wait. I look around at the faces. They all express the same concentration and anticipation. I also see an inkling of disappointment for we all know the road ahead will be much more difficult now. Fritz and I share a brief look at each other. I know what he is thinking. It is the same thing I am thinking. A new directive from BdU would have us remain surfaced and engage the contact with our AA defenses. After "May" I feel this crew deserves a fighting chance to return home alive, and I will not needlessly endanger this boat. No, we hide.

"Surface splashes…wasserbomben!" from the sonar room.

And so it begins. We must wait for the first pattern of depth charges to blow before making any counter-active maneuvers. We are all praying those Tommies have guessed wrong about our position. At this depth, not even our silhouette is visible to them from the air. They must estimate in three dimensions, our course, speed, and depth. The odds are in our favor that those "Roast Beefs" can't get them all correct. However, they don't need to be exact either. All they must do is get close, and the depth charges will tear this boat apart. The first one goes off! Then the second! Third and fourth! The sound of each detonation pierces the air like a scream. The boat shudders and rattles, but not violently. They've missed! The charges were set too shallow. This is our chance.

"Right full rudder, come to course two-seven-zero," I command.

I intend to run west to escape the coming patrols. We will continue on this westerly heading for the remainder of the day and return to our base course surfacing only after nightfall. Hopefully, the enemy will continue to search along our previous track. We will have to travel very slowly, though. No more than three knots in order to conserve battery power and defeat the enemy's hydrophones.

26 June 1943

0853 GMT

Almost two hours have passed since we dove. Apparently we made it down before the enemy had a chance to plot us. There is no doubt though that they called in a contact report. I have secured the crew from battle stations, ordered damage inspections, and relieved the night watch. Cruising underwater is an almost relaxing experience. There is no pitching and rolling. No loud diesels running. Aside from the low-frequency buzzing of equipment and a few creaks and groans from the ship, it is very quiet.

Taking over the navigator's position is Lieutenant Otto Weiss. Although new to the boat, he brings with him the experience of two previous war patrols and that of surviving the sinking of his last boat. So far, he seems to be a capable young man. I have briefed him on my plan for evasion and he has made the necessary adjustments to his navigational calculations. It seems we have thumbed our noses at death yet again. If luck is on our side, we should be at our patrol station in about ten days.

28 June 1943

2018 GMT

"Surface contact, merchant, bearing one-one-zero, five-thousand meters!"

I quickly spin towards the relative bearing called by the lookout. Though my binoculars I clearly see the smoke plume on the horizon, followed by the shadowy outline of a large freighter. We have been running on the surface for two hours and picking up intermittent radar returns from about the same position as this new visual contact. The weather is bad; low clouds, steady rain, and a hazy five-km visibility.

"Lookouts below, clear the bridge," I shout.

I am the last man down the ladder, and as soon as I reach the conn, I order the boat to periscope depth. It is unusual to encounter a lone merchant this far out to sea. They don't normally brave the open waters without some sort of escort. At the very least, I must identify this vessel. If the situation warrants, I will attack. The British have been known to disguise men-of-war as merchants, so I will not be lured in so easily. I raise the scope and immediately regain visual contact.

"Fritz, take a look."

Fritz takes over the periscope smartly. After a few moments, he provides his assessment while lowering the scope.

"Looks like a C3, Kaleun."

"Ja, a lone C3 way out here?"

Both of our minds are racing, running possible scenarios.

"Could be a Q-Ship, Kaleun," he suggests.

It could be. It could also be a very brave, or very arrogant, captain making a run on his own. It is no secret the Kriegsmarine has suffered tremendous losses in the Atlantic. After last month's tally of over forty boats lost, it is widely known the battle for the Atlantic is over. Could this just be a case of pure foolishness?

"Sound battle stations, Fritz. I mean to close the range for a closer look."

Once again, the klaxon sounds and the crew spring into action.

"Come left, one-eight-zero, ahead slow."

I raise the periscope as I give the order. This time the ship is closer, maybe 3,000 meters, directly astern, with a heading of about 170 degrees. I order Lieutenant Mayer, now manning the weapons station, to start his target track. I feel the range is closing too rapidly, so I also order engines ahead one-third. I can see no other contacts visually. Sonar reports only the single contact, now 2,500 meters. I must attack. I decide it will be a stern shot, two torpedoes. I order a turn to 090 and begin running the pre-firing checks.

"Ready tubes five and six. Range...two thousand. Target speed, seven knots. Bearing...two-one-zero...heading...one-nine-zero."

Each number I send in is repeated and verified by Mayer as he enters them into the computer. The seas are heavy, so he announces we will use the magnetic pistols rather than contact and will run them at a greater depth. I give him an estimate on the target's keel based on how low she is riding in the water.

"Open outer doors tubes five and six," I order.

"Doors open, tubes five and six ready to fire."

This is the moment. I feel like the spear hunter, arm cocked back ready to throw. Breath held. Heart pounding faster and faster. Here it comes.

"Tubes five and six...torpedoes *los!*"

The tubes are fired in succession, ten seconds apart. One minute and forty-seven seconds later there is a brilliant geyser of water amidship the target. She practically jumps from the water. A yell comes from the sonar room at almost the same time.

"Torpedo hit!"

Before any of us can begin celebrating, a second explosion is observed just aft of the first. Two shots, two hits. I take a brief moment to marvel at how far we have come in such a short time. It seems like it was just yesterday when we were lucky to have one in four actually hit the target. As I step aside from the periscope to allow my officers a view, I see the entire crew is jubilant. We must not yet get complacent, though. There will be time for celebrations later. I order a turn to the north. Although the target is settling by the stern, it may not be completely crippled. It is my intention to get into firing position and remain there to be sure the ship is finished. We may have to help it along. I am proud of this crew. If they continue to perform like this, we should have a successful patrol.

4 July 1943

2004 GMT

The boat is at periscope depth, and the crew are at their action stations. A merchant has been sighted to the south, and once again I am faced with the decision. We have settled into the monotonous everyday cycle of life aboard ship. Just as the morning sky reveals itself, we dive. When we are certain the sun has set, we surface. As we near our patrol area, I order drills for every watch. I want this crew prepared for all contingencies. From the performance I have seen so far, I am certain they will make me proud. It seems adding another 8,000 tons of enemy shipping to our record really bolstered morale. I only hope we continue to enjoy such success. I raise the periscope and acquire the target visually. After a few back-and-forths between the scope and the identification manuals, I decide that I am looking at a Victory Ship. Had it been moving in a straight line, I may have had a hard time with identification. However, this ship is zigzagging, providing me with excellent contrasts between the purple glow of the setting sky and the ship's outline. The vessel is also clearly flying the flag of the United States. I scan the horizon 360 degrees for any other contacts. Again, it seems we have found a lone merchant outside the normal lanes.

"Fritz, have a look."

Fritz spins his cover to the rear and presses into the periscope sight. I advise him of my choice for identification. He checks my decision against what he is seeing and responds.

"Ja, Victory Ship...seven thousand tons."

I am about to take the scope back when he shouts.

"New contact, sir...bearing three-three-zero...six-thousand meters."

I shove him out of the way and question what my eyes are seeing, a second merchant contact, much smaller, about seven or eight hundred meters astern of the Victory. It must have been hidden from sight during the zigzag.

I call out, "Come to course two-six-zero, all ahead full."

The boat turns perpendicular to the target's track. I intend to attack as close to a 90-degree angle as possible. The weather is good tonight. The sea is calm and there is a low, broken cloud layer. While the Victory Ship is our primary target, I have Fritz identify the smaller contact. The Victory is undoubtedly armed with some deck guns. Attacking it on the surface is out of the question. But this smaller target might be worth a surface engagement if we can sink the Victory.

Fritz proclaims, "It must be a British small merchant, but I can't be completely sure from the current angle."

I take another look. I can't be sure, either. At this point, I know whatever it is; it is traveling with an enemy-flagged vessel. As far as I am concerned, it is our secondary target. I announce my intentions to the crew and turn to the XO.

"Fritz, take over...this one is yours."

"Jawohl, Herr Kaleun, I have the conn."

It is important for the executive officer to be as proficient at attacking targets as the captain. One day, he will command his own boat, and will need the practical experience. If something should happen to me while on patrol, that day may come sooner than expected. I stand back near the observation scope and enjoy the opportunity to watch it all come together. Being able to watch from a distance every now and then also allows me to pick up on possible crew deficiencies. So, I don't mind sitting this one out.

Fritz begins calling out target data. Tonight, Chief Carlewitz is manning the TDC. Due to the significance of the target, we all concur that contact exploders should be used. Each fish must count. We do not want any premature detonations as the magnetic ones are known to do. Now, I sit back and enjoy the spectacle. It is like attending a well-choreographed ballet. Every movement made by every man is precise and deliberate. This dance has been rehearsed time and time again. Everyone knows his place. Everyone knows his cues.

At 2,000 meters Fritz orders a spread of four torpedoes fired at the Victory. Each second now ticks by like a minute. Then we hear that horrible sound. The torpedoes have found their mark. One...two...three... Three of the four have struck near the target's stern. The explosions resonate through the boat and rattle our teeth. Then, a single large explosion rocks us. Over the jubilant yelling, we all hear the death throws of our evening kill — bulkheads collapsing and steel twisting, making it sound like slow death itself. I would say a prayer for her crew, but our job is only half-complete.

I yell, "Surface the boat, man battle stations, surface!"

Again the hustle begins as the crew prepares the boat for surface action. All seem very eager and without fear.

"Gentlemen," I call out, "Mr. Friederichs will command the gun attack."

I turn to Fritz. "As soon as your gun crew is ready, begin firing rounds into her hull. Do your best to place them close to the waterline and don't parallel her course. I'll be on the bridge, now good hunting."

He replies with enthusiasm, "Jawohl, Herr Kaleun."

The moment we surface, the gun crews scramble topside. I wait for the last man to clear the hatch and then make my way to the bridge. I immediately see the small merchant, maybe 2,000 tons. She is starting to turn away from us. Fritz is giving orders to the deck gun crew as they ready their weapon for duty. Suddenly, I catch a flash of light from the direction of our prey. The merchant is armed and engaging us! It is too late to choose another recourse. We must press the attack. Fritz's crew begins firing. I turn and order Senior Chief Otto Kals to have his AA batteries open up on the target as well. When I look back toward the target, I see Fritz is on his mark and pumping rounds right into her side. By my count his crew seems to be getting off a shot every three-to-five seconds. I am pleased. Lucky for us, it seems those merchant sailors are very poor marksmen. The closest shot they have made landed at least 100 meters away. We continue to close the distance. My estimation is we are about 1,500 meters from the target. I call down for a small course correction to the starboard; about ten degrees should do it. As our range closes to 1,000 meters, I see a large explosion from the target's deck and can make out small figures jumping into the sea. That's it! She is finished. I allow Fritz to fire five more rounds before calling cease-fire.

I yell down to the foredeck, "Well done, Fritz."

"Danke, Kapitän."

I turn the boat's operations over to my subordinate, again with instructions to attempt positive identification of our two unlucky friends before bringing the boat back on course. He acknowledges and I head below. As I pass through the conning tower, I notice that the entire engagement has taken place in less then one hour's time. Forgive me, but

129

somehow I find that fleeting thought amusing. And so ends another day at sea. I wonder what tomorrow will bring.

9 July 1943

1711 GMT

We have been patrolling our assigned sector without success. I sent a request to headquarters for a new hunting ground along with our latest patrol report. I just had the reply decoded and I now understand the saying, "Be careful what you wish for." Along with the normal housecleaning information came our new orders. BdU has directed us to patrol grid square CJ76. This puts us in the Mediterranean, off the North African coast. It means we must pass through the Straits of Gibraltar, the most highly patrolled body of water in the war. The entire area for a thousand kilometers is saturated with Allied ASW assets, both air and sea. I am suddenly consumed by the horrible thought that I must announce this order to the men. Their spirits are so high right now. This is a merry crew. Posting this order will strike down on morale like a hammer. Having premonitions of death and failure or a complete loss in the belief that one will see home again will surely lead to poor performance. I need these men to stay sharp. It is a difficult task, but I must find the right words to say when I read the orders. I must instill faith in the men. Not faith in God, nor faith in the Fatherland, not even faith in the boat. Their faith must be in me. They must believe without question that I will get them home alive and they will help me do it. I must do this even though I know these orders are a death sentence. So, I summon all my officers to the conn and address the entire crew.

"Men of *U-103*, you have performed your jobs admirably. You have earned my respect and brought honor to the boat again. We have all been through much together. We have accepted and overcome many challenges. As far as I am concerned, this is the finest crew of the Kriegsmarine. We must now face yet another test of our abilities. We have been ordered to disrupt enemy shipping between the North African coast and Sardinia. You all know what this means. We must run the gauntlet...Gibraltar. I have led you all to many victories. More importantly, I've brought you all home safely. And I shall do so again. You do your jobs and do not fear. I shall sneak us through their patrols and attack their convoys where they feel the safest. It is we who have the advantage. The element of surprise is on our side. We will carry out our duty and we will return home again as victors. I am counting on each of you. Do not let me down. Carry on."

And with that, I retire back to my cabin with the knowledge that I have just fed my crew a good helping of *scheiße*. I don't think we stand a chance in hell.

21 July 1943

1013 GMT

My first skipper told me, the secret of the successful U-boat commander is patience. Over the years, I have found this to be gospel. Success hinges on one's ability to estimate the situation honestly and to wait until conditions favor victory. It is simply foolish to conduct yourself in any other manner. So here we sit, fifty kilometers west of the Straits. I have kept us here for two days now. I have let large task groups and convoys pass unmolested. I also refuse to break radio silence for reporting. I am certain the enemy must have come up with some way of detecting our signals. For now, I collect as much information as possible. Once in the clear, I will send it all in. We surface around 0100 each night and idle the diesels to recharge the batteries. During this time, I listen to the latest weather broadcasts. I am waiting for a patch of bad weather to move through the Straits. Rough seas would help reduce the effectiveness of the enemy's hydrophones. I intend to wait for this added advantage to materialize before I attempt to traverse the Straits.

For two days now the weather has not been cooperative, but according to the latest forecast, there will be a front moving through the area tomorrow. Frontal passage is usually preceded by a squall line. When this line of thunderstorms moves in, we will dive to fifty meters and pass through undetected. The risk is that if the front passes too quickly, we will be caught with our proverbial pants down. I will inform the crew of the good news and have them prepare the boat. Once we get underway, we must remain rigged for silent running the entire way and will most likely need maximum use of the batteries. It is the best plan I can come up with. I believe this is our only chance for success. Now if only it will all fall into place.

22 July 1943

1901 GMT

The weather has turned as anticipated. This is our chance. I order the boat ahead slow and dive to fifty meters. Before, our descent sonar reported the usual warship contacts lurking about. Much to my delight, once below the surface, sonar can hear nothing but surface noise now. If we can't hear them, they definitely can't hear us. So long as they don't try an active sweep, we should be all right. Even if they do, they may not be able to get a decent return ping. I have ordered all off-duty personnel dispersed throughout the ship to provide extra help for damage control should the need arise. No matter what happens now, we are committed. There is no turning back. Whatever happens, we must press forward.

"Fritz," I summon.

"Ja, Kaleun?"

"I want you to plot a secondary course from our current position to the submarine facilities at Toulon. It is our contingency should we become damaged in any way while passing through the gap. I want you to keep it updated every thirty minutes, understand?"

"Ja...why don't you get some rest?"

"Bah, could you sleep?"

"Nein...I don't suppose I could."

"It will be a long day, dear Fritz."

A long day indeed. It will take us at least twenty-four hours to pass through by my calculations. Twenty-four hours of uncertainty and we will be through. I feel our chances are getting better by the minute.

25 July 1943

1400 GMT

We made it through the gauntlet.

Our current position still has us a few days out of our patrol area. A convoy has been detected by sonar. I have maneuvered the boat into a position north of its easterly track. By first count through the periscope, it appears to be screened by no less than four destroyers, but our position is so perfect, they will not know we are here until after we fire.

"Udo, I want a four torpedo spread," I say.

The young Lieutenant is once again manning the weapons station.

"Jawohl, Kapitän."

"Kapitän," comes the hushed voice of sonarman Wilhelm Barsch. "Very large screws at the head of the convoy, sir. Slow turns, bearing zero-six-zero."

I raise the periscope and look in amazement at what I see at bearing 060. Leading the convoy main body is a large, two-stack passenger liner clearly flying British colors. Oh, what a bounty we have been served. I don't understand why such a capital ship isn't screened in the center of the convoy, but who am I to question the intelligence of our adversaries? Also visible now are two C2 transports and many smaller commercial vessels.

"Fritz, come take a look, quickly."

After a brief look, my astonished XO lowers the scope and turns to me.

"A liner, Kapitän, twenty-thousand tons of her!"

I smile and nod my head in agreement.

"Ja, twenty-thousand tons, at least. Must be a troop transport. Pass the word."

In less than a minute, the entire boat knows what we are facing. They all know this is the largest target we have ever encountered, and that we must make each shot the very best we have ever made. All we have left are our four bow torpedoes and four bow reloads. However, we won't have the luxury of waiting around this area for the tubes to be reloaded. So, a single spread will be all we can do. I raise the scope for another check.

"Target course, zero-nine-zero...bearing zero-five-zero...range three-thousand meters...speed four knots."

I lower the scope.

"Ready tubes one through four, spread angle five degrees, depth one meter, and set your pistols to contact exploders."

Udo reads back each command before relaying them to the forward torpedo room, all the while working the target data in the computer. Now, I will wait a few minutes more to allow both the range and shooting angle to decrease a bit before firing. I just hope we aren't detected before then. I scan the control room. The anticipation is maddening. I savor each moment as I watch the second hand on the clock tick. It is during this time one's senses become heightened. I notice the foul, acrid order that permeates a ship when it has been submerged for a long period of time. I notice a small leak near the port engine telegraph. I feel the rumble of the convoy against our hull.

"Open outer doors one through four."

With that order, I raise the scope for one last look.

I call out, "Target bearing...zero-three-five...range...two-five-zero-zero. Standby tubes one through four...torpedoes...*los!*

I lower the scope and with a loud hiss throughout the boat, our fish are away. The temptation to watch through the scope is great, but discipline must prevail. I don't dare raise the scope until I am certain we have scored a hit. One minute has passed already. I look to Fritz. He is concentrating on his stopwatch. All is silent until Sonarman Barsch calls out.

"Impact...one...two. Now three...four, sir. All four have struck, Kapitän."

I raise the periscope and see the liner in distress. Fires have broken out all over. Secondary explosions begin popping fore and aft. In a matter of moments, her bow begins to settle. I can also make out hundreds of figures making their way into the sea. Poor devils. I allow Fritz, Udo, and Chief Carlewitz a quick look through the scope before ordering it lowered.

"Mr. Friederichs, come to course two-seven-zero, make your depth nine-zero meters...let's get out of here."

25 July 1943

1713 GMT

"They are pinging us Kapitän!"

Damn! So soon? We have been creeping away from the attack at 130 meters and rigged for silence. How in hell could they have detected us?

"Sound battle stations," I order. "Chief, stand ready on the decoys."

"Jawohl, Kaleun."

Sonarman Barsch calls out, "High-speed screws, bearing one-six-zero. Splashes...wasserbomben!"

"Helm, right full rudder, come to course three-six-zero, ahead standard. Chief, launch the decoy."

"Decoy away."

Again, I will wait until the first pattern of charges detonates before making any changes in depth.

"Kapitän, I'm hearing screws, sir. Many...possibly more destroyers. One close, bearing one-four-zero. Another bearing...one-six-zero...a third bearing...two-four-five...more pinging, sir."

With a pop and crack, we have been jolted by the first round of charges. I can hear men yelling from the forward torpedo room and order Fritz to investigate. More explosions, this time farther away, but still close enough to shake us like we were a toy. I fear our luck has been pressed too hard, and may have run out.

"Can you estimate the depth of those charges?" I ask sonar.

"Best estimate is one hundred meters sir."

"Make our depth five-zero meters," I order.

If they fall for the decoy, they will continue setting their charges too deep. I hope the others joining in will take their cues from the first destroyer. I also order ahead slow to try and draw more attention to the decoy. Fritz calls from the forward torpedo room.

"Kaleun, we have damage to the forward torpedo tubes and are taking water...will have it shored up in a few minutes."

"Keep me advised, Fritz."

Now I can't chance a deep dive. We will have to stay in the green arc.

Crack!

Ah, more depth charges. They are getting even farther away now. The boat didn't even shake this time. Perhaps we have given them the slip? It is now that the battery gauge catches my eye. We are down to 50%. That may not be enough power to get us out of this mess. I think our only hope is for the decoy to allow us to extend the range beyond their sonar.

25 July 1943

1807 GMT

"Kapitän, propeller noises are fading rapidly, sir," is the call from sonar.

I think we have made it.

"Fritz, give me some good news, my friend."

As he approaches he replies, "Nothing good to report, sir. The temporary patches are holding, but they won't hold for long at this depth. The wasserbombs must've gone off pretty close. Tube two is in very bad shape, I think the outer door got pranged a bit. I wouldn't recommend staying at this depth, sir."

"What do you think, Fritz?"

"I think we must implement our contingency plan, sir. We've scored a big prize already and I don't believe we can continue in this condition. I recommend we head for southern France."

"Agreed. Set the new course, but maintain our current depth for another hour. You have the conn. I'm going to have a look around my boat."

"Jawohl, Herr Kaleun."

I am about to head through the aft hatch when a scream comes from the sonar room. "Alarm! Wasserbomben close aboard! They just popped in above us, sir!"

"Ahead flank, right full rudder!"

"Sir, contact warships bearing one-eight-zero and two-one-zero, they are pinging us and are closing!"

They have us.

"More splashes, sir!"

My promise to the men. I fear I have failed them and their families. Through the lethal barrage thoughts flood my mind. What was it all worth? Did our lives have meaning? Did we serve a purpose greater than just the Third Reich? Will our families know we died with honor? The boat is shaking violently now, worse with every detonation. The boat cries out and shudders in pain. A flash of searing heat. And now all is quiet, and cold, and black.

Joined Subsim/WPL several months before SHII was released. I'm talking *way* back. There was even this weirdo Texan as BdU of Wolfpack League then! Ya know, Neal is so old…that when he was born the Dead Sea was only sick! What woulda happened if Horace Lawson Hunley had heeded Neal's advice? And look where he ended up…oh wait…

Rick "StdDev" Bartholomew

Things They Don't Tell You at New London

Bob 'Dex' Armstrong

In high school, I watched a TV program called *Silent Service*. An old retired codger named Rear Admiral Thomas M. Dykers opened the show with, "Tonight, we bring you another thrilling episode of *Silent Service* stories, of warfare under the sea." I did not know that for the first time, I was being radiated by minute television-transmitted bull manure particles, that I would come to know as 'sea stories'. *Silent Service* sold BS by the yard, beginning with the words, "Thrilling adventure...."

There were things they didn't tell you. Things like being the guy who threw trash over the side in heavy seas ... rigging doubler plates in a rolling boat ... reversing the blank flanges in the air lines to the fuel ballast tanks when they were empty, and you lined them up as main ballast tanks ... and several other thrilling adventures that the devil and the USN combined talents to bring you.

Let's begin with rigging blank flanges for main ballast tank use. (For you boat sailors; 'blank flange', alias 'spectacle flange' alias 'Dutchman plates').... For those of you not initiated in the smoke boat fraternity, imagine a dental procedure where you had to remove a wildcat's molars through his rectum while inside a washing machine in the rinse cycle.

A spectacle flange was a half-inch thick piece of metal that looked like a mask for a 'blind-in-one-eye Lone Ranger'. On one side it had a hole ringed by stud holes. The other side was solid with a ring of stud holes. They were in all fuel ballast tank lines (600# MBT blow).

When the fuel in a tank was used up and the tank was filled with compensating seawater, the hand of the Almighty would descend on some undeserving E-3 bastards — always two — fun must be shared. They gave you each box wrenches and words of encouragement, "Go get em', bucko!", "Have a good time, sweetheart...", "Don't be out too late dearest, you know how daddy and I worry about you...."

You always politely responded with an all-inclusive, "Screw you and the horses you rode in on."

They gave you things like dog chains... A thick window washer's belt with fifteen feet of heavy chain attached. At the end, there was a clamp called 'the dog'. The dog connected with a T-track that snaked its way down the full length of the topside deck. The purpose of the device was to keep a man from being washed overboard and out to sea. Any baboon with six in-line brain cells could take a look at the length of chain and recognize immediately that when the overwhelming swell hit you, the chain would position you between the limber holes and the tank tops, where you would enjoy the sensation of being repeatedly smacked with a coal shovel.

To move up and down the deck at sea, you had to grab the chain and pull the clamp along the track. This was known as 'Walking the dog', 'Strolling with Fido', 'Taking Spike for a walk'. If you were E-3 or below, these terms held no humor. E-4 and up, they were sidesplitting gut busters.

Once you had your box wrenches and dog chain, they cracked the after battery hatch and opened it (If the engines were running and it was cold topside, the compartment below became an arctic cyclone with grown men yelling, "Put the iron back in the pneumonia hole!").

It was always wet when you dragged your dog back to the superstructure access lid. You stayed clear of the exhaust lines if steam was coming off them. Contact with a hot exhaust was one of those once in a lifetime experiences where you learned everything you ever wanted to know in a fraction of a second.

From then on, it was simple. While operating in a space a little larger than an Oldsmobile glove box, you and the poor sonuvabitch you were paired up with had to pull the ring of stud bolts, put the nuts and bolts in a coffee can that you zipped up in your jacket, reverse the flange to the open or 'Lone Ranger Seeing-Eye' position, then replace the stud bolts. After this, you repeated the whole fun and games procedure on the opposite 600# MBT blow line. While you were down in the devil's crawlspace, large portions of the Atlantic Ocean were sloshing in and out of the limber holes and raining down on you from the free flooding deck above. This was about the most thrilling it got...other than unscheduled major leaks.

Then, there were doubler plates. I never liked the whole concept of doubler plates. You only put the damn things up when your fellow 'above the surface' Navy brothers were going to shoot or drop something on you that might possibly enter your place of residence, or poke some kind of unwelcome hole in it. Even as E-3s, we instinctively knew that this kind of activity had originally germinated from a bad idea.

For those of you who have been fortunate enough to have never been next to a doubler plate, imagine the first cousin of a railroad locomotive wheel. Now imagine the Three Stooges trying to fasten this contraption to the inside of a rotating cement mixer. There, you have it, fun and games in the Silent Service. And the service became rapidly unsilent if you ever dropped one of those sonuvabitches on any of the five appendages living in your boot.

Another thing, no one at New London ever mentioned the inboard vent for #2 sanitary tank in relation to non-qual sleeping arrangements.

They also failed to mention the vertical spud lockers that masqueraded as showers on diesel boats.

No one told us that failure to secure one little gate valve and one little kick throw could actually reverse-percolate previously digested Navy chow through your morning coffee.

They conveniently left out the fact that anything that came from Fleet Supply or the forward hold on Orion, came with free roaches.

Nobody said anything about bug juice and panther piss.

And worst of all, no one mentioned that Chief Petty Officers couldn't take a joke.

I discovered this site when I bought Sub Command and signed up straight away. I was using a dial-up connection in the local library to read it. I was attracted to the site because it offered tips on how to play the SC game. This site really makes the game live by injecting a lot of real life into the game and also background into the game design. I try and visit daily and I have read Subsim in places as far apart as Cuba and the Maldives.

Christopher "Linton" Morgan-Jones

Sub Club Inter-Continental Meetings

Laura "Sharkstooth" Sands

London - October 2003

In April of 2003, Sub Club member Panz from Carlisle, England came to visit me in Chicago. (Sub Club is the social arm of the Subsim Empire). We had such a great time that he returned with his family in August. They extended an invitation for me to visit them in October and it sounded wonderful, so we started planning. At this time, I was the Flotilla Commander (FC) of the 1st Flotilla at the Wolfpack League (a branch of the Sub Club, which, as was previously mentioned, Subsim's buddy club), which is primarily a UK Flotilla (time zone wise). I wanted to take full advantage of my trip across the pond and meet everyone that I possibly could. Deerhunter lives in London and bravely offered to organize the details, so I put the word out. It was looking like the meeting was going to be bigger than just a handful of friends and we couldn't imagine Neal being absent from this, so we invited him. To our great surprise and delight, he agreed to join us, and the stage was set for the first Intercontinental Sub Club Meeting.

We rented a cottage in Tunbridge Wells for a central meeting place. Panz and I were driving down from Carlisle, which is in the northwest corner of England, just before the Scottish border. We stopped in Nottingham to pick up Nosferato, and then unbeknownst to me, we went to Birmingham to pick up Club favorite and Wolfpack Radio DJ Nemesis, whom I thought wasn't going to be able to make it to the meeting. What a fantastic surprise that was.

We all met up in Tunbridge Wells at a little pub. I had to pinch myself to believe it was real. The Dutch Bunch was there, Drebbel, Bluebeard and Dargo, along with so many unfamiliar faces. Strangers really, yet their voices were so very familiar because we had all regularly talked in voice on Yahoo IM. I recognized Neal's famous Texas drawl from hearing it on WPL Radio, and McBeck's distinct way of pronouncing my name —"Shargie." Among the members present was "Crow", one of the Wolfpack League's biggest supporters.

The next day, we took a bus to the tube and before I knew it, we were at the entrance for the *HMS Belfast* in London. It was a good thing I wore my trademark red heels because a few of the guys waiting for us at the Belfast recognized me by them.

It was a fantastic day, touring the *Belfast*, dining at a great restaurant afterwards and then all meeting back up at the cottage in Tunbridge Wells. The next day was spent having just darn good conversation and getting ready to leave. We all decided it was a great meeting, but just way too short.

⊕⊕⊕⊕⊕⊕⊕⊕

Galveston, TX - October 2004

It wasn't long before we started planning for another meeting, this time at Neal's place in Texas. That man sure can plan a party. He had a jam-packed week set up for us, and it was a blast. It started for me with Deerhunter and Nosferato flying in from England and meeting me at O'Hare Int'l on a connecting flight to Houston. We picked up our rental car — a snappy little silver convertible, and then hooked up with Neal, whom we were supposed to follow back to his house. The details from that trip alone could fill a novel, but I won't go into it except to say that we finally made it and all in one piece too. The first day was mostly just spent picking people up from the airport and getting everyone situated in their hotel rooms. Our rooms were in Galveston so we went there and settled in for the evening.

The week was a whirlwind of a real Texas BBQ, visiting the Battleship *Texas*, testing our skills at a shooting range, dining at a wonderful seafood restaurant, shopping, Tex-Mex cooking, sipping Margaritas at the pier, and just relaxing and renewing friendships. Neal surprised us with a special guest: Kelly Asay, President of Tesseraction Games and the mastermind behind the great naval action game, Enigma. He answered a lot of questions about Enigma and he sure knows his Cabo Wabo.

Neal had a special event planned. The guys could spend the night aboard the *USS Cavalla* or the *USS Stewart*, their choice. The *USS Cavalla* SSK-244 is a Gato class submarine built in the summer of 1943 and converted to an SSK hunter-killer during the Cold War. The *USS Stewart*

(DE-238) is a Destroyer Escort that was decommissioned nearly six decades ago.

Captain Zeb Alford, who was an XO on *Cavalla*, aid to Admiral Rickover, and captain of the *USS Shark* and *USS Sam Houston* gave our group a wonderful speech and presentation. We sub enthusiasts learned a great deal from this American hero.

The guys who chose to sleep aboard the *Cavalla* had a bit of an extra special surprise. It seems Drebbel, Bluebeard, and WildViper were planning a midnight raid. They sneaked in and set off all the sub's alarms in the middle of the night, and, just because they are so thoughtful, they filmed it. The look on everyone's face was absolutely priceless. Needless to say, KP had some deserving inmates that month.

The last surprise Neal sprang on the group was an exclusive, hands-on preview of Silent Hunter III. The game was still in development, but Florin Boitor, the executive producer, sent Neal a beta copy to demo for the group. The game was great, even at that stage we knew it would be a smash hit. We were the first subsimmers to sink ships with SH3!

The week ended and we all promised to do this once again in 2005. This time, the Dutch Bunch was going to host it in Amsterdam.

Amsterdam - September 2005

The Dutch Bunch was well prepared for this meeting. They had two separate itineraries. One was simply if you could only make the weekend meeting, and the other was a week long, hardcore "I want to be a Dutchman" tour of Holland. There was no way in the world I was going to miss that week.

The trip started with Neal and myself flying into London, Heathrow. Nosferato picked us up and we headed over to Deerhunter's place to collect him. On the road again, we headed to Dover, picked up Neal's friend Richard, and caught the ferry over to France. We drove up to Ghent, where we spent the night in a bed and breakfast. Drebbel had driven down from Den Haag and met up with us. The next day was spent leisurely seeing the beautiful sights of Belgium.

We headed up towards Monnickendam, where our cottage was located. Bluebeard and Jolanda had a lot of people bunking with them, and everyone got unpacked and situated in their quarters. It would be next to impossible to list everything that we did during that week, but a few of the items on our schedule were: A canal tour through Amsterdam led by the inimitable Abraham, visiting the Dutch sailing ship *Batavia*, touring the Dutch Naval Museum, shopping (*Ed: for red shoes, no doubt*), and a trip to Arnhem (*A Bridge Too Far*) and viewing its Memorials.

Subsim invited Silent Hunter III executive producer Florin Boitor to be our special guest. He flew in from Romania on Thursday and we peppered him with every SH3 question imaginable. We learned everything imaginable about SH3. Florin was a peerless gentleman and fun to party with.

One day we split into two groups. One group drove up into Germany to visit the *U-2540*, now called *Wilhelm Bauer*, at the German maritime museum in Bremerhaven. The other toured around the Nederlands, and thanks to Abraham, saw the beautiful countryside, getting a look at their dykes and lines of defense against flooding.

Saturday we picked up a few people that were coming in for the weekend meeting. We had Panz and U-48 coming over from England, Iceman from Scotland, Crow from Belgium, and Doc and Dan from Germany. Add that to the people already at the meeting, and we were quite the international bunch.

We all go to these meetings for different reasons. I love seeing all the sights, visiting unique places, but what really keeps me coming back are the fantastic people I have met that can show you their country. Seeing their hometown through their eyes is so wonderful. Not once have I felt like a tourist, but more like a part of the family. It will keep me coming back as long as we keep having the meetings.

The Dreadnought Era

David Millichope

Introduction

In February 1906 a new player entered the world of naval politics: *HMS Dreadnought*. She was to define the naval era for a period of no more than thirty years. Her brief stay occurred during a global power struggle that stretched from the beginning of Britain's relative decline as the world's first superpower, circa 1870 (the emergence of other major industrialized nations) until 1989 (the collapse of the Soviet Union and emergence of America as the world's only superpower). This struggle polarizes around the eruption of two massive military conflicts we now refer to as the First and Second World Wars. *HMS Dreadnought* was so significant she lent her name to a whole generation of capital ships that held centre stage in naval strategy and thinking. By the time of the second major conflict (1939-45) her day had all but gone, her pre-eminence replaced by aircraft carriers. Roughly speaking, we can say that the dreadnought era ran from 1906 untill 1941, but in real terms probably extended no further than the mid 1930s.

Dreadnoughts : the measure of a nation's power.

Much as America and Russia counted nuclear warheads between 1950 and 1989, so did the industrial nations of the world count its Dreadnoughts between 1906 and 1941. Here was measurable proof of a nation's standing. They were the biggest manmade machines of war ever constructed. They were probably also Britain's last gasp as a superpower. Between 1906 and the end of the Great War she constructed a fleet she could barely afford economically. However, as Jutland and the final demise of the High Seas Fleet at Scapa Flow testifies, the effort was justified by the importance of its role in the defeat of Germany.

The evolution of the Dreadnought Battleship

Between 1860 and 1880 navies had fleets of ships that were ironclad type vessels with mixed calibre guns and were expected to operate in coastal waters in support of marine operations. During the period roughly defined by 1880/90 they evolved into ocean going vessels capable of engaging other warships in fleet actions at sea. The effective operational range of the gunnery duels was about 6000 yards. In 1905 the battle of Tsushima was conducted largely with ships of this type. *Dreadnought's* launch in 1906 redefined the ship into a faster moving vessel which concentrated its firepower into single large calibre guns capable of hitting the enemy at ranges beyond even the horizon. Between 1906 and 1914 these longer-range guns were being serviced by complex systems of central fire control that allowed reasonably accurate gunnery duels to take place at ranges beyond 20,000 yards if visibility permitted. By 1914/18 this vessel had evolved into a warship type we would recognise as the "battleship"—later made famous by such epic names as *Bismarck*, *Missouri*, and *Yamato*. What we must remember is that these famous names wielded less power and influence than their less famous WWI antecedents. The reason for this was the primitive nature of naval aviation during the Great War. Unlike *Bismarck* and *Yamato*, the WWI Dreadnought did not have its operational effectiveness rendered impotent by marauding hordes of aircraft that carried ship busting bombs and torpedoes. It must also be said that the WWI dreadnoughts did not have it entirely all their own way. They were vulnerable to both mines and torpedoes. Nevertheless, this did not stop them from operating in massive battle lines, the like of which the world had never seen and probably will not see again. In essence, the tactics of 1914-18 were still rooted in the days of Nelson. The dreadnoughts were still expected to form into concentrated lines and deliver firepower. The classic set piece battle was expected to be gunnery exchanges between rival battle lines of Dreadnought ships.

Jutland: one of History's most decisive battles

One of the first naval war games I ever played (way back in the 1970s) made a very important point that has stayed with me ever since. Specifically, that naval engagements are largely meaningless unless seen in the context of a campaign.

On this point, Midway and Jutland are two battles that make interesting comparison when seen in the context of their relative campaigns. Midway (1942) is usually seen as one of the great decisive naval battles of all time; Jutland (Skaggerat, 1916) is forever remembered as a short and inconclusive encounter loaded with the frustrations of what should have been. It seems poetic justice that the war which is associated with stalemate and attrition had Jutland as its defining naval moment, whereas WWII with all its dynamic twists and turns has Midway as its iconic "turning point." Yet which one in itself truly decided anything? I would argue that, in the context of its campaign, Jutland was a genuinely more decisive naval engagement than its more famously celebrated cousin.

Midway was a battle skilfully and bravely fought by the American fleet against superior numbers that radically altered the naval balance in the Pacific. Before Midway the American Fleet was on the defensive, desperately trying to stem the shock of Japanese success and expansion. After Midway the balance was altered forever. From that point on it is clear that the Japanese could never win their Pacific war with America. However, the important point to bear in mind is that America could have lost Midway six times over and still won the Pacific war. For Japan, the war was already lost the moment Nagumo's aircraft began offloading their devastating payload at Pearl Harbor. For America it had already been won, years ago, in her industrial heartlands. Her industrial capacity was truly awesome in comparison with every nation state that existed at the time. The bottom line is that Japan had little chance of winning the Pacific war from the moment she gave America the sense of purpose needed to wage an industrial war to the death. This she did at Pearl Harbor.

Industrial War – the players

The two world wars of the early 20th century were the first industrial wars involving a global power struggle of the newly industrialised nations. The problem for the two expansionist nations, Japan and Germany, was that they had arrived into this new game about sixty or seventy years behind the Anglo-Saxon nations. For Japan this was a fatal flaw in its Pacific ambitions because she was hopelessly outclassed by America's economic capacity to wage industrial war. I would respectfully submit that Midway was a decisive tactical battle won by a naval power (America) that had already effectively won its campaign.

Turning to the other expansionist nation, Germany, her relatively late industrial start had become less of a problem because she was already economically outstripping her chief European industrial rival – Great Britain. However, Britain was not at the time a continental power. She invested her economic strength almost exclusively in her worldwide navy.

Jutland was a tactical battle that in itself significantly influenced the paths of its two protagonists. Its consequences maintained the naval domination of the weaker industrial power (Britain) and, as I will argue, proved to be the decisive turning point of WWI.

The Strategic Significance of Jutland

Famously, Germany claimed Jutland as a victory on the basis that its numerically inferior navy had inflicted greater losses on its enemy. In terms of propaganda this was indeed very useful to the German war cause and catastrophic (though not fatal) for British perceptions of their war effort. However, the British losses did not strategically alter the naval balance. What it did achieve, in the wider context of the naval campaign, was to decisively feed the conviction amongst the German Naval Command that its High Seas Fleet could never break Britain's distant blockade.

Thus, in terms of its influence on strategic campaign thinking, Jutland proved utterly decisive. This decisiveness becomes evident in two ways. Firstly, the continuing blockade of Germany trapped her in a deadly timetable. She could not afford the war to continue indefinitely because her economic system would begin to fail under the strain of economic blockade long before that of the Allies. It was the seismic cracks created by these pressures that led to the request for armistice.

Jutland's decisiveness was also evident in a second way: the German drift towards the adoption of an unrestricted U-boat campaign. This shift in naval strategy ultimately (and probably inevitably) led to the entry of America, the world's greatest industrial power, on the side of the Allies. As in WWII her economic might made victory in Western Europe inevitable provided the European allies could hold out long enough. Unable to make decisive progress on the Western front, facing the build up of almost inexhaustible supplies of fresh troops from America, suffering economic privations and revolt on the Home front, Germany sued for peace in 1918.

If we see the two world wars as part of the same industrial struggle for world ascendancy then the surrender of German naval power has significant consequences for the resumption of the military struggle in 1939. Britain and France had continued their relative industrial decline prior to the resumption of European hostilities in 1939. Significantly, Britain's naval strength had become less influential. However, the terms of the Versailles Treaty, (acquired through the consequences of Jutland) had led to the effective elimination of nearly all of Germany's naval strength. Given that Germany was stronger industrially in 1939 than she had been in 1914 we can see how the consequences of Jutland run right through this mighty struggle.

Between 1939 and 1945 Britain's weaker naval position is still sufficient to hold off the neutered German navy. The battle of the Atlantic proves decisive and America's industrial might yet again comes to bear on Germany.

Prelude to Jutland

Britain and Germany had been in a naval arms race ever since the launching of the battleship *Dreadnought*. Almost overnight this ship made all the capital ships of the world's fleets obsolete. In theory this put Britain, at a stroke, on a par with all other major navies. For a nation accustomed to, and dependant on, maritime supremacy this could have been a defining point in decline. However, at the time, Britain's naval planning was driven by the mercurial genius of Jackie Fisher, and he was not a man to allow decline.

The race was on, and given Fisher's drive, ambition, and ability to squeeze budgets out of Parliament, Britain's ship building capacity was able to outstrip that of Germany. By the outbreak of war she had a decisive numerical advantage in capital ships over the German High Seas Fleet, and it was these ships whose head count decreed naval supremacy. As in all wars previously fought by Britain since the late 18th century she immediately mounted a blockade of her major adversary.

Traditionally this had been a close blockade, with ships stationed closely offshore to the enemy harbours preventing the breakout of the enemy fleets. In 1914 this was far too risky because of the threat of enemy torpedoes and mines that could easily be deployed from the enemy bases. The answer was a distant blockade. In essence, the British Isles are like a huge breakwater, blocking continental Germany's route to the Atlantic and world trade. The concept of distant blockade was to station the British fleet at a number of bases in the northeast of the British Isles so they could rapidly concentrate at the northern exits of the North Sea to cut off any attempted breakout of German forces. The English Channel was sufficiently secure because of the huge complex of minefields covering these narrow approaches.

The lighter scouting forces were stationed at Harwich, the most southerly of these bases. The Grand Fleet, the sledgehammer force of battleships, was stationed at the most northerly base at Scapa Flow. Between them lay the bases of Cromarty (Invergordon) and Firth (Rosyth), where the fast moving battle cruisers (under Beatty) were stationed as powerful rapid response forces to counter the threat of the German battlecruiser squadrons. If the German High Seas Fleet were to put to sea in support of their scouting forces, then the Grand Fleet was in a position to likewise support Beatty when he was threatened.

Germany's strategic problem was how to break this blockade with a numerically weaker fleet.

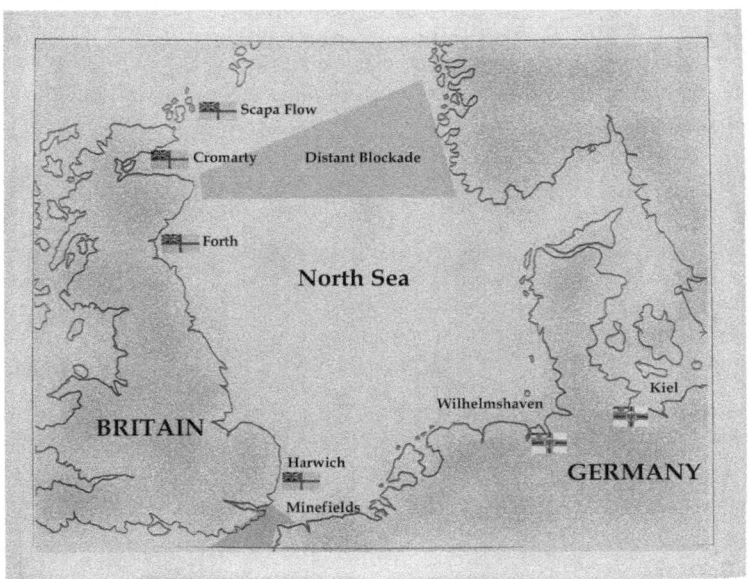

Jutland was an attempt on the part of the German High Seas Fleet to trap and destroy major elements of the British Grand Fleet. In theory this would eventually bring the two dreadnought navies to near parity of numbers and enable the High Seas Fleet to mount a head on confrontation against British naval supremacy. The result would be a reverse Trafalgar, in the sense that Britain would lose control of the oceans. Britain's blockade of Germany would be lifted, and she herself would be starved into an armistice. With Britain's armies eliminated from the western front it was expected that France would be unable to hold out indefinitely and German victory in the west would follow. Germany could then turn the full weight of her armies against Russia, whose defeat would establish German hegemony over Europe. The collapse of Britain's naval supremacy was thus seen as the start of this domino effect.

⊕⊕⊕⊕⊕⊕⊕⊕

Jutland – the Plan

The German battlecruisers would mount a demonstration of force off Norway. This would entice major British forces in the form of Beatty's battle squadrons, who would come irresistibly rushing over to engage them. Waiting for them would be the entire German High Seas Fleet. A submarine trap was also laid close to the Scapa Flow base with the intention of destroying as many of the Grand Fleet as possible who were likely to be ordered to sea in support of Beatty. Reconnaissance by Zeppelins would keep the German's furnished with vital intelligence information as to the British movements.

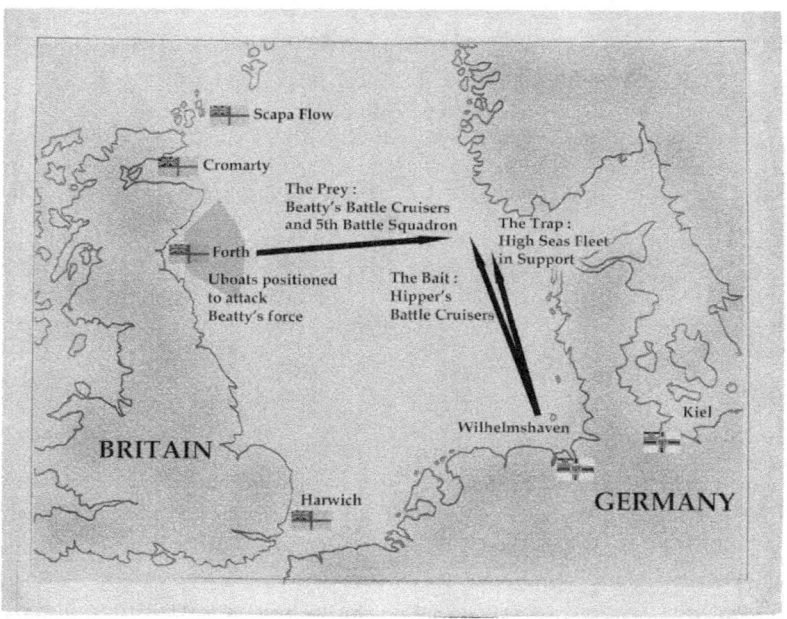

However, bad weather rendered the German Zeppelins useless so the Germans were boxing blind from the start. Crucially, the British were well aware of the German intentions and dispositions because they had been able to read the German code since early in the war. Unbeknown to the Germans, the British Grand Fleet was able to put to sea even before the German High Seas Fleet. Consequently, far from being a trap for the British forces, the engagement turned into an ambush for the entire High Seas Fleet. For the British Grand Fleet the long anticipated Trafalgar was upon it.

Jutland – the engagements

One cannot help but feel sorry for the British CIC, Jellicoe, because there was not much he did wrong. Unfortunately, events were to turn this beautifully laid and executed ambush into a frustratingly missed opportunity. Beatty's squadrons did encounter Hipper's battlecruisers. In turn Hipper lead Beatty onto the High Seas Fleet. It, also in turn, accepted the bait and set off in pursuit of Beatty's squadrons, who duly led it into the arms of the numerically much stronger British Grand Fleet. To make matters even more perfect for the British expectations, Jellicoe brilliantly deployed his dreadnoughts to achieve the classically deadly "crossing the enemy's T" at just the right moment. Annihilation for the German High Seas Fleet beckoned.

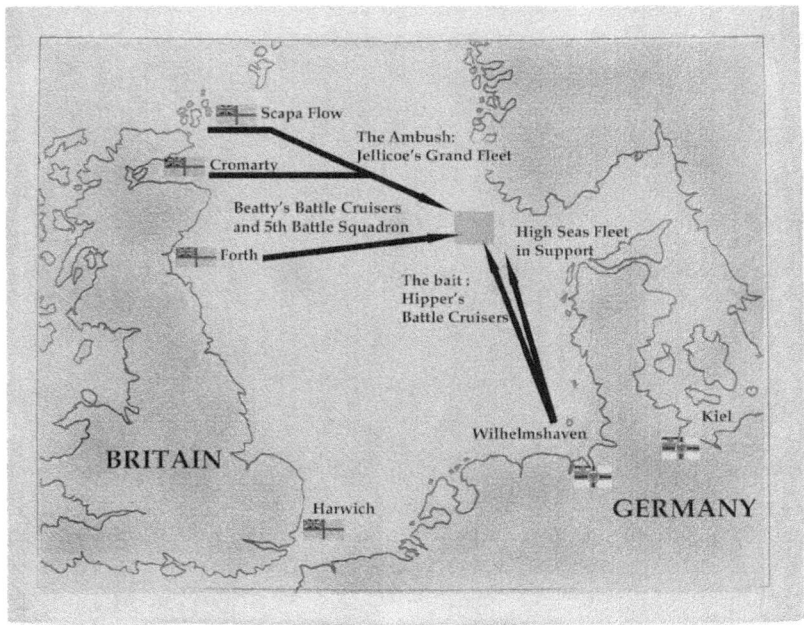

So what went wrong? It has to be said that at one important level, communications were mainly to blame. Jellicoe had received the vital information of the German intentions, but he was given very misleading timing. The origin of this can be traced back to the arrogant behaviour of a vital liaison officer at the Admiralty. The end result was that Jellicoe believed the High seas Fleet to be further away than they actually were. He therefore ordered the Grand Fleet to steam at a much more leisurely pace to the place of expected action. Had he shown more urgency, he would have had several more hours of daylight in which to affect a decision with the enemy. Instead, when eventually they did make contact, the High seas Fleet was able to escape into the safety of darkness somewhat earlier than they should have.

However, even at this juncture all was not lost. The Grand Fleet still lay between the High Seas Fleet and the safety of its base. Jellicoe could have confidently expected to bring his superior force to bear against the enemy on the following morning when daylight returned. All that he needed was for his subordinate commanders to keep him informed of any attempts of the enemy to break through the cordon he set. But this is exactly what they did not do, and during the night the High Seas Fleet was able to boldly thrust through the rear of Jellicoe's cordon of ships and reach the safety of a passage through to their base at Wilhelmshaven.

In effect this meant that Jellicoe's one and only opportunity had lain with his brief engagement with the High Seas Fleet in the late afternoon and evening of May 31st. This was the occasion when the Grand Fleet was "crossing the T" of the Germans and threatening annihilation. We must now turn to the decisions of the two commanders-in-chief to see why this annihilation did not occur. Firstly, the German CIC, Scheer, executed what is now generally regarded as a brilliant "about turn" manoeuvre, away from the imposing array of British naval firepower about to rain down on him. The Germans had practiced this manoeuvre for just such an eventuality. So we must say that one factor in the escape of the High Seas Fleet was down to good training and brilliant execution. Now comes the most controversial decision of the entire battle, executed by Jellicoe. This became the rod that was used to beat Jellicoe in the years following Jutland. It is probably largely to blame for Jellicoe's name not joining the celebrated ranks of all time British naval heroes. It is also, paradoxically, the decision that is absolutely at the crux of the overall naval campaign that was to win the war for the Allies. Instead of pursuing the High Seas Fleet into the mist, he *turned away*. That ensured that the Germans were able to claim a famous tactical victory. Most of the initial action had involved the battlecruisers of either side, and in this respect the British had suffered three spectacular losses when the battlecruisers *Princess Royal, Queen Mary,* and *Invincible* blew up and sank with virtually all hands within minutes.

Germany lost one battlecruiser, *Lutzow* and an old pre-dreadnought battleship, *Pommern*. Comparisons with cruiser and destroyer losses were equally unfavourable to the British.

All of this was to be trumpeted to great effect in the years immediately after the engagement. It broke the "spell of Trafalgar;" it gave the Germans an absolute propaganda gift. However, all of this would have paled into insignificance if the two dreadnought battle fleets had been able to slug it out to the death. The dilemma that preyed on Jellicoe's mind was this: If he pursued the High Seas Fleet, he did indeed maintain the possibility of another British Trafalgar. On the other hand, Jellicoe understood the campaign implications of significant British naval losses. As someone once said, "If the British Fleet is destroyed, the Germans will be in London within months. If the German Fleet is destroyed we will still be no closer to marching into Berlin."

So what was the issue that made him turn away? The answer is in the threat of torpedoes and mines. Both of these could be laid in the wake of a German retreat and both had been shown to be very effective against surface ships. Early in the war *U-9* had single-handedly sunk three British cruisers off the coast of Belgium and the dreadnought *Audacious* had succumbed to a mine in the Irish Sea – the only British dreadnought loss to enemy action of the entire war. A successful attack against massed dreadnoughts was any admiral's nightmare. In the context of the campaign, the risk of losses to torpedoes and mines weighed far more heavily than the opportunity of enemy destruction.

Once viewed in this light, it becomes apparent that Jellicoe was fighting the new war of commerce, and not the old war of battle lines. It is sometimes easy to forget that the Great War was initially seen as a "limited war" very much in the style of the limited wars that had been fought between European states for many centuries. Hence the notion that it would all "be over by Christmas". Only as the war progressed did it become apparent that the old ways had given way to a total war, which had become industrialised. The yearning for a Trafalgar type victory was still very strong in the psyche of the British public and to be fair it was also strong amongst many of the naval personnel of the two respective fleets. However, it was becoming apparent to many of the commanders that the wider campaign issues of controlling the commerce of the seas were now the main purpose of the fleet. Two largely alien concepts came to replace the old notion of set piece battles: the "distant blockade" of Germany by Britain, and the strangulation of Britain's imports of food and war by Germany's U-boats. In this respect Jellicoe had a strong grasp of what was required of him. By preservation of her fleet and its numerical superiority, Britain needed to maintain the status quo in which she controlled the seas, and this meant she controlled all ocean going trade that was denied to the enemy. Germany, on the other hand, had to break this grip. Her failure to do this was ultimately a major factor in her defeat.

As early as 1914, Jellicoe had put in writing to the Admiralty what he intended to do in the event of an enemy fleet breaking off an engagement and heading away from him. He did not intend to subject the fleet to the risk of a torpedo attack launched by the enemy's torpedo boats screening its retreat. In that respect he was doing what he had said he would always do. Here lies the origin of the myth that "Beatty delivered the High Seas Fleet and Jellicoe threw it away." Beatty was a consummate man of public relations. He had an easy charm and swagger that appealed to the press and public alike. He was seen as a second Nelson, full of dash and courage, ready at all times to lead his battlecruisers, cavalry style, to cut out the black heart of the Hun. In the years following Jutland his case grew with the telling and Jellicoe's reputation suffered accordingly. With time and a clearer perspective, history has been able to correct this imbalance and Jellicoe now gets something closer to the respect that was his due for his time as CIC of the British Grand Fleet.

The seeds of Germany's defeat – the Kaiser's diktat

If we can argue that the strategic influence of Jutland was decisive despite its inconclusive tactical outcome, then we can also come to accept that one of the most complete naval victories of all time does not even have a name, let alone much of a mention in the history books. This naval "victory" was the internment of the German High Seas Fleet by the Royal Navy and her allies in November 1918. Although internment does not have the same meaning as surrender, it is inconceivable that the Versailles Treaty of 1919 would not have ultimately included unconditional surrender of the

fleet as one of its clauses. It is all the more remarkable for knowing that the German High Seas Fleet constituted numerically the second largest navy in the world at the time. At a stroke almost the entire fleet was removed from contention. Had hostilities been resumed, following a possible rejection of peace terms in 1919, Germany would have been fighting a war in which her fleet had been removed from the campaign as effectively as if they had been annihilated in battle. There is no parallel in history. Midway and Leyte Gulf are both crushing defeats, but not annihilations. Trafalgar and Tsushima are near annihilations, but they are not complete. Neither do the total ship count nor tonnage losses come close to those of 1918. The unconditional surrender of Germany and Japan in 1945 involved the surrender of two navies whose forces were all but spent in battle. The Italian navy surrender of 1943 also involved much smaller numbers.

Of course, because of the perverse nature of battle code, the internment of the High Seas Fleet is not usually viewed as any kind of victory. Indeed some contemporary officers have documented in letters and memoirs of how they felt it was Britain's shame to have the undefeated fleet delivered into their hands, implying it to be almost tantamount to a defeat. There can be sympathy with this view, and recognition of how hollow the "victory" must have seemed. Nevertheless, when viewed outside a militaristic viewpoint, the elimination of virtually an entire battle fleet by the steady attrition of naval blockade is in real terms a victory of one military force over another. Not until the "winning" of the Cold War in 1989 do we have to again accord "victory" to such an apparent non-event.

So how did this drama unfold?

The first seed of defeat starts with the Kaiser's order to "not risk the High Seas Fleet." This dictate had a suffocating effect on the German Naval High Command throughout the entire Great War. With hindsight this timid approach to the use of the world's second most powerful fleet appears incomprehensible. Having antagonised the British by its creation, to the point of making Britain's entry into the war inevitable, it seems doubly foolish to then not use it. But then the Kaiser was a foolish man.

In December 1914 the numerical gap between the capital ships of the two fleets was effectively reduced to about two. The three British battlecruisers *Invincible*, *Inflexible* and *Princess Royal* had been dispatched to the South Atlantic on an avenging mission to destroy Von Spee's Asiatic Squadron – the victors of the Battle of the Coronel. *Audacious*, a dreadnought battleship, had hit a mine and sunk in the Irish Sea. This event was never officially announced by the British to cover up its loss. Four dreadnought battleships were temporarily out of contention whilst they were being refitted. Had the two fleets met head on at this point in the war the outcome would have been far from clear. In view of what we know now of the superiority of design of the German ships, the result would have been too close to call. All of which demonstrates that the High Seas Fleet could have been deployed with more conviction and effectiveness at this time. As the events were to unfold, Germany was doubly blessed in December 1914. Not only, as we have seen, was her numerical inferiority at its least critical, but she was also presented with an opportunity for which the German Naval High Command continually strove.

The missed opportunity

In the wider context and with the benefit of hindsight, it is quite possible that Germany missed her best opportunity of winning the Great War at this time. The key to this was British sea power. Remove that and Britain would have to sue for peace. We now know that the elimination of Britain from "The Allies" would almost certainly have led to the defeat of France and Russia. If we measure naval supremacy by dreadnought head count then Germany had a deficit of about ten to overcome. Only by removing this numerical advantage did she have the freedom of action to challenge British control of the seas head on. Consequently, the aim of German naval strategy was to isolate and destroy a significant portion of the Grand Fleet. On December 16th, 1914 the main body of the High Seas Fleet was on a collision course with a British task force, which would have numbered ten capital ships. Incredibly, this was the very scenario Germany had planned for, and had those ships met, we may well be reading history books that spoke of the great German victory at the Dogger Bank that turned the naval balance and eventually the war for Germany. Who knows? Maybe we would be reading it in German.

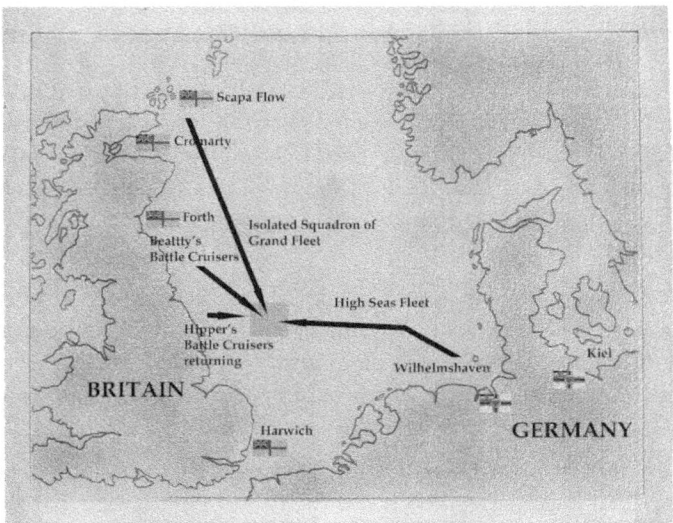

This battle was not, of course, the battle of the Dogger Bank as fought in 1915. Almost certainly, that battle would never have happened if our earlier Dogger Bank battle had occurred. Let's examine the events that nearly led to it, and the reasons why it never occurred.

By December 1914, the Germans had already experimented with hit and run raids on the English East coast. These were, in themselves, of no real military significance, but their nuisance and propaganda value was out of all proportion to their real effects. The German Naval High Command hatched a plan to use Hipper's battlecruisers to raid several of the English coastal towns in the North East — Scarborough, Whitby and Hartlepool. They felt sure this was enough to entice some British forces south from their bases in Scotland, probably Beatty's battlecruisers. The accompanying light forces were to lay mines in British coastal waters. It was hoped that these would cause significant damage to the British battlecruisers and maybe the Grand Fleet as they passed over them in pursuit of the German force. The High Seas Fleet was to act in support of Hipper's battlecruisers should any British forces manage to engage the German battlecruisers.

What actually happened was a perfect example of the usefulness and dangers of intelligence information that so often determines success and failure in naval encounters. The British had access to the German code, and for much of the war were able to track the German orders. This was the first occasion they were able to decipher information and it provided them with advance warning of Hipper's battlecruiser movements. The question was how big a force they should assign to deal with them. Critically the code information did not alert the British to the presence of virtually the entire High Seas Fleet east of the Dogger Bank. Indeed, the Admiralty conveyed to the British CIC that they believed the German High Seas Fleet was not going to be operational. The fatal decision was taken to send only a single battle squadron of six modern dreadnoughts with supporting cruisers and destroyers. These would rendezvous with Beatty's battlecruisers and then deal with Hipper's returning force somewhere in the vicinity of the Dogger Bank.

Jellicoe protested on the basis that the Grand Fleet should always act in concentration. If, for any reason, the intelligence information was incorrect, then the British always ran the risk of running into the High Seas Fleet. He was overruled. Thankfully for Britain and her control of the seas, events did not allow him the dubious pleasure of saying, "I told you so."

This is where we now speculate on what might have been if Ingenhol, the German CIC, had been given more freedom of action than a timorous Kaiser was willing to give him. The Kaiser's standing orders were to avoid risking the fleet in a major surface action anywhere outside the Heligoland Bight, unless circumstances appeared that were favourable to the High Seas

Fleet. Ingenhol reasoned that dispatching the fleet to the Dogger Bank "in support" of Hipper's battlecruisers was just about in the spirit of the Kaiser's orders.

However, he received news of skirmishes between his advance scouting forces and some British cruisers and destroyers. He reasoned quite correctly that these ships must be screening a larger more powerful force which, as far as he knew, could be the whole of the Grand Fleet. Mindful of the Kaiser's orders, he was in an invidious position. If he risked the fleet and his worst fears were realised, he could face a damaging naval encounter and possibly an ignominious dismissal from his post. He had no option but to withdraw the High Seas Fleet to the safety of its base, leaving Hipper's force to the mercy of whatever British forces were out there. Had he been free to continue his current course, he would have made contact with the main British forces. Ingenhol's fourteen dreadnought battleships, eight pre-dreadnoughts, nine cruisers and fifty-four destroyers would have run into the British force whose capital strength consisted of six dreadnought battleships and four battlecruisers. Hipper's four battlecruisers would also be closing from the west.

Unless the British made a remarkable escape, Ingenhol could have been in a position to destroy sufficient British capital ships to significantly reduce the British numerical advantage that gave her command of the open seas. We can surmise that future generations of German capital ships would have proudly borne the name *SMS Ingenhol*. In the event even Hipper's battlecruisers, left precariously to their fate, were able to slip past the British forces and history was able to record only skirmishes between the scouting cruisers and destroyers.

As a footnote, the whole event is largely remembered for the damage inflicted on the English coastal towns of Scarborough, Whitby and

Hartlepool by Hipper's battlecruisers. The British were more intent on trapping Hipper's forces returning from their sortie than defending the towns in the first place. Hipper's bombardments killed 105 men, women, and children and wounded 525. With the possible exception of some minor defences at Hartlepool, the towns were of no military significance. Already by the end of 1914 the nations of Europe were waking up to a new ruthless way of waging war.

The seeds of German Defeat – The Jutland Syndrome

The second seed of ultimate defeat was sown at Jutland. We have touched on this previously. The German High Seas Fleet enjoyed the better of the early exchanges between the battlecruisers, but then ran into one of the biggest naval ambushes in history. Although the High Seas Fleet escaped its fate and claimed Jutland as its victory, in private Scheer conceded that it should never be attempted again for fear of being wiped out. Despite the euphoria of its small losses, the High Seas Fleet had in fact been badly mauled. Its survival had been partly due to the resilient design of its ships. However, many of these ships were in no fit state to sortie out into the North Sea for many months. The experience deepened the feeling of hopelessness that a German victory by fleet action was not achievable.

The seeds of German Defeat – the U-boat campaign

The Jutland experience led inexorably to the third seed of ultimate defeat – the U-boat campaign. Following the stalemate on both land and sea, the idea began to take root that Germany's best hope of victory lay in the U-boat weapon. Confident predictions were made that six months of unrestricted U-boat attacks on all ships going to and from the British Isles would force Britain to the peace table. To the opponents of an unrestricted U-boat campaign, the risks were palpably obvious. The experiences of 1914-16 had illustrated how easy it was to sink ships with Americans on board and how this lead to a worsening of relations with the United States. Proponents of the unrestricted campaign would point to American outrage at Britain's practice of stopping and searching American ships in pursuit of its own blockade. But as one American journalist would point out, the comparison was as odious as suggesting that a man trampling all over your flowerbed could be equated with someone bursting open your door and shooting dead your sister. Anyway, it would all be academic if the British were eliminated from the war before America was antagonised sufficiently to declare war.

They were nearly right. Initially losses were horrendous and the British were for a while looking into the abyss. Only the introduction of the convoy system brought the slide to a timely end and Britain was able to survive by the skin of her teeth. The U-boat campaign continued, but could no longer achieve its aim. It did, however, do what its opponents said it would always do – bring America into the war. Germany now had to win the war sooner rather than later .The failure at sea meant this had to be on land on the Western front. The collapse and surrender of Russia gave Germany her opportunity, but after unprecedented early successes at the beginning of 1918 the offensive ground to a halt. The Allies were bolstered by ever-increasing numbers of fresh American troops and remorselessly the door began to swing open for the main body of the British and French armies. Staring at military defeat on land, and a worsening of morale on the Home Front due to major food shortages caused by the naval blockade, the German leadership began to look for a way out. There was only one way, of course, and that was to sue for peace.

And here was the final seed of ultimate defeat that ironically came directly from within the High Seas Fleet itself.

The Last Sortie

With an unfavourable peace treaty beckoning, the German Naval High Command hatched a desperate plan to temper the severity of the peace terms. Basically, they were to deliver one final bold naval assault. Its purpose was to inflict sufficient damage on the British Grand Fleet to weaken its bargaining position. Light forces were to mount several hit and run raids on various coastal targets around the Thames Estuary. It was expected that the British would have to respond by bringing the Grand Fleet down from their Scottish bases over cordons of waiting U-boats and mines positioned in its path. Simultaneously, the High Seas Fleet would be dispatched then to engage the Grand Fleet, which would have been disrupted and hopefully weakened by the U-boats and mines.

The plan actually had a reasonable chance of success. All it had to do was inflict significant damage on the Grand Fleet. It was not even necessary to win the engagement. However, insidious damage had already been done to the High Seas Fleet. This was not in material losses, but in the morale of the crew. Firstly, the long period of inactivity had taken its toll. Secondly, many of the best crews had been removed to the U-boat service. Thirdly, those who remained were suffering the privations of food shortages familiar to the Home Front. With talk of the final sortie being a suicide mission, the collapse of authority in Russia still a fresh memory, the unthinkable began to happen. Revolution was being muttered.

The German Naval Mutinies

The High Seas Fleet put to sea, but several elements of crew on two of the battleships refused to respond to orders. Scheer reacted by issuing an ultimatum, lining up several boats and torpedo boats that threatened to torpedo the mutinous ships unless they surrendered. After what must have been a very tense standoff, the mutinous crews capitulated, but under the circumstances Scheer was forced to return the fleet back to port and abandon the mission. Things only got worse. News soon spread around the bases of Wilhelmshaven and Kiel. In sympathy with the incarcerated mutineers a general revolt broke out amongst the shore staff. Inspired by the Soviet model, red flags were raised and Workers and Soldiers Councils set up. Law and order inexorably slipped away and traditional authority broke down.

As with Russia, the mutinies of the navy spread to the Home Front generally. In some cases republics were even declared. With an Empire he no longer controlled, the Kaiser abdicated and the German High Command sued for peace. An armistice was signed in a railway carriage in the forest of Compiege. The war had been like no other. It had been bitter and bloody. The Allies were in no mood to be magnanimous, demanding stringent conditions before they were willing to accept an armistice. The High Seas Fleet, still largely intact, was not going to be allowed to remain as a fighting force.

After some prolonged haggling amongst the Allies, who were far from united in what they wanted to do with the world's second most powerful fleet, the following conditions were added to the Armistice terms. Germany would disarm, and take the bulk of its most modern ships to the Firth of Forth, where its final destination would be decided. These ships were to be ten dreadnought battleships, six battlecruisers, eight light cruisers and fifty destroyers. Despite protests from some British naval staff who wanted outright surrender, it was decided that the ships would be interned and nominally commanded by skeleton German crews

Internment of the High Seas fleet

It must have been a remarkable sight. Through four years of war the two fleets had barely caught sight of each other. Now they were both to meet in open water, and sail together into the Firth of Forth. What surprised all onlookers, particularly the Germans, was that virtually the entire British Grand Fleet and major elements of the French and United States navies came out to accept their surrender and escort them to their place of internment. To the German crews, particularly the officers, this was to be a source of some satisfaction. Clearly the British were affording them a backhanded compliment in that they feared and respected the power of the High Seas Fleet even under conditions of surrender. As if playing into this sentiment, or maybe in disbelief that such a surrender could actually be taking place, Beatty continued throughout the whole proceedings to take no chances with any German surprises.

The formal acceptance of internment was strained. It is generally agreed that Beatty, now CIC of the Grand Fleet, behaved somewhat unmagnanimously, perhaps resentful of "his" hollow victory. No doubt he also carried memories from the past four years of the many traps and German escapes. To Beatty, this war should have been his opportunity to become immortalised as a new Nelson. Now the best he could do was accept a limp and almost contemptuous surrender from an enemy that had eluded him for so long. The fact that he was accepting the most complete victory in the history of naval warfare must have barely registered with him. For the British there was a final touch of the inappropriate. The three men who probably had the most to do with ensuring this final victory, however dismissively obtained, were not represented at the formal surrender. Jackie Fisher, John Jellicoe, and Winston Churchill had all in some way been demoted from their frontline duties.

Ideally, they should have been interned in neutral ports, but no neutral nation had sufficient forces to police them. The risk of a breakout back to Germany was too high for this to be practical. In the end Scapa Flow was deemed the only place where adequate supervision of the interned fleet could be guaranteed. This was to be a source of some resentment to the interned fleet, which had been led to believe that their final destination was neutral ports.

The End of the High Seas Fleet

And so on to the final act. One midsummer's day in 1919, as the peace negotiations dragged on at Versailles, a most extraordinary act occurred at Scapa Flow. A boat full of sightseeing schoolchildren was to experience something they would never forget. One of the most formidable fleets in the world, of its own volition, chose to scuttle itself inside the principal base of its adversary. Sixteen Dreadnought capital ships and even greater numbers of smaller ships slipped almost silently beneath the cold waters of the desolate Scottish base. It had cost a fortune to build, brought a major European power into conflict with its national aspirations, and indirectly lost it the war. During its brief existence it had been virtually useless.

Now very little of it remains. Most of the ships were the subjects of a series of remarkable salvage operations between the two world wars. The handful that are left have been variously broken into and several of them lie topside down in the mud.

There remains one final footnote. Pre-1945 shipwrecks are amongst the few locations where you can find the high-grade radiation free metal required for the delicate scientific instruments used in such places as nuclear installations and lunar vehicles. Ever since the atomic explosions at Hiroshima and Nagasaki in 1945 the earth's atmosphere has carried extra traces of radiation. Because steel processing requires huge intakes of air, all steel produced since 1945 has always incorporated some radiation. Scapa Flow is probably the best source of radiation free steel in the world. It has never been officially admitted, but almost certainly some of the High Seas Fleet is now on the moon.

Suggestions For the Ex-submariner Who Misses "the Good Old Days on the Boat"

- Sleep on the shelf in your closet. Replace the closet door with a curtain. Two to three hours after you fall asleep, have your wife whip open the curtain, shine a flashlight in your eyes, and mumble "Sorry, wrong rack."
- Repeat back everything anyone says to you.
- Spend as much time as possible indoors and avoid sunlight. Only view the world through the peephole on your front door.
- Renovate your bathroom. Build a wall across the middle of your bathtub and move the showerhead down to chest level. Shower once a week. Use no more than two gallons of water per shower.
- Sit in your car for six hours a day with your hands on the wheel and the motor running, but don't go anywhere. Install 200 extra oil temperature gauges. Take logs on all gages and indicators every thirty minutes.
- Watch only unknown movies with no major stars on TV and then, only at night. Have your family vote on which movie to watch, then watch a different one.
- Have the paperboy give you a haircut.
- Take hourly readings on your electric and water meters.
- Eat only food that you get out of a can or have to add water to.
- Make up your family menu a week ahead of time without looking in your food cabinets or refrigerator.
- Set your alarm clock to go off at random times during the night. When it goes off, jump out of bed and get dressed as fast as you can, then run to your kitchen with the garden hose while wearing a scuba mask.
- Once a month take every major appliance completely apart and then put them back together. Ensure you have parts left over.
- Use eighteen scoops of coffee per pot and allow it to sit for five or six hours before drinking. Never wash any coffee cups.

- Have a fluorescent lamp installed on the bottom of your coffee table and lie under it to read books.
- Check your refrigerator compressor for "sound shorts."
- Put a complicated lock on your basement door and wear the key on a lanyard around your neck.
- Every so often, yell "Emergency Deep," run into the kitchen, and sweep all pots/pans/dishes off of the counter onto the floor. Then, yell at your wife for not having the place "stowed for sea."
- Put on the headphones from your stereo (don't plug them in). Go and stand in front of your stove. Say (to nobody in particular) "Stove manned and ready." Stand there for three or four hours. Say (once again to nobody in particular) "Stove secured." Roll up the headphone cord and put them away.
- At night, replace all light bulbs in the living room with red bulbs.
- Find out how long it will take to do a job. Give yourself half the time it should take then have someone scream at you for not working fast enough.
- Yell "Torpedo Evasion" and run through the house knocking over everything that isn't bolted down.
- Yell "Man Overboard" and throw the cat in the pool.
- Sit up from 1130 to 0530 in front of your stove to insure it doesn't turn on by accident.

I bought SHII off the shelf and ran into immediate problems with crashing. After reloading and experimenting for several months I finally realized there must be someone who knows how to fix it. A forum search brought me here, where I learned of the official patch and mods which not only fixed the problem, but actually made the game fun to play. The first thing I noticed about Subsim was how many people there were here who were willing and able to help, to mod and to have a good time. A lot of the faces have changed, but those characteristics haven't. I will never forget the infamous 'Drebbel Trial' while we were waiting for SHIII.

"Sailor Steve" Bradfield

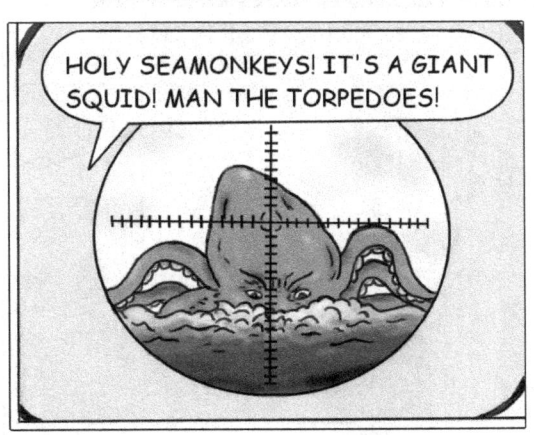

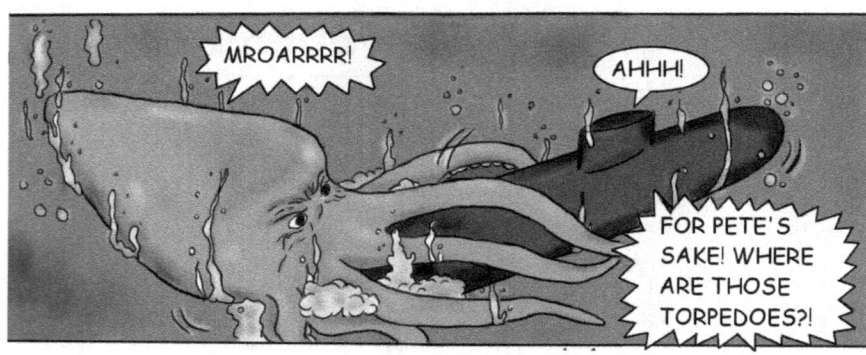

In Memoriam: Erich Topp (U-552)
Theodore P. Savas

I have always been fascinated by the sea and machines with the ability to travel underwater. This interest melded with my passion for WWII, and the result has been more than two decades of reading and learning about German U-boats. My other lifelong interest is the American Civil War. Unfortunately, the participants of that conflict were all long dead before I was born. One of the benefits of an interest in WWII is the ability to speak, meet, and/or correspond with people who participated in the events and shaped them, each in his own small way, into what we read about today. One of the men I had the pleasure of meeting on many occasions was Admiral Erich Topp.

We carried out a lengthy correspondence and met many times over the years. For the purposes of this essay, I will focus on three specific events: our first meeting, a book-signing event we did together, and the last time we met.

I had the pleasure of meeting Topp when he was visiting the States in January 1996. The genesis of that meeting was rather unusual. I had just finished reading Eric Rust's wonderful book *Naval Officers Under Hitler: The Story of Crew 34* (Praeger, 1991). One of the footnotes mentioned a recent letter and interview with Topp conducted by Rust. I am still not sure what prompted my decision, but I picked up the phone and called Baylor University in Waco, Texas. According to the book, Dr. Rust served there in the early 1990s as a lecturer. As luck would have it, Dr. Rust was still at that institution working as an associate professor.

Dr. Rust was friendly and helpful. "I just spoke with Topp," he told me. "He happens to be here in Texas visiting his daughter. I don't think he would mind if you called him. Would you like the number?"

I scribbled down the ten digits, thanked him for his help, and hung up. My wife was shocked. "You are going to call a German submarine commander on the phone and chat?"

"Well, sure, why not?" I laughed. "All he can do is hang up on me. But I've heard he is a decent man, and who knows where this will lead?" Indeed, I could never have envisioned the journey upon which I was about to embark.

A pleasant female voice answered the phone. "Is Admiral Topp available?" I inquired. She asked me to hold.

"Topp here." The voice was clear and strong with a heavy German accent.

"Admiral Topp, thank you for taking my call," I began. "My name is…" We hit it off well and the questions I had for him piled up in my mind, one behind the other. The conversation consumed perhaps fifteen minutes before I finally hung up. Topp was on his way to visit Elizabeth Stern, an "old friend" who lived in Southern California. He had known her during the war. Of far more interest to me was that he invited me to fly down and interview him there the following week.

I caught a Southwest Airlines flight at dawn on January 27, 1996, picked up a rental car when I landed, and drove into the desert to meet Topp. I admit that when I pulled into the driveway I was a tad nervous about the whole thing. Just as I remember the last time I saw him in Germany many years later, I recall clearly the first time I set eyes on him. As soon as I turned off the car engine a tall man in a dark blue suit with a red bow tie walked outside and around a bush, tall, erect, and with shoulders square. I recognized him immediately.

He offered a broad smile. "And you must be Mr. Savas," he said, extending his hand.

I grasped it and returned the smile. "Admiral, it is a pleasure to meet you." His gaze was striking in its intensity, his eyes a warm and brilliant blue. The hand I was shaking belonged to an 81-year-old man, but he looked and acted as though he was 55.

We walked inside, where he introduced me to "Mrs. Stern." I believe she was about 76 years old at the time, a widow living alone in an upscale, gated California community. She had prepared brunch, complete with tea, coffee, pastries, and more.

Topp settled into an armchair while I did the same next to him. With his permission I taped our conversation — hours and hours of dialogue. He continued to call me "Mr. Savas." When I told him to call me 'Ted', he responded, "Well, then you will call me Erich."

"Admiral, I appreciate that, but I don't think it would be proper. If you don't mind, I will call you Admiral, and to you I am Ted." He smiled his understanding.

"When is your return flight?" he asked.

"They have flights every hour to San Jose, so I can fly anytime," I answered. "My wife is pregnant and our second child is due any day."

"Ah, then we will have to make sure you are close to a telephone. You don't want to miss the big day! Do you know if it is a boy or a girl?" he inquired.

"We don't know," I replied. "But my first child was a girl — Alexandra."

"Then we should hope your new baby has a periscope!" he joked with a twinkle in his eye and a broad smile on his face. We both laughed.

My questions filled the gamut, from his early years and early heroes and role models to U-boat training, recollections of different commanders, logistics, training, and more. Before we knew it, it was lunchtime. When I broached the subject of leaving, Topp waved off the suggestion and invited me to stay. We drove through the gated community to a small cafeteria-style restaurant. Another hour of questions, answers, and commentary followed. When we drove back I told him it was unfair of me to take time away from his visit with Mrs. Stern," but he waved that off as well and invited me back inside.

When I mumbled my apologies as I stepped through the door, a smiling Mrs. Stern patted me on the shoulder and responded graciously, "You will please stay as long as you wish." She poured us coffee and we settled down, this time at the small kitchen table.

It was during this third round of conversation that I learned of their unique relationship. As best I recall (as I write, I don't have access to my taped conversations), Mrs. Stern was a cabaret dancer in 1941. She led me to her living room where pictures of a stunningly lovely young dancer lined one wall. Traces of her beauty were still evident more than half a century later. And then she told me the story of how she and Topp first met.

"I was dancing [I believe it was in St. Nazaire] and U-boat men were tearing apart the place — throwing chairs, making a ruckus, yelling. And then, in walked Erich Topp!" She clapped her hands and smiled at the Admiral, who was standing next to Mrs. Stern as she related the story. "He did not have to say a word — he just glared at them. They stood as best they could — many were very drunk — and gathered themselves together.

He ordered them to clean up the place and they did. He was very handsome — the perfect picture of a U-boat commander!"

I turned to look at Topp, who shrugged his shoulders. He had a rather sheepish look on his face. "Ja, Ja," he muttered, almost like an embarrassed schoolboy.

Images of the opening scene in *Das Boot* immediately sprang to mind. When I mentioned it, Topp nodded, shrugged, and replied, "Something like that, maybe not quite as wild. No gun shots!" All three of us laughed.

"Well, we saw each other once in a while over the course of perhaps one year," she said. Topp nodded as she spoke. "He was of course gone often and I never knew whether he would come back," she continued. Her voice grew softer. "And then I was sent elsewhere, out of St. Nazaire, and we lost track of each other. For some reason I believed he was killed during the war, and always believed that."

A confluence of events fifty years later conspired to bring them back together. In 1995, Topp's wife of 50 years died. Around that same time, Mrs. Stern learned of a BBC U-boat special that featured commentary by Admiral Topp. I believe her daughter brought it to her attention. Her U-boat commander was still alive! She made inquiries and eventually contacted him. If memory serves, his visit to the Southern California desert was the first time they had seen one another since 1942.

Mrs. Stern brought out a photo album that had photos of Topp I had never seen. One was especially handsome — a formal smiling portrait without his cap — and I commented on it and several others. (Two weeks later they appeared in my mailbox signed by Topp; either he or Mrs. Stern had had them reproduced for me.) It was nearly time for dinner. Mrs. Stern insisted I stay and I gladly took her up on the offer. Twelve hours after I arrived, I was driving back to the airport for a late night flight to San Jose. Right before I left, Topp signed my copy of his memoir *Odyssey of a U-boat Commander: Recollections of Erich Topp* (Praeger, 1992). This is what he wrote: "Especially for Ted Savas. Waiting with him for his second child, and with best wishes for his wife Carol. January 27, 1996." My son Demetrious Theodore (DT) was born a few days later.

Sadly, the reunion of the pair of young lovers wasn't destined for a happy ending. Topp and I traded letters regularly (several each month) for many months. And then his replies stopped. Weeks and weeks passed. "Something is wrong," I told my wife. Something in fact was very wrong. Topp owned a lovely home by the sea in Spain. Later in 1996, while he and Mrs. Stern were visiting there, the couple was involved in a head-on collision on a dangerous highway near his home. The crash either killed Mrs. Stern outright or she died shortly after the accident. Topp was critically injured and forced to endure many months in the hospital. He recovered, but the accident took its toll on him, both physically and emotionally. "I am sorry you did not receive word from me for many months now," his next letter began, "but for 5 ½ months my right hand has been out of action."

A year later and *Silent Hunters: German U-boat Commanders of WWII* (Campbell, 1977; Naval Institute Press, 2004) was in development. This was a book project I put together as editor with several of the world's leading U-boat scholars, each of whom provided invaluable contributions. I explained the idea and general thesis of the project to Topp and asked if he would write a foreword. He agreed to write whatever would help the project, but told me he had something else in mind. Shortly thereafter, a manuscript essay in German arrived in my mailbox. The lengthy work detailed his relationship with Engelbert "Bertl" Endrass, a fellow commander (*U-567*), Knight's Cross holder, and Topp's best friend. Their close relationship is the stuff of legend in U-boat circles.

Topp wrote the manuscript in snippets in his Captain's cabin during his 15th war patrol in 1942. It opens with these words: "I wrote these pages in the lonely wastes of the Atlantic when all hope had vanished that Engelbert Endrass might return alive from his last patrol." Someone told me rumors had circulated for years that Topp had written this piece, but he had never made it available for publication. I don't know whether that is true or not, but his offer of it in 1997 made it widely available to the general public as the opening essay in *Silent Hunters*. Baylor University's Eric Rust was also on board for the project, and graciously agreed to translate and annotate the chapter.

Shortly after the book's publication, Topp flew to San Francisco to take part in a weekend-long "Celebrate History" military conference on February 13-15, 1998. Hundreds of people passed through the event, including U-boat reenactors who were thrilled with the opportunity to meet one of history's most successful tonnage aces.

As best as I recall, Topp flew in the day before the event. Tom Broadfoot of Broadfoot Publishing Company was with me when we picked him up at the airport. The next day Topp and I met up with Admiral Ralph Metcalf, another friend of mine from Saratoga, California. Metcalf was the former commander of the submarines *USS Pogy* (SS-266) and *USS Loggerhead* (SS-374) in the Pacific Theater during World War II.

We toured the San Francisco Maritime National Park along the harbor and enjoyed lunch overlooking the bay. To our disappointment, *USS Pampanito* (SS-383), at Pier 45 at Fisherman's Wharf, was closed that day for maintenance.

Being an Admiral has it privileges, and Metcalf wasn't about to take no for an answer. He pulled aside a park employee, explained who he was visiting with, and arranged a VIP tour of the boat.

As we walked down into the submarine, someone working inside recognized Topp and we spent the next hour standing in the control room while Topp and Metcalf regaled a very receptive audience with war stories. When one of the boat's "crewmen" advised us that the conning tower was accessible (it is normally off limits to visitors), Metcalf made a beeline for the ladder. Topp held back, his leg and hip still bothering him from his prior accident.

"Admiral Metcalf, please be careful," I pleaded. "I promised Mrs. Metcalf I would bring you home in one piece."

Metcalf, who was (and still is, as of the date of this writing) hard of hearing, replied, "Mrs. Metcalf is here?"

"No Admiral, she told me to watch out for you, and I just don't want you to get hurt on that ladder." I quickly added "sir" to the end of the sentence when I saw fire stir in his eyes. It sounded stupid, but it sure felt right.

Metcalf tilted his head to one side and shook it. "I was climbing these ladders before you were born, son! Out of my way!"

With that, Metcalf scampered up the ladder and into the tower more quickly and easily than I imagined possible (he was a 1932 graduate of the Naval College). I followed him up and we stood inside the cramped tower. "You shot right up those steps, Admiral. How did that feel to do that again?"

"Felt damn good!" he shot back. "Damn good!"

I don't think a student of history ever felt more thankful than I did at that moment. Below me, smiling up through the hatch, was Admiral Erich Topp. Standing next to me was Admiral Metcalf, another successful submarine commander from the Pacific Theater. I asked them every question I could think of, and they shot back answers based upon personal knowledge rather than something they had read in a book.

After we left the *Pampanito* we walked along the dock taking in the ships in the harbor. While Topp was engaged in conversation with someone from the submarine, Metcalf and I discussed his successful attack against the Japanese army cargo ship *Awe Maru* off Suruga Bay, Honshu, Japan. He pointed out into the bay to offer another example of a larger ship. It was a dreary day, with patchy fog that made it difficult at times to see too far into the distance.

"See that ship out there?" he pointed. "That's big. About 10,000 tons."

I cleared my throat. "Ah, Admiral. That's not a ship. That's a small island. But it does look like a ship from here, sir, with the fog and all."

"What?" he asked in surprise. "You don't think I know what a ship looks like?" I bit my tongue while he squinted and looked again. After a few seconds he rubbed his eyes and shook his head. "Looks like you're right. It's a damn island." We both chuckled.

A few minutes later Topp joined us. Metcalf and I filled him in on the conversation we were having about the attack against the *Awe Maru*. When I asked Topp about one of the attacks he had made against a large tanker off the Eastern seaboard of the America, he scanned the harbor and pointed into the distance. "See that ship there?"

"Where?" both Metcalf and I asked.

"There, right there!" he pointed.

"Topp, I'm almost deaf, but you must be blind!" Metcalf joked, sticking his elbow out and tapping it into the Admiral's side. "Can't you see that's an island?"

"No, no . . ." Topp answered, shaking his head. The former German ace narrowed his eyes and took a second look before muttering in German, "What am I thinking!" Metcalf and I had a good laugh, and when we filled in Topp on the joke, we all three laughed heartily.

That evening we drove from the hotel back to the San Francisco harbor, where the Liberty Ship *S.S. Jeremiah O'Brien* is docked. After touring the ship with a group of people from the conference, we sat at tables out on the deck. It didn't take long for the buzz to spread that Admiral Topp was on board. An older man sitting at the next table stared at the Admiral for several minutes before finally getting up and walking over. On his head was a *USS Niblack* (DD 424) hat.

"You're Erich Topp?" he asked. His eyes were cold and narrow, as if he was suspicious of the answer.

Topp looked up, smiled, and was about to reply when he spotted the hat the man was wearing and the cold look upon his face. The Admiral shot a quick glance in my direction before standing up. "Yes, I am Erich Topp," he said extending his hand. "And what is your name?"

The German hand remained unclasped for several seconds until the American stranger, reluctantly, reached out and gripped it. He introduced himself by name and then added, "My best friend was on the *Reuben James*. I watched his ship go down after you torpedoed it. I never saw him again." He continued holding Topp's hand, refusing to let it go. The meeting and conversation had taken a sudden — and chilling — turn.

Topp slowly shook his head. "That was a terrible day, and war is a terrible thing," he began. "I am very sorry for your friend's loss. I am sure he was a good man. When I saw that rescue operations were underway, I did not fire again." He paused before continuing. "We all lost good friends during that dark time."

The *Niblack* veteran took a deep breath and it looked as if he was choking back tears. I don't think this was the answer or the sort of man he had been waiting half a century to meet and ask. He let go of the Admiral's hand and mumbled something I didn't catch. Topp offered him a seat next to him, and together they sat down. I excused myself so the two could talk privately.

I don't know exactly what was said and I never asked. But I can happily relate that fifteen minutes later the man removed his *Niblack* hat and placed it on Topp's head. They put their arms around one another's shoulders and someone snapped a picture. It was a memorable moment and one I shall never forget. Two men, one of them harboring deep resentment against the other for more than fifty years, finally met his "enemy" face to face. The meeting exorcised demons that night, perhaps in both men.

The next day Topp and I had an early breakfast and I walked him back to his hotel room to freshen up. He was scheduled later that morning for a joint question and answer session with Admiral Metcalf. It was misty and spitting a light rain when I picked up Topp to guide him back to the main conference area. Something extra special was in store for him.

During our short walk, he slipped his left arm through my right arm and we made our way down a few slippery steps and outside into a small courtyard between the buildings. A navy whistle sounded. Someone yelled "Attention!" in German and a group of U-boat reenactors decked out in their "leathers" stood crisply at attention.

Topp stopped in his tracks and leaned slightly forward, peering through the misty rain as if his eyes were deceiving him. He looked at me, smiled slightly, let go of my arm, and in a single moment transformed into a commanding officer once more. Without missing a beat he conducted an inspection, adjusting a collar here, a cap there, exchanging a few words with each man as he made his way down the line.

When he finished his inspection, the naval escort walked Topp inside, picked up Admiral Metcalf and myself in the lobby, and escorted us into the main room toward the stage. "Castor Mourns Pollux," a haunting original score written for the occasion by composer Patrick Brennan, a friend of mine from Chicago, filled the room, which was jammed with conference attendees anxious to hear the men speak. It was a memorable event, where two former warriors explained the differences in conducting submarine warfare in the Pacific and Atlantic oceans. A riveting hour of question and answer followed.

Later that day Topp and I signed copies of *Silent Hunters* for customers anxious to secure a book with his signature and inscription. I remember at one point he looked up to see the line stretching off into the distance and around a corner booth. "I would never have guessed such a thing — such interest!" he exclaimed over and over. "I just can't believe it."

That evening, after we finished dinner and coffee, he told me he was tired and wanted to retire early. I escorted him back to his building. Within a few minutes, several people acknowledged his presence and asked for an autograph. He accommodated each one, in turn, with superb graciousness. When we were alone, he grabbed my arm as we stopped in front of the elevator. The reception he was receiving in San Francisco (of all places) stunned him: the interest in his life and accomplishments in Europe in general, and in Germany in particular, was nothing like it was in the United States. Here, people everywhere stopped to talk with him, the lines to see him were long, and the interest in his career palpable. "Intense" is not too strong of an adjective for the reception he was experiencing. And it really shocked him.

"I wrote my book to put all this behind me — to tell my story and move on with my life." I still recall how forcefully he said those words, his eyes burning as he tried to explain himself. "Do you understand what I am saying?"

"Yes, Admiral. I know exactly what you are telling me," I replied. "But I am sorry to say that what you are asking is impossible. Like it or not, those few years have defined you for life." No matter what else he had accomplished, he would always be the commander of *U-552* (the Red Devil Boat), Engelbert Endrass's legendary best friend, the war's third-leading tonnage ace, and a good man whose outspoken condemnation of Admiral Karl Doenitz and the Nazi regime he had served deeply divided the U-boat community. I told him this. He listened and nodded. He already knew.

The fire in his eyes slowly cooled. "More than ever I realize now I will never be able to put any of this behind me. The interest is growing stronger each month, more and more letters and calls from around the world, especially from America. They are always pleased to learn I am still alive so they can try to understand from me, personally, what it was all about. It is my responsibility to explain it as well as I can."

We were about to enter the elevator when a rousing cheer went up from a pub across the hall. Someone was shouting something in German — a toast to victory. I looked at Topp and nodded my head toward the bar. "Shall we see what all the excitement is about?"

He lifted his hands as high as his shoulders, palms facing in, and answered, "Okay!"

Inside we found the U-boat reenactors crowded around a booth with several pitchers of beer on the table. The lights were low and the men were having a grand time. One of them spotted us and leaped to his feet.

"It's the Old Man!" he yelled. "Attention!"

Topp strode in as the men struggled to their feet. "No, no! Please, sit down, sit down! You are having too good a time to stand up!" I know he was slightly embarrassed by the attention, although he loved center stage — even if he denied it.

"Please join us!" implored several of the reenactors. Topp looked at me, nodded, and we slid into the booth. Two mugs were quickly located and filled. Everyone raised a cup to drink toasts to Endrass, Topp, and to fallen sailors everywhere, of every nation.

At one point during the conversation Topp turned to me, leaned over a few inches, and clinked my mug with his own. "To the Sheherazade."

Topp was referring to the notorious nightclub in Paris frequented by U-boat crews while on leave. We had discussed it in depth just an hour earlier. "To the Sheherazade," I replied.

The hour we spent in the pub was unforgettable, but it was also a bit eerie. Sitting next to Topp surrounded by young men dressed in authentic U-boat uniforms while drinking beer in a dark pub-like atmosphere conjures up unusual thoughts in a man steeped with a sense of history. It is impossible to go back in time, but we approached it that evening.

The next day Topp gave a rousing lunch talk about the Battle of the Atlantic, we signed more books, and later that afternoon sat with Metcalf and enjoyed a final cup of coffee together. My wife Carol arrived with our daughter and brought with her my favorite art print: Michael Wooten's Type VII, a lovely hand-shaded U-boat print signed by ten different U-boat commanders. I removed the back and Topp penned a lengthy inscription. The art hangs above my computer as I type.

It had been a weekend to remember. Before we parted, Topp handed me a bottle of good red wine tucked inside a gift bag. "One of the U-boat men here at this conference gave me this," he said. "I cannot carry it back on the plane and it won't fit in my luggage. Is it all right if I give it to you?" I accepted.

I saw Topp again in 1999 in Germany with my wife. We met up in Cuxhaven for a reunion with other members of *U-552*. Together we toured the *Wilhelm Bauer (U-2540)*, a Type XXI boat anchored in Bremerhaven harbor. Topp commanded a Type XXI electro-boat *(U-2513)* during the war's closing months and wrote the tactical manual on the boat. He said his brief tenure as a XXI commander was the only time during the war he violated an order. During the conflict's closing days, orders to sail for Norway arrived. He was to leave with several other boats at the same hour. Topp told me the directive was foolish and little more than suicide. "Allied intelligence and spies would know we were all leaving together. We offered a rich target too good to pass up!" He asked his chief engineer to find a reason to delay the patrol by a day or two. Allied aircraft sank or seriously damaged most of the other boats, but Topp's made it safely to Norway, where he surrendered his U-boat and was imprisoned. Although he alluded to one particularly horrendous event during his time in captivity, he refused to elaborate.

We enjoyed a five-hour train ride back to his home in Remagen, talking most of the way about religion, philosophy, European art, architecture, and much more. Topp designed his own home overlooking the Rhine. The views are spectacular. Our two days there were consumed touring cathedrals, scanning his photo albums, and eating good food and drinking outstanding wine. One particular opportunity remains with me: looking out his window through the end of *U-552's* attack periscope, which one of his men cut off for him after the war. He also showed me his decorations, which he kept in the basement unprotected. I implored him to put them into a safe-deposit box to protect them, to no avail. Sadly, shortly thereafter someone took advantage of his friendliness to steal these priceless historical artifacts. To my knowledge, they have never been located.

The last time I saw Admiral Topp was nine days before Christmas 2003. I flew with my daughter Alexandra (Alex for short, who was then 13) to Germany to meet up with Juergen Oesten (*U-61, U-106, U-861*). I first met Jurgen in 1999. We have since corresponded frequently via email (and do to this day). I was in Germany on behalf of Ubisoft and the Silent Hunter III team from Romania — Florin Boitor and Tiberious Astianax "Maddog" Lazar. I had persuaded Admiral Topp to act as technical advisor for Silent Hunter II, and had enlisted Oesten for the same role for Silent Hunter III.

We spent most of two days sitting in the Oesten's lovely home, interviewing him and asking questions, discussing everything from how he managed to sneak his Monsun boat home on fumes in the spring of 1945, to his fascination with playing Aces of the Deep! We were discussing his attack against the *HMS Malaya* when Oesten took out a piece of paper and pen and mapped out the attack, together with the geometry supporting it. When he finished, I reached out and asked if he might sign the paper for me, so that I might keep it. He did so, and today I have it framed in my library. Lazar later shook his head and said to me, "I was thinking the same thing, but you beat me to it!"

We left Oesten's home on December 15 and took the train to Bonn, where we spent the night. Topp's son Michael advised me to call his father before we arrived to jog his failing memory. I had written him that we were coming to Germany, but he had not replied before we left. I dialed his number and the Admiral answered. I explained who I was and that we would like to stop by to see him.

He asked my name several times. "I am sorry. I am sure I know you, but you see, my memory is kaput. You say you have been here before? Please come by and when I see you, I will remember you." Time had finally caught up with the U-boat ace.

Alex and I took the train to the small station in Remagen and caught a taxi for the house in the hills. It was already late in the afternoon. When we pulled in front of the familiar home, I asked the driver to stay until we were inside. I was unsure what sort of reception awaited us.

Admiral Topp opened the door sooner than I expected. He shook my hand, told me he didn't recognize me, but invited us in nonetheless. I rather reluctantly waved the taxi away and we stepped across the threshold. We stood in his foyer, which boasted a lovely three-story winding metal stairway on the right, and the trials of Odysseus painted on wall panels on left.

"I am sorry about my memory," he began again, patting himself on the front of his head as he shook it slowly. "Please tell me again your name and how I know you."

And so I did exactly that, explaining how we had met, the *Silent Hunters* book project, and much more. Something clicked after a minute or so, and the Admiral lifted his hands in the manner I had come to know so well. A smile spread on his face.

"Savas! My Greek friend and author from America! Yes, of course! Ah, I have no memory, it is kaput, but of course I know you!" He shook my hand again and put his arm around Alex's shoulders. "And I remember you, too, from San Francisco."

He looked very tired, and I offered to leave if we were bothering him. "No, no. I always enjoy visitors and you are an old friend who has come a long way. Come in and sit down."

The once warm and bright home was now largely closed off and cold. The great room overlooking one of the most historic rivers in the world was no longer readily accessible. The last few years had not been kind to Topp. His world had contracted until it encompassed little more than his small kitchen, dining room, foyer, bathroom, and bedroom.

We sat at his table in the dining room. He offered a glass of red table wine, which I gladly accepted. Alex settled for soda water. We talked of many things, but never discussed World War II even a single time. He spent most of his days sitting in his small dining room responding to correspondence, he explained. He showed me a stack of recently opened mail, and another tall stack waiting to be opened. He picked up one letter from an American asking if he could fly to Germany and assist Topp in signing a large number of art prints (which of course the organizer would then sell for profit). Some still prey on the old and infirm; the request disgusted me, but suited the man asking for the favor.

The Admiral was concerned I would be disappointed because I couldn't ask him questions about the war and his role in it. "That's not why I am here, sir," I told him. "We are here to spend some time with an old friend, and that is more than enough."

A look of mild surprise crossed his face when he fully understood my answer. His features appeared to brighten at the thought. I think he appreciated my response and knew it to be heartfelt. Most of the people over the years who wanted to spend time with him wanted something from Topp the U-boat commander: information, an autograph, a memento — something. I was once one of those people. Few wanted to discuss the simple pleasures of art, wine, food, or the weather with Erich Topp, a former architect, father, husband, diplomat, and friend — and a man who happened to also be an aging naval veteran and one of the leading tonnage aces in history.

After perhaps thirty minutes, the Admiral offered to cook us something for dinner, but we finally agreed to all go out to a riverboat restaurant docked on the Rhine. He had taken my wife and I there in 1999. I offered to call a taxi, but he said his BMW was in the garage and we could take that.

I opened the old-fashioned garage door (the sort that lifts from the front and tilts backward) and urged him to let me drive. Topp waved me off and backed the car out. My daughter climbed into the backseat. "Admiral," I began while standing outside the driver's door. "I can drive us down into Remagen. It is dark now and difficult to see at night."

"No, I can drive. I know the way," he answered. "It is my car, you are my guest, and I will drive us." He proceeded to back out into the street too quickly, where an oncoming car nearly smashed into the passenger side of his BMW.

Alex shook her head at me; her eyes wide open in fright. I walked over to the driver's door and opened it: "Admiral, I know you used to command U-boats all over the Atlantic, but I'm driving your car down these winding streets to Remagen. Please sit on the other side."

Topp looked at me with a surprised and rather serious countenance for a few seconds before finally sighing. "Okay, okay, you have pulled rank on me. I know you are now in charge."

Once we found a place to park by the large boat-restaurant, it was clear to me it was closed. "Do you like Chinese food?" he asked. We did. Topp navigated us into town, but was unable to locate the restaurant. "I am sorry," he repeated over and over. "I thought it was right down this street." I finally stopped and asked directions. We were only two blocks away.

We enjoyed a nice dinner that night. The hostess used my camera to snap a few pictures of Topp, Alex, and myself. I remember how cold it was when we left the restaurant. We made it back to his home fine, and stepped back inside.

"Would you like another glass of wine?" he asked. I gladly accepted. We discussed a book someone had sent him. "It is on my bed. Let me get it," he said.

"Admiral, I can get it for you, if that's okay." He said it was, and I made my way across the foyer to his room. I found the book and turned to leave when I spotted a painting of Engelbert Endrass staring at me from the wall. Looking back, I remember feeling sad about seeing it. Topp's life was a colorful pastiche of memories, but his recollections were rapidly fading beyond the point of recall. The portrait of his long-lost best friend likely served to keep at least that portion of his life alive to him.

When we finally drained the last drops of wine from the bottle, I knew our time together was rapidly drawing to a close. "Admiral, I have to call the taxi company to come and get me." He offered to drive me to the station, but when Alex exclaimed, "No!" we both laughed in response.

Fifteen minutes later the taxi pulled up in front of the house on Siebengebirgsweg. I slipped on my jacket and we stood in the foyer to say our good-byes. I gripped his hand and told him thank you for supporting *Silent Hunters*, for providing the moving tribute to his friend Endrass, and for all of his support over the last seven years. He bowed slightly, reached down and hugged Alex, and opened the door for us to step outside. Even though it was quite cold, he remained in the doorway, his hand up and waving as we climbed into the taxi and pulled away from the curb. I still remember him as I last saw him, silhouetted against the soft foyer light waving his farewell. My daughter, who loves older people, was crying softly. "I don't think we will see him again," she whispered.

"I had a notion that he had been hiding his decline in order to maintain some of his customary lifestyle, but your account reveals that he is indeed an old man now in the winter of his life waiting for the end," Eric Rust wrote to me after hearing of our visit. Dr. Rust knew Topp well. He had translated his memoirs from German into English, assisted in translating and footnoting the Endrass piece for *Silent Hunters*, and co-authored with Topp a book-length manuscript on Oscar Kusch that Savas Beatie LLC will publish in the near future. "Your description of Topp reminds me, ironically, of his piece on Endrass," Dr. Rust continued, "where he expresses admiration for those who are cut down at the peak of their glory and don't have to face the indignity of agony and decline."

We exchanged a few letters over the next several months. Topp agreed to write a foreword for a book I was editing called *Hunt and Kill: U-505 and the U-boat War in the Atlantic*. His last published contribution during his lifetime appeared in the summer of 2004.

⊕⊕⊕⊕⊕⊕⊕⊕

Admiral Topp died the day after Christmas 2005. A friend notified me a few days later that he saw a message about it on a website. I confirmed the sad, but not wholly unexpected news with Erich's son Michael. I remember glancing up on the library shelf near my computer. For several years, a bottle of good red wine in a gift bag has been perched there. After Topp gave it to me in San Francisco, I resolved to drink it when he died and toast him on his next journey. It remains unopened. For some reason I can't bring myself to drink it.

Someone passed on an e-mail to me that I think is a fitting conclusion for this essay. Although Topp told me he didn't believe in an afterlife in the religious sense, let us all hope he is mistaken:

"He will be now on his last patrol meeting his beloved Pollux, Bertl Endrass and great friend Teddy Suhren when St. Peter will direct him 'to a special heaven for U-boat commanders where they can continue their old ways of singing, drinking, and merrymaking'."

May it always be so.

Sileo in pacis

The History of Subsims
Brian H. Danielson

Submarine simulations were present before the beginning of the personal computer. The first "electronic subsim" may have been "Seawolf", "Sea Devil", or "Submarine", one of those arcade games found in many bowling alleys and truck stops. Similar in size and layout to a pinball machine, this subsim consisted of a periscope and handles. As you peered through the scope at the twilight horizon, ship silhouettes would slowly cross. You timed them and pressed the torpedo launch button on the right handle with your thumb. A small red light would streak forward and intercept the ship, lighting up the sky and emitting explosive sounds. You could play as long as your father's patience and supply of nickels didn't run out.

Today, many years later, submarine simulations have grown in scope and realism. The graphics allow you to suspend reality with ease and feel the tension and excitement our heroes lived. Instead of timing a torpedo run by eye, today's subsim skipper has an array of screens, equipment, and options at hand. In some cases, the complexity of the best subsims rivals reality.

Here we examine the pioneer programs of subsim history. By current standards, many of these subsims are antiques. But all things must have their genesis and these early efforts gave rise to a growing niche of computer games. One should consider; without GATO and Sub Battle, we might never have enjoyed Aces of the Deep and Silent Hunter.

GATO

This was the first sub simulator for a personal computer. The graphics were very limited, CGA with stick figures for ships, of which there were five enemy types. Although it simulated Gato class subs, there were only four bow torpedo tubes available. All functions were generated by the keyboard (mouse? what's a mouse?). The mission area was very limited — a group of islands in the Pacific, which was subdivided into twenty "quadrants." You pressed the "M" key for new missions and they were transmitted by Morse code. You were instructed to keep in mind "the enemy may break Allied code at some point. Some messages may be enemy fakes designed to trap you." How's that for early efforts at realism? Resource management was one of the game's subtler features. Your primary strategic objective was to complete as many missions as possible with the supplies you carried before returning to the quadrant where your subtender waited.

Gato was the Pong of submarine simulations. When the first players encountered Gato, you can imagine they thought it was pretty significant. They had to start somewhere.

Publisher: Spectrum Holobyte
Circa 1983
Requirements: 4.77 MHz, CGA, 128K RAM, 5.25" disk

SILENT SERVICE

Back in its day, this was considered the standard by many. It was available on several types of PCs and game stations. It first supported CGA and then was updated to be used on EGA, but the graphics were still at CGA resolution. Even so, the visuals were a big improvement over the earlier Gato. Silent Service allowed the player to visually determine the angle on the bow of the target and factor it into the solution. The destroyer search algorithm was the best of the first generation subsims. The player commands American subs in a

hunt for the Imperial Japanese Navy. This sim was revamped later and released as Silent Service II.

Publisher: Microprose
Circa: 1985
Requirements: IBM PC, 128K RAM, CGA, 5.25" disk

SUB BATTLE

This subsim from Epyx was one of the early superstars. It first supported CGA and Hercules graphics and was later updated for EGA (and really used EGA). Enemy ships were all black in color, but had good and clearly recognizable forms. The user interface was beautifully executed. Control and status information was provided on one screen. A window within the screen featured specific views such as radar, sonar, periscope, etc. It had four difficulty levels. In level four, one could see the rolling sea from the periscope.

The programmers get an A+ for overall scope. Sub Battle supported single missions or entire WWII campaigns in the Pacific or Atlantic. For the German side, Type II, Type VIIC, and Type XXI U-boats were available. For the U.S., your choice was between S-class, *Gato* class, and *Tench* class, with the mission determining the actual type used. All subs had reasonably correct specifications.

Sub Battle had an impressive list of features. The year the mission takes place determines the level of equipment available. Aircraft are present. If the sub was near the shore or a carrier, occasional patrol aircraft could cause a bad day at sea. Ships in an attacked convoy would change speeds and courses in an attempt to escape destruction. Damage a ship and its speed diminished. Radio was used to give reports and orders, including downed pilot locations for pickup. Missions also include shore parties and minelaying. Many types of surface ships were featured, including battleships and carriers. This subsim went so far as to include the sun and moon, even with changing phases. A rough spot to note: The time "slow down" didn't work properly. After slowing down from accelerated time, a detected convoy might be too far out of range to plan an attack, whereas if you had been running in real time, you could have easily made the attack. But overall, Sub Battle shined.

Note of interest: game designer Mike Jones, who later went on to craft Aces of the Deep, worked on this game.

Publisher: Epyx
Circa: 1986
Requirements: IBM PC, EGA, 512K RAM, 5.25" disk

UP PERISCOPE!

Another WWII subsim from the American perspective, Up Periscope! supported single missions or patrols. Several historic scenarios were also available as well as the option of general patrols.

Up Periscope! originated a few new ideas, but they weren't well-implemented. First, the concept of variable visibility (possibly due to fog) for daytime was introduced. Second, the user had to make several periscope, radar, or sonar contacts at different times to generate an accurate shooting solution. A minimum of two marks was required to fire using the TDC in manual. This added to the realism and difficulty, but the drawback was once you fired at a specific ship in a convoy, you couldn't transfer the settings to the next ship for a second shot. By the time you developed another solution the convoy would slip out of position or the destroyers would be all over you.

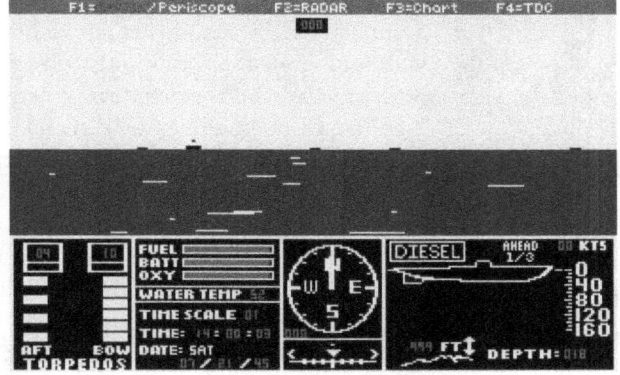

Torpedo implementation had some interesting characteristics. Torpedo gyro angles were a factor. The closer you pointed your bow to the projected track, the greater your chances of scoring a hit (this wasn't implemented in any other subsims to this time). Strangely, to fire the aft torpedoes at a target, you needed to point the bow toward the target.

Graphics weren't bad, considering it was CGA resolution. Ships were rendered in white and black with black masts. Aircraft weren't included, but the skill level of the enemy ships could be adjusted, affecting their evasive zigzagging, aggressiveness, and depth charge accuracy. Up Periscope also included a 120-page booklet titled Submarine Action in the South Pacific. Filled with facts, illustrations, and historical data, this booklet doubled the value of the sim.

Publisher: Actionsoft
Circa: 1986
Requirements: IBM PC, CGA, 256K RAM, 5.25" disk

CODENAME: ICEMAN

Sierra pioneered the graphic adventure game with King's Quest in the early eighties. Codename: ICEMAN was a graphic adventure game that included some submarine play. Not a true subsim, ICEMAN entertained nonetheless. The nature of a graphic adventure game meant the player must explore the environment, "looking," "talking," and "getting" every object and person his character comes across. You couldn't leave the island unless you discovered and used all the items hidden in the setting. Johnny Westland, the game character, was a Navy commander on leave in Tahiti.

You began the game with Johnny lolling on the beach. During the Tahiti episode, you would play volleyball, administer CPR to a drowning victim, dance with and romance a club girl, and eventually receive news you must return to active duty. You did all these things by directing Johnny with the arrow keys and endlessly typing in "LOOK AT WALL," "GET EARRING," or "SIT DOWN." I nearly went bonkers trying to use the phone to get a transport to the airport — the other party didn't like what I said and kept hanging up. Until I found the exact words, I was stuck in the bungalow bedroom, redialing, redialing, redialing....

Patience was a big necessity with ICEMAN. Once in the futurist sub, the captain gave you directions. If you failed to respond correctly or promptly, game over. The aim of all this adventuring was to locate and rescue an ambassador being held by terrorists. The submarine part of this game wasn't very realistic, but a player could enjoy the "action" if he didn't mind lots of typing.

Publisher: Sierra
Circa: 1986
Requirements: IBM PC, 8 MHz, 512K RAM, VGA, 5.25"/3.5" disks

688 ATTACK SUB

In 1988, a company named Electronic Arts presented subsim skippers with 688 Attack Sub. Designed by John Ratcliff and Paul Grace, 688 Attack Sub was the predecessor to Seawolf SSN-21 and Jane's 688(I) Hunter/Killer. Ten missions chronicled your part in a Cold War that grows progressively hotter. In most of these missions, you command either a U.S. 688 class sub, or a Soviet Alfa. The Alfa was the USSR's titanium-hulled hotrod, capable of an estimated 45+ knots and ~2000 feet.

The goal of each mission was dependent on the type of sub chosen. In some missions, the 688 was requested to protect a convoy, whereas the Alfa would attempt to sink it. Other missions involved only sub vs. sub scenarios. There were many types of vessels included in the sim, but within any particular mission, the same set of vessels was used. There were a limited number of initial setups for a mission, so the action, while not entirely random, had some variability.

Performance was best on a 286 or faster PC. EGA graphics gave good results, with VGA preferred. A mouse greatly benefited the interface use. There were several screens that one must flip through (i.e. from the navigation console to the weapons room) and this presented some difficulty in executing commands quickly (defined function keys helped solve this problem).

Several other features worth noting: Multiple thermal layers were provided on most missions, giving hiding places if used correctly. The implementation of these layers was very well done, making the act of hiding from enemy ships and submarines a real art. Wire guided torpedoes were included. The passive sonar had a signature processor that allowed the player to determine ship/sub types by both sound and graph. Last, your sub came with a retractable sonar array.

688 Attack Sub had one very special feature. It allowed two players, on different PCs, to play each other via a modem (or null-modem cable). Typically, one player chose to be the Alfa, while the other player was the

688 sub. Such games tended to be incredibly intense since players don't always behave in a predictable manner. 688 Attack Sub, therefore, heralded one of the most useful and important features a subsim can possess; the ability for friends to pit their acumen against each other.

This sim also has a few problems. Foremost on the list: explosion sounds locked up interactive commands. For example, if you were diving and near the bottom, and torpedo explosions were heard, you couldn't stop the dive until the sounds were over, which probably meant a damaged sub as a result of hitting the bottom. Also, modem play would occasionally have strange results (player 'A' sunk while player 'B' continues tracking player 'A'). Overall, though, with the mouse interface, multiple screens, good graphics, and modem play, 688 Attack Sub can be viewed as a milestone in subsim development.

Publisher: Electronic Arts
Circa: 1988
Requirements: IBM PC/XT/AT, 384K RAM, 256 VGA, 5.25" disk

RED STORM RISING

Red Storm Rising was based on the plot of the book by the same name. It ran well on 8088 turbo systems with monochrome Hercules graphics, EGA or VGA (color is recommended). A faster PC helped speed up the game slightly. A joystick can be used in conjunction with the keyboard, but keyboard entry alone was very satisfactory. There were so many keyboard controls that a keyboard overlay was provided to aid in remembering the functions of the keys.

While this simulator had several single mission selections, the main challenge was the entire WWIII campaign. There were several skill levels and several levels of technology (determined by year of conflict: 1984, 1988, 1992, and 1996). Several classes of subs were available, including *Permit*, *Sturgeon*, 688, improved 688, and *Seawolf* (for 1996 only).

There were a vast number of campaign missions and, if repeated, they were always random in location as well as types of ships involved, a truly dynamic campaign. Subsim skippers are never satisfied with "canned" missions and this contributed to the success RSR enjoyed. Some missions included only submarines, others included only a surface convoy, others included a surface task force (with carriers), and others with both ships and subs. Helicopters with sonar dipping equipment and sonar buoys added much to the turmoil.

With the most challenging settings, events got so complex, and events occurred so quickly, that having two people at the keyboard was a great help, and for beginners, almost a necessity (plus it was fun). Trying to identify ships with signature analysis, running from incoming torpedoes, launching noise makers and decoys, dealing with several of your own wire-guided fish, and launching Harpoon and Tomahawk missiles made for a very busy captain.

There weren't many known bugs or problems with this simulator. Since much modern submarine technology is still classified, it is difficult to specify any technical errors, but most would say that the simulation 'feels about right' (especially in the highest skill level).

Red Storm Rising demonstrated one other characteristic: it was still sought after and played by cyber skippers after its release, the greatest legacy any subsim could hope to aspire to.

Publisher: Microprose
Circa: 1988
Requirements: IBM PC/XT/AT, 384K RAM, 16 VGA, 5.25" disk

DAS BOOT

Das Boot has aircraft, enemy and friendly submarines, interactive anti-aircraft gun and deck gun firing controls, fantastic ideas for radio (including the capability to decode enemy transmissions), and more. Yet few programs with so many great ideas came together with such disaster. This was due, in part, to the lack of scope of the simulation. For example, one of the missions is the *Das Boot* mission. It began in the Gibraltar Strait. You were ordered east through the strait. You were quickly attacked by aircraft, and not permitted to dive. It was foggy and you can't see anything but aircraft. If you were a good shot, you could shoot down all the planes. If you did so, you won, and that was that.

This simulator was the only one that used true three dimensional polygon filling graphics. This type of graphics was usually used with flight simulators. The advantage of this technique was that you got a real three-dimensional effect. However, the game was unable to show object detail. With fast jet aircraft simulators, lack of detail is fine, since you only see the enemy for a split second. But for submarine simulators that operate at such slow speeds, lack of detail is very annoying. In Das Boot, it was impossible to determine the identification of a vessel by sight. You must have the XO "ID" the vessel to know what type it was.

Other features were the use of time compression, different external views, enemy use of HF/DF against you ("maintain radio silence to avoid detection"), Metox, and different torpedo types. With the groundbreaking graphics, Das Boot had a lot of promise, but the lack of career missions pretty much sank it.

Publisher: Three Sixty/EA
Circa: 1990
Requirements: IBM PC, DOS 2.11+, 512K RAM, 256 VGA, 5.25" disk

HUNT FOR RED OCTOBER

Taken from Tom Clancy's career-launching novel, The Hunt For Red October attempted to follow the general plot of the book. This meant that you were to head west and escape all the ships of the USSR as well as appear ready to defect to the United States. The graphics were EGA only. The joystick was used, but very difficult to control. The graphics and screen layouts were very poor in general. HFRO didn't enjoy any lasting success as a subsim.

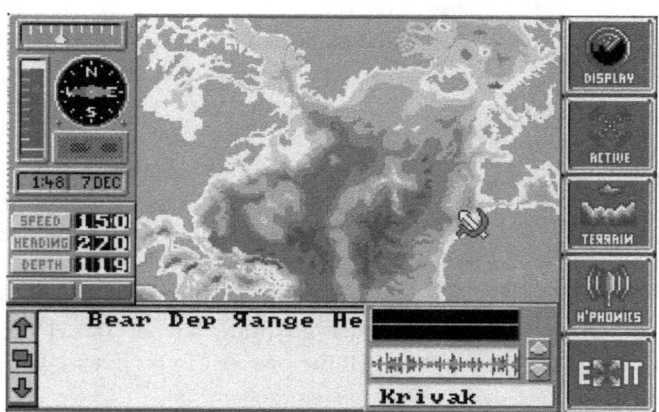

Publisher: Grandslam Entertainment in 1991, Software Toolworks in 1989
Requirements: 768k of RAM, 3.5" inch disk

⊕⊕⊕⊕⊕⊕⊕⊕

Submarine simulations have come a really long way since the simple gameplay of Gato. The realism has become ever more accurate. Graphics and sounds have evolved to the point of photo-realistic quality and will continue to improve. Almost all subsims here included historical information with the playing manuals, proof that the creators were really passionate about the nature of the programs. When you slide the Silent Hunter 4 DVD into your PC, pause and reflect on the early eighties. Stick graphics, buzzing PC speakers, and boot disks...these were the Model T subsims indeed!

....You don't look at minor plumbing leaks the way most people do.

Submarine Silencer

Donald Ross, Ph.D.

Introduction

The writer was active as a submarine silencer and anti-submarine-warfare (ASW) analyst for about twenty years, from 1954 to 1974. During that period the U.S. Navy stressed the ASW role for submarines, for which quietness was an important requirement.

During both World War I and II, the primary mission for submarines had been to attack and sink surface ships. Following WWII, our only potential enemy was the Soviet Union, which had only a few oceangoing ships. With no significant anti-surface-ship mission for U.S. submarines, our need for a submarine force was being questioned. But the Soviets had a huge submarine fleet. To counter this threat we would need quiet submarines designed specifically for the anti-submarine role. In 1949, at the instigation of the submariners themselves, CNO directed that each fleet assign one division of four submarines "to the sole task of solving the problem of using submarines to detect and destroy enemy submarines." Submarines now had an important mission. Submariners could give a collective sigh of relief. A new category called SSKs (K for Killer) was soon designated.

Throughout the Cold War period we led the Soviets in the area of submarine quieting. Since our subs were quieter, we assumed that they were always trying to catch up. However, our lead in this aspect didn't mean that we led them in all submarine technology. The Soviets still had anti-surface-ship missions for their subs and quietness was not their highest priority. Apparently they saw our aircraft carriers as their major threat, and so built a class of submarines with sufficient speed to tail our carriers and thus to be in position to attack them if hostilities began. Some of their subs could make more than 33 knots when ours couldn't go near that speed, and some could operate at significantly greater depths than ours. Perhaps they saw us as being "behind" and trying to catch up. I do believe that our subs were superior to theirs, but the difference was more in personnel training and in safety rather than in technology.

Failed Attempt to Quiet *Nautilus*

In 1952, when design decisions were being made for the first nuclear submarine, noise reduction by the use of vibration isolation was in its infancy. Our Navy was just starting to apply techniques learned from the Germans to some of our new SSK conversions, such as the *Cavalla* (SSK 244). Capt. Hyman Rickover decided that his most important consideration for the first nuclear powered submarine had to be safety. If *Nautilus* were to fail, nuclear power would be set back for decades. His staff reported that the safety of the new silencing techniques was as yet unproven. So, not only did he refuse to apply them to the *Nautilus* design, but also he omitted all noise considerations in the overall design.

The result was a very noisy sub, even at the slow speeds when submarines most need to be quiet. When she first went to sea, *Nautilus* was plagued by a dominant low-frequency tone. This was traced to a tank opening and was then eliminated by changing the tank design. This left the noisy machinery systems. The suggestions made by Herman Straub and myself when we rode *Nautilus* in March 1956 (see "Civilian Submariner") didn't quiet the sub but only served to reduce its detectability when operating at high speeds.

Sometime in 1956, when *Nautilus* had proven itself and the decision had been made to build a nuclear submarine fleet, Rickover decided to let the noise control people have a go at quieting *Nautilus*, but within limits. *Nautilus* had two machinery systems, port and starboard. For many functions there was a third unit as a spare. Rickover ruled that the spares weren't to be touched.

To evaluate the effectiveness of the various noise treatments installed during the next overhaul, before and after over-the-side noise surveys were conducted. For these tests, the sub was operated from shore power and each machine was run one at a time. I participated in both tests with the Electric Boat group led by Bob Collier and Bill Ezell. The tests were run between midnight and seven a.m., usually on a Sunday morning. The rest of the yard was shut down to avoid noise interference. It was eerie being in an almost deserted shipyard.

Comparison of the before and after over-the-side noise surveys showed mixed results. The sound reduction measures were effective for machine systems entirely internal to the submarine, but had no effect at all on seawater pumps. Also, the reactor coolant pumps couldn't be touched. Since at slow sub speeds the coolant and seawater pumps were the noisiest items, the resultant overall reduction was negligible.

Submarine Silencing Contractor

My understanding of submarine noise sources began in the mid 1950s when I worked at Bell Labs developing a new low-frequency passive sonar for submarines and had spent many weeks carrying out detailed self-noise measurements first on the *Cavalla* (SSK 244) and later on the *Nautilus* (SSN 571).

In January 1958, I resigned from Bell Labs to join the small acoustics consulting firm of Bolt Beranek and Newman (BBN) in Cambridge, Mass. Here I contributed directly to submarine silencing, but didn't ride any submarines. At BBN I was in the engineering division headed by Dr. Ira Dyer. This group conducted most of the firm's military R&D. The projects I worked on were sponsored by the Office of Naval Research (ONR) and the Ship Silencing Branch of BuShips (Code 345).

In those years, working as a Navy contractor was both pleasant and efficient. Code 345 was headed by a naval officer, usually one with an engineering degree. It was staffed by about a dozen civilians and a junior officer. In my first years at BBN, Cdr. Patrick Leehey, a math Ph.D., was in charge. He was replaced in 1963 by Cdr. J.R. Baylis. During my entire time at BBN, Bob Sherwood was the chief civilian. Projects were originated either by the Code 345 staff or by proposals from one of their stable of contractors. Our projects for the Acoustics Branch of ONR were handled smoothly by Aubrey Pryce and Marvin Lasky. These were relationships characterized by mutual trust and respect. The Navy trusted their contractors to perform honestly, and we trusted that the Navy would pay us fairly for our efforts. It was an effective partnership.

Analysis of Noise Surveys of First Nuclear Submarines

By 1959, over-the-side noise surveys had been conducted on six nuclear submarines. Each survey was documented by a collection of one-third-octave-band analyses, one sheet for each machine for each operating condition. BuShips Code 345 was sure that these tests had an important story to tell. They contracted with us at BBN to analyze them. The shipyards sent us all of their reports, which when piled on top of each other approached four feet in height. Our problem was how to analyze so much data. As this was in the days before computers and scanners, we started out by tracing each page of data on to a transparent sheet. We then used a light table to find those graphs that were similar enough to be represented by a single curve. In this manner we slowly reduced the number of curves to a manageable number.

What we found was relatively simple, and proved to be of great importance to controlling the noise radiated by our nucs when operating at slow speeds. We found that the dominant parameter was what we called the "rated" or "name-plate" horsepower (hp) of a machine corresponding to its operating speed (rpm). By this we meant the maximum horsepower for that rpm. Thus, if a pump were rated at 25 hp for a speed of 1750 rpm, then we used 25 hp as the rated power for that rpm no matter how low the actual power may have been. Only if the speed were reduced did we reduce the rated power. Thus, if a machine was run at half speed, we reduced the rated power by a factor of eight, corresponding to the reduction in power.

The analysis of the noise surveys for the first six nucs made clear what had to be done to quiet future submarines for slow-speed operations. On diesel-electric subs, most auxiliaries were run by DC motors powered by a large battery. Speeds were controllable over a wide range by the use of voltage control rheostats. If a certain speed excited a resonance, that speed would be avoided. The few pieces of equipment that required precision 60 Hz power were fed by small AC generators. Nuclear subs, on the other hand, had relatively small batteries, being used like an automobile battery primarily for start-up operations and emergencies. Almost all auxiliaries on nuclear submarines are powered by 60 Hz AC generated by the ship's service turbogenerators (SSTG). When designing the first nucs, power ratings for the various propulsion auxiliaries were chosen based on the maximum power required for maximum ship speed. For the slower ship speeds, for which power requirements were much lower, flows to some fluid systems were throttled, while other types of machines simply didn't work so hard.

Our analyses showed that constant-speed units made at least as much noise when delivering reduced powers as they did when working at their rated power. Reducing power without reducing rpm didn't result in less noise. There were two obvious solutions to the problem. One was to provide separate low-power systems for slow sub operating speeds. The second suggestion was to replace the various single-speed motors by two- or even three-speed ones. The lower rpms, with their reduced power ratings and reduced noise, were to be used for slow-speeds. This is what has been done for all nuclear submarines starting with the *Thresher* (SSN 593).

While this, my first major project while at BBN, made a major contribution to the quieting of U.S. nuclear submarines, other projects on which I worked, though equally successful technically, were less significant.

Vibration Damping Treatments

My office at BBN was next to that of Dr. Ed Kerwin. In 1958, Ed was involved in developing damping treatments that would reduce structure-borne noise by suppressing flexural vibrations of plates. His work, which was applicable to light structures in air such as airplane fuselages and automobile bodies, was based on science that had been brought to BBN by Prof. Lothar Cremer in the mid 1950s. Cremer and his co-workers in Germany had recognized that although beam-like and plate-like structures can experience several types of vibrations, flexural (bending) vibrations are the most easily excited and are therefore the most important radiators of sound. Their work during WWII had been done for application to submarine silencing. After the war, Cremer returned to Berlin to lead a group applying this knowledge to the design of noise-isolating walls and floors for apartment buildings. It was this application that had led Leo Beranek, CEO of BBN, to invite him to spend a year with the architectural acoustics group at BBN. This recognition of the role of flexural vibrations made enabled BBN scientists to become leaders in the field of noise control in the 1950s and 1960s.

The Germans realized the critical role of damping in reducing sound radiation from flexural resonances. Their approach, led by H. Oberst, was to find, or develop, lossy materials that could be glued onto vibrating surfaces. For instance, mastic undercoat used on automobiles and railway cars reduces their resonant responses and makes them sound less *tinny*. Many rubbers are characterized by high internal damping and much of the German research concentrated on the development of rubber-like (visco-elastic) materials that could be sprayed on or otherwise readily attached to metals. The problem with these single-layer homogeneous damping treatments was that to be really effective they had to be relatively thick and heavy.

Kerwin was the first to realize that the use of a thin metal cover on top of a thin layer of damping material would cause the latter to experience shear and that such shearing action would be more efficient in producing damping. He derived an expression for the loss factor of *constrained-layer damping treatments* and verified the results experimentally. Kerwin's discovery led to a rash of applications. All told, BBN obtained ten patents in this area. Under license, the Minnesota Mining and Manufacturing Company (3M) produced damping tapes composed of aluminum foil and a sticky, lossy material like that used in Scotch Tape. These tapes are still in production and are widely used in the aircraft industry.

Thinking that there might be applications to submarines, I became involved in this exciting development. In his first paper describing constrained-layer damping, Ed had made a number of simplifications in his derivation of the equations. I re-derived them without the approximations and extended them to multiple-layer damping treatments. We were joined in this work by Dr. Eric Ungar, who applied shear damping to beams and other massive structures.

Project TEACUP

Bob Sherwood and others at BuShips were aware of the work on damping by BBN and also of related work by several of their other contractors. Some of these men were especially enthusiastic about the potential for quieting submarines by applying damping to the hull. Despite the lack of any full-scale tests, the decision was made to damp the engine-room hull sections of the *Thresher* (SSN 593) class submarines using a rugged constrained-layer treatment.

The need for a full-scale evaluation of the untried damping treatment was clear. BuShips arranged for the Mare Island Naval Shipyard (MINSY), builders of the *Permit* (SSN 594), to construct the engine-room hull section two months before it was needed and to make it available for tests. The shipyard capped one end with a heavy bulkhead and attached hook eyes to the other end. They turned the open section up and then suspended this hull section in the water next to an available pier. Even in the planning stage it was named "Project Teacup." BuShips contracted with several of its labs and contractors to carry out the evaluation of the proposed applied damping treatment under the leadership of Bob Sherwood. In January 1960, Ed Kerwin and I led the BBN contingent of several people to Vallejo at the north end of San Francisco Bay. The test program, which lasted six weeks, was characterized by full cooperation and high morale among the participants.

The test procedure was to excite the hull and the frames in several locations with vibration generators fed by swept-frequency oscillators which covered the frequency range from as low as 10 Hz to as high as 4 KHz. Vibration pickups on the hull measured the vibrational resonances. The resultant underwater sound field was measured by about thirty hydrophones strung from a set of booms. The first set of tests was made with the bare hull suspended in air by a crane. The hull was found to be very alive, with many measurable resonances. The Teacup was then lowered into the water. As expected, the effect of water loading was small. The hull was still very much alive with numerous resonances. Noise levels were recorded for all of the hydrophones.

We then took a few days off while yard workers installed the standard damping treatment. Vibration measurements showed that the treatment was effective in that all but a few resonances were greatly reduced. However, below about 2 KHz, sound levels in the water were hardly affected. Only for the higher frequencies were there measurable changes. To verify this surprising result, we then had the yard install a second layer of damping treatment on top of the first. This second layer suppressed the resonances even more, but again the sound in the water below about 2 KHz hardly changed. As the purpose of the treatment had been to reduce the low-frequency noises radiated from machines, BuShips cancelled engine-room hull damping treatments for future submarines.

The task of writing up the Teacup experiment fell to us at BBN. One of Cremer's German students, Manfred Heckl, was with us that year. He derived the equations for the radiation of sound by water-loaded plates, finding that below the coincidence frequency, which is about 3 KHz for submarine hulls, damping should have no effect at all on sound radiation. We might not have been so surprised by this result if we had been aware of an earlier work by K. Goesele which had implied the same thing. However, Goesele's paper had been published in a German journal and hadn't been known to us.

It wasn't until many years later that I realized that BuShips had probably made the wrong decision in removing the engine-room damping treatments. At that time the U.S. was emphasizing long-range passive sonars operating at frequencies for which damping would have been ineffective. The Soviets, on the other hand, we learned years later, were using short-range passive sonars that operated at higher frequencies than those covered by Project TEACUP, frequencies for which applied damping would actually have been quite effective.

Piston Slap and the *Tang* Class Fiasco

During the early period of the nuclear submarine building program the Navy hadn't yet given up completely on diesel-electric propulsion, especially for quiet submarines. The SSK conversions showed that such subs could operate very quietly when on battery at slow speeds. The problem was the frequent requirement to run the noisy diesels to charge the batteries. The Navy called on a leading marine engineer from MIT to advise them on building a quieter diesel for installation in its new *Tang* class boats. He concluded that the problem was that all of the cylinders of standard in-line diesels exerted forces on the crankshaft from the same direction, causing a fundamental unbalance. He reasoned that if the cylinders could surround the crankshaft, then this fundamental unbalance would be removed and the engine should be quieter. Based on his recommendation, a new 16-cylinder diesel was designed having a vertical crankshaft, with four banks of four cylinders each surrounding the crankshaft. Because of the vertical stacking, these diesels were called "pancake" engines. Five sets were procured, for four *Tang* class subs and for the experimental sub *Albacore* (AGSS 569).

Noise tests of the *Tang* (SS 563) showed that the pancake diesels were actually 6 db noisier than the standard in-line engines on other diesel subs. Not only were the engines noisier; but also, while mean times between cylinder-liner failures of standard diesels were measured in months, for these engines they were measured in hours. The situation was so bad that the engines on the *Tang* class had to be removed and replaced by Fairbanks Morse in-line diesels. The pancakes were sent to the Portsmouth Naval Shipyard, where they provided spare parts for the *Albacore*'s engines. *Albacore* was able to increase their engines' useful lives by operating them at reduced speeds.

While at Bell Labs in the mid-1950s, I was well aware of the *Tang* class fiasco. It had not been a surprise. We knew that submarine diesels produced spectra containing as many as a hundred harmonics of the fundamental rotational frequency. Such a spectrum implies that the source of the noise had to be a repeated sharp pulse. It was thought that the pulses might be from the firing explosions. In fact, some studies were conducted in which the timing of the explosions was altered, but with inconclusive results. In the mid-1950s, an interesting experiment was carried out at the Applied Science Lab. at the Brooklyn Navy Yard. The airborne noise of a small diesel generator was measured as it was gradually dismantled. Starting with it running as a powered diesel generator, the fuel supply was then disconnected and the generator run as a motor. At the same speed, the spectrum was virtually unchanged, showing that the explosions had little effect. Then the engine was dismantled in steps. There was little change in the noise until the pistons were finally removed. Clearly the noise was dominantly related to the pistons.

This was the status of my knowledge of diesel noise at the time I joined BBN and was invited by Prof. Ted Litovitz of Catholic University to join a study of Soviet acoustics literature sponsored by the C.I.A. The Agency had this idea that one could learn about Soviet capabilities to quiet their submarines by studying their unclassified publications. Litovitz was contracted to form a diversified group of acousticians who would study translations of Russian books and journals. We would meet several times a year to discuss our reading assignments. Our sponsor took notes on what we were learning. What a project! Scientists need to keep up with the literature in their fields and here we were being paid to do what we should have been doing anyway. As the only one in the group with a practical background in submarine noise, I was assigned the engineering items. From this study I became familiar with the names and agencies of several very capable Soviet experts on ship noise, men and women who showed as much knowledge of structureborne sound and airborne noise control as the Germans under Cremer or any of us in the U.S. However, this didn't mean that they were actually going to apply this knowledge to quieting their submarines.

One day in 1960 I was asked to read a translation of the book *Noise of Marine Diesel Engines* by V.I. Zinchenko, originally published in Russia in 1957. This book, which contained detailed airborne noise measurements of some fifty Russian marine diesels, had separate chapters on each of six different noise mechanisms. When I read the chapter on piston slap noise, I knew immediately that here was the explanation for the underwater noises of submarine diesels, as well as the key to designing quieter diesels and other reciprocating machines. Piston slap explained everything I knew about the noises from reciprocating machines. It was one of the most exciting days of my life.

Piston slap refers to the impact of a piston against a cylinder wall as a result of sidewise motion of the piston across the cylinder clearance space due to reversals in the direction of the cross component of the connecting-rod force.

I was so excited about piston slap that I immediately wrote a short proposal for a project and hand delivered it to my sponsors in Code 345. They approved it on the spot and I started work the next day. In our study of the noise from the first six nucs, we had noticed that reciprocating compressors and diesels had similar spectra and showed similar dependencies on power and speed. At Bell Labs, we had been were very much aware that the signals received by SOSUS stations from U.S. submarines' diesels were strongest during the warm-up period, the first ten minutes after starting. Both of these results were consistent with piston slap. As to the *Tang* pancake-engine fiasco, Zinchenko showed that piston slap vibrations and noise depend on rotational speed as well as on rated power. In order to fit the pancake engines into the allotted space, the designers had almost doubled their operating rpm. Not only did the resultant radiated noise increase by 6 db, as Zinchenko would have predicted, but also his results implied that the rate of fatigue failures of the cylinder liners would increase by an even greater factor.

Once we understood that piston slap was the dominant noise mechanism for marine reciprocating machines, it was immediately clear how they could be quieted. In the early years of the twentieth century, piston slap had been the dominant noise of early cars. Automotive literature of the 1920s has many articles on the subject. Modern car engines have their cylinders displaced slightly so that the pistons ride on one wall. Two-stroke marine diesels used on large merchant ships are very quiet as they employ articulated connecting rods and so their pistons don't experience any cross forces. We could have designed quiet diesels and compressors for U.S. submarines, but the Navy had switched to all nuclear power, and centrifugal compressors had replaced reciprocating ones. However, several European countries did develop quieter diesels for their subs.

Lofargram Analyst

The interpretation of submarine noise measurements was revolutionized in the 1960s by the application of the Lofar spectrum analyzer. Prior to the introduction of Lofar, spectra from ships and submarines were usually analyzed in one-third octave bands, which however yielded very little information about the nature of specific sources. Detailed information was needed about narrow-band tonal components. To obtain a narrow-band spectrum in those days a sample of perhaps several minutes would be recorded on an endless tape and a narrow filter swept through the desired frequency range. Since the time required for this process equaled the frequency band to be scanned divided by the square of the effective filter bandwidth, it took at least two minutes to scan a 500 Hz band with a filter bandwidth of about 2 Hz. Frequency instabilities occurring during this period simply increased the apparent tonal bandwidth. Lofar, invented by Larned Meacham at Bell Labs in 1951 as a detection tool, speeded up this process by using frequency multiplication. With a multiplication of 100, the 500 Hz band of the previous example became 50 KHz, and the filter bandwidth 200 Hz, leading to a required sweep time of only about one second. The filter outputs were recorded on teledeltos heat-sensitive paper, which advanced a small amount for each sweep, thereby revealing important details of short-term instabilities.

The radiated noise spectrum of a submarine contains a large number of tones, from a variety of different sources. The analyst attempting to determine dominant noise sources needs to learn not only the frequency of each tone, but also its frequency stability and the presence of modulation frequencies. While Lofar displays yield only qualitative amplitude information, they do display tonal instabilities, enabling the analyst to tie whole families of tones to a single source. As a member of the submarine Lofar project at Bell Labs from 1953 to 1957, I was among the first to use Lofar this way. I firmly believed that every observed feature had a physical explanation, and that it was my task to find it. In most cases I was successful. By placing numerous hydrophones in the superstructures of the diesel sub *Cavalla* (SSK 244) and the nuclear sub *Nautilus* (SSN 571) we had collected recordings of the noises from a wide variety of machines and systems. The resulting knowledge was useful not only for the purposes of sonar classification, but also for the submarine silencing program.

One of the more important results of Lofar analysis was the ability to sort between tones influenced by the propeller(s) and those from other systems. As an example, we compared the diesel spectra of U.S. and foreign subs. The U.S. used diesel-electric propulsion systems in which the diesels charged a battery and the propellers were driven by DC electric motors. Tones generated by our diesels were very stable and showed no influence whatsoever from the propellers. Foreign subs on the other hand used slower speed diesels directly coupled to the propellers. Tones generated by their diesels were much less stable, being modulated by propeller shaft and blade frequencies as well as being influenced by motions of the stern planes.

Tonal stability often is a clue to a physical source. All propellers located in the stern of a vessel radiate tones at multiples of the blade frequency, this being the product of the number of blades times the rotational speed in rpm. Due to one blade often making more noise than the others, these tones are frequently modulated by the shaft frequency. For submarines operating at speeds and depths sufficient to suppress cavitation, propeller-generated tones are relatively stable, responding only to movements of the planes. On the other hand, submarines experiencing propeller cavitation produce diffuse tones because of the random nature of the cavitation process. In the case of surface ships, the tones are even more diffuse due to the pitching and rolling motions of the ship.

Having been an early user of Lofargrams relevant to submarine noise analysis, I was called on to train many Navy lab staffs as they too adopted this technology. On several occasions I was asked to help the Navy's acoustics intelligence analysts interpret signals from new classes of Soviet submarines. Perhaps the most exciting was in 1957, while I was still at Bell Labs. I was called in to interpret Lofargrams that had been made of sounds from a new Soviet submarine. The Russians had claimed that they had built a nuclear sub. Not knowing that plans for the *Nautilus* had been given to the Soviets by a spy, our Navy denied their claim because it was simply too soon for them to have developed it on their own. However, after a day of studying the Lofargrams, I concluded that the spectra were indeed from a twin-screw, geared, steam-turbine drive and were therefore consistent with nuclear propulsion. The next day our Navy confirmed that the Soviets had indeed produced their first nuc, named the *November* class.

The *Tullibee* and Sonar Self-Noise Reduction

No discussion of submarine silencing would be complete without reference to the parallel development of the powerful BQQ-2 sonar system consisting of a low-frequency passive array (BQR-7) and a bow-mounted, fifteen-foot-diameter passive-active spherical array (BQS-6). The prototype of this system was installed in the *USS Tullibee* (SSN 597), the first nuclear submarine specifically designed to be the quietest submarine that could be built within the state of the art at that time (1958-1960). The idea first tried on *Tullibee* was to make the bow region as quiet as possible by moving the noisy bow planes to the sail and the torpedo tubes to a midships location below the control room.

The *Tullibee* was unique in several respects. Not only did it boast the Navy's most powerful sonar, but also it was the only nuclear submarine powered by a turbo-electric drive, thus avoiding the need for noisy gears. However, the restricted power of the DC electric motor limited the ship's speed, and thereby its ability to transit fast enough when taking station or when closing on a target. The gear noise problem was soon solved, by a combination of using many small teeth and precision machining, enabling the Navy to adopt geared-turbine propulsion for all future nucs.

Studies of sonar self-noise conducted on *Tullibee* and subs of the *Thresher* class found that there were three major sources: ocean ambient noise, turbulent flow over the bow, and propeller and machinery noises radiated from the stern and received by the sonar by reflection from the ocean surface. The aim of the silencing program was to reduce the flow and machinery noise components to the point where ambient noise would be dominant. The surface-reflected component was reduced by installing an acoustic baffle above the sonar sphere. It was found that the ribs in the steel bow domes were playing a major role in causing flow noise. To eliminate the ribs, fiberglass domes were developed and these were installed beginning with the *USS Drum* (SSN 677).

Reducing the sonar self-noise increased the ability of our subs to detect target subs while remaining undetected. It also enabled our sonars to be used as part of a system to measure the radiated noise of uncooperative targets. For this function, the spherical sonar needed to be calibrated.

The task of calibrating sub sonars was assigned to the Ship Acoustics Department (SAD) of the Naval Ships Research and Development Center (NSRDC) in Carderock, MD. As the Department Head from July 1967 to December 1971, I was intimately involved in this program. Sonar calibration isn't a simple matter, it is quite prone to error. I participated in the series of trials conducted by SAD on the *Tullibee* in April 1969 in which our calibration techniques were checked and perfected.

"Mechanics of Underwater Noise"

Around 1972, ONR's Aubrey Pryce and Marvin Lasky became aware that the set of summary reports on underwater acoustics that Pryce had co-authored in the 1950s had become obsolete and needed to be replaced. I talked with them about my doing a three-volume update on radiated noise, ambient noise, and ocean sound propagation, which I ambitiously projected as a three-year project. By this time, I was working at the Roslyn VA office of TetraTech, Inc. ONR gave us a one-year contract for the volume on underwater radiated noise. I quickly produced a twenty-chapter outline for what I hoped would be used as a self-education text as well as a reference for workers in the field. My approach was to stress physical explanations of the basic mechanisms by which noise is generated, transmitted by structures, and radiated into the sea. While other comprehensive reports stressed collections of actual data and so had been classified, my intention was to give the reader a basic understanding of the dominant mechanisms that would be neither classified nor subject to becoming obsolete. The book was to be called *Mechanics of Underwater Noise* (MUN).

By the end of the first contract year, I had written a detailed outline and had developed a card file summarizing hundreds of pertinent articles, but I had only started on the first three chapters. I had also moved to La Jolla and was now with TetraTech's San Diego office. As the contract for the book was cost plus fixed fee, I was able to receive an extension. Progress in the second year was also slow and again ONR gave us an extension. Finally, in the third year, I gave up. In January 1976, I went to TetraTech's headquarters in Pasadena and told Vice President Don Stern that I couldn't write the book and that we would have to refund the monies we had received from ONR. Knowing that non-performance is unacceptable, I expected a really hard time. But Stern surprised me. After patiently listening to my story, he said that he understood that I couldn't write *the* book. "But," he said, "Donald, do you think you could write *a* book?" That did it. I understood his point immediately and returned to San Diego to work on a ten-chapter book, incorporating the other ten topics as sections rather than as full chapters.

Mechanics of Underwater Noise was published by Pergamon Press in the summer of 1976. It has been used as a reference text in many university and commercial courses ever since. It was reprinted in 1987 by Peninsula Publishing of Los Altos, CA. It seems somehow to have become *the* book on its subject, and even today is still readily available through Amazon.com.

While I certainly made my share of direct contributions to submarine silencing, I truly believe that my most significant contribution to this field was in the training of hundreds of engineers who attended my numerous lectures and short courses and/or have used MUN as a reference.

Back in the Day
Life aboard a U.S. Nuclear Submarine
Tim Grab

"Request Permission to Come Aboard"

It was a sunny spring day in May of 1993 when I reported aboard the Improved *Los Angeles* class submarine (a.k.a "688(I)") *USS Annapolis*, SSN-760, at the Electric Boat shipyard in Groton, Connecticut. The sight that greeted me when I arrived at the pier was one of organized chaos. Hoses and ductwork snaked across the decks and in the overheads. Toolboxes, welding machines, and other instruments of the shipyard trade, both topside and belowdecks, conspired to trip up the unwary *nub* (nub: a non-submarine-qualified crewmember), or cause him to knock himself in the head.

It was just such a banged-up brain-pan (mine) that I was presented to the new hospital corpsman, freshly reported only a day or two after I did, as his first case. However, instead of bashing my skull on a piece of shipyard gear, I hit my head on a slightly protruding bolt in the crew's mess while helping to stow emergency gear after a casualty drill. It was the first time in my life that I actually saw stars as a result of bumping my head.

"Doc" examined the resulting gash, pronounced me clear of a concussion, and congratulated me on my first "battle scar." This, and several less stellar experiences, rapidly educated me on how to navigate inside the confines of the submarine.

For a nub, the first few days after reporting aboard can be a bit overwhelming. He must become familiar with the layout of the ship, meet the 120 crewmembers (and at least *try* to remember some of their names), and gain a minimal familiarity with the locations of basic accommodations (including the heads, or bathrooms — very important) and emergency equipment. It becomes an even more unusual challenge when navigating a ship that is cluttered with shipyard equipment and encountering passageways blocked off due to shipyard work. I didn't have such a hard time of it, and for that I have my shipmates to thank, particularly the guys in my division, the Nav Ets — as well as a few shipmates with whom I had gone to some of the Navy electronics schools, who had reported aboard before me and were already familiar with the layout of the boat.

Annapolis was at Electric Boat for her PSA, or Post-Shakedown Availability. This is a shipyard period in which a new submarine, after her commissioning, launch, and first year of operations with the fleet, receives all of the upgrades and equipment that weren't part of its original construction contract. (One example of such equipment is the anechoic hull tile coating. Prior to PSA, the hull is simply painted black with white markings on the hatches and other openings, which explains why some photos of submarines show the white markings and others do not.)

⊕⊕⊕⊕⊕⊕⊕⊕

To crank or not to crank

The only thing I really dreaded about reporting to the boat was the prospect of "cranking," or "mess cranking."

Cranking is galley duty, and those who perform this duty are called "cranks." It is a duty assigned to most new non-submarine-qualified enlisted crewmembers, and most get it shortly after reporting to the boat. It is the dirty (but absolutely necessary) work of the galley. Retrieval of food items from their storage areas (which can be anywhere in the forward compartment, including the decks in berthing), basic food preparation such as peeling and chopping, washing dishes, clearing and cleaning tables, scrubbing decks, serving crewmembers during mealtimes, and preparing trash and garbage for disposal. Cranks usually worked eight or ten-hour shifts, in addition to spending several hours studying their submarine qualifications. As the saying goes, "Ain't no slack in a fast attack!"

A non-rated crewmember, or striker (one who hasn't received specialized training before reporting to the boat) could expect up to ninety days, or even more, of cranking. On the other hand, a rated crewmember, or one who has received training for specific duties (such as I was, as are most submarine sailors) could expect about sixty days in the galley. The length of time a nub spends in the galley depends on the needs of the galley, the needs of the nub's division, and the whim of the COB, or Chief of the Boat. The COB is the senior enlisted man on the submarine and, among his many responsibilities, assigns nubs to crank in the galley.

My friend and shipmate Matt Lentz, a fire-control technician, recalls cranking (I added some items in brackets to clarify some of the navy-speak and protect the innocent): "Cranking sucked (and I'm sure it still does). No other way to put it. You still had to meet your division responsibilities... You'd be there at 4 a.m. and sometimes not leave until after dark, but the cooks, other than the duty cook, would leave whenever. The one good thing was not standing duty. When you were done [with cranking], sometimes they'd hook you up with goodies. For example, when I finished my time, the last MSC [mess chief] that was there didn't release me until Tuesday, but he made sure I wasn't on the [duty] schedule that weekend. It was a 3-day weekend, and I would have had duty one of those days. I never saw a cook helping with a weapons load, but I had to do their dishes...

"The work wasn't too bad. Doing the dishes sucked. At sea, you'd have to scrub the TDU [trash disposal unit] room, which could get pretty nasty.

"We usually worked eight to ten-hour shifts. You'd have an early crew at 4 a.m. (in-port) and a normal crew at a normal time (8 a.m., I guess; we had to be in for quarters). The early crew finished at 2 p.m. I think it was so you had an overlap at lunch, when the whole crew was on the boat. I forget how it was run at sea; I only did that for a few days. I think we just did eight-hour shifts [at sea]."

Matt commented on trying to keep up with qualification studies while working in the galley: "Cranking when I was trying to qualify wasn't really that big a deal. I was always ahead on points, and the numbers [point requirements] were halved when you were cranking. If anything, you could get some hook-ups — an extra chocolate-chip cookie can go a long way."

Regarding cranking after qualification, Matt remembers: "I actually did most of my time after I had my dolphins. After the COB screwed up his scheduling, he sent a lot of qualified guys back [to the galley to crank]. It started with Billy [a torpedoman], who had more cause to bitch about it than anyone else. He probably did about 120 days [total cranking] between his two boats. When the fact was pointed out to the COB that the SSORM [Standard Submarine Organization and Regulations Manual] said that qualified personnel shall not serve as FSAs [Food Service Attendants, the fancy Navy name for cranks], the COB said 'Well, sometimes in the Navy, you have to ignore the regs.' If I had ever gone to Captain's Mast, I wonder if I could have said *that*."

As I said, a rated nub could spend about sixty days cranking — unless his division happened to be short-staffed. In this situation, the nub is usually sent straight to his division to begin qualifying for his primary watchstation. Of course, he'll be expected by his shipmates to qualify rapidly at his watchstation, so the pressure's on! If the nub doesn't qualify in a timely manner, the senior watchstanders might have to stand port-and-starboard watches, with two crewmembers alternating shifts at one watchstation, six hours on and six hours off. This puts extra stress on the senior watchstanders and makes them unhappy, and when they're unhappy, so is the nub (or, put another way, if they don't sleep, the nub doesn't sleep either). It's in the nub's best interest to qualify quickly.

It was in this situation that I found myself when I reported on board. My training was in ESM (Electronic Surveillance Measures, which is basically a fancy radar detector used at periscope depth). There were only two guys left on the *Annapolis* that were qualified to stand the ESM watch, and one of them, Joe, an Electronics Technician 2nd class — was due to rotate to shore duty not long after I reported aboard. Joe was also the only technician aboard who was qualified to maintain the highly over-sophisticated BRD-7 Radio Direction Finding (RDF) Set, so a new technician was needed to replace him. Since I was going to be on the boat for at least four years, I was the obvious the choice to be the new BRD-7 tech, which meant three more months of tech school. Although I was a little embarrassed to be a freshly reported nub who got such a sweet deal as another tech school almost immediately after reporting to the boat (many qualified crewmembers wanted, but weren't assigned to, extra schooling during extended in-port periods), I wasn't about to complain, nor would it have done me any good. So, after being on the boat a mere three months, it was back to school for three more months.

My sweet deal was cut short with only three weeks left to complete the BRD-7 course. *Annapolis* had been out of the shipyard for several months and gone on several Atlantic training missions, and was due to deploy to the North Atlantic for some meat-and-potatoes submarining. ET2/SS Joe's shore rotation was fast approaching, and we needed to get me qualified to stand the ESM watch ASAP. I was pulled from my school and told to prepare for deployment. This was my first taste of "the needs of the Navy."

While I had been away at school, blissfully ignorant of what was happening on the boat, our old COB rotated back to shore duty, and a new one reported aboard. I think he was eager to make an impression on the command, because when he noticed that I was a non-qual who hadn't cranked, an evil gleam appeared in his eye, and he tried to put me in the

215

galley. However, my stint in the galley wasn't to be. My LCPO (Leading Chief Petty Officer, or the boss of my division) squashed the assignment immediately, showing the new COB the watchbill and the qualification list, which showed that we would have only one ESM watchstander for the next deployment. The writing was on the wall, and the COB couldn't ignore it. His plan for me had been thwarted by circumstance, and he couldn't send me cranking. From that point forward, even after I qualified in submarines, the COB seemed to give me dirty looks (or maybe it was just my perception) and gunned for me every time he needed a fresh crank. Fate always seemed to intervene on my behalf, and I never served a day in the galley.

"Are you a nuke?"

There was an additional semi-embarrassment that I had to deal with when I reported aboard, which concerned my paygrade, or rank. At the time, sailors who received advanced training for certain duties on board submarines, such as Sonar, Radio, and ESM, were enlisted under the Submarine Advanced Electronics Field (SAEF) program. There was a decent enlistment bonus, along with performance incentives such as paygrade promotions, which meant that if you achieved a high-grade average when you completed certain schools, you advanced faster through the ranks than other rated sailors. I was enlisted under this program, and within nine months, I had gone from E-1 (Seaman Recruit, the lowest enlisted paygrade) to E-4 (Electronics Technician Petty Officer Third Class, or ET3).

Because of some magic Navy math, I became eligible to take the test for E-5 (ET2) even earlier than I expected (I think you had to wait at least a year after making E-4 to take the test for E-5). I, of course, couldn't shirk my duty and *not* take the test ... so I sat down in the SUBASE New London gymnasium with hundreds of other sailors and took the advancement test. Keep in mind that at this time, I hadn't yet reported to the *USS Annapolis*; I was still in Basic Electronics Rate Training, or BERT. I thought nothing of actually passing the test, and I figured that even if I did, I wouldn't be advanced to E-5. Imagine my surprise when, in the school following BERT ("C" School, in which students were trained on the actual equipment used in the fleet), I received word that I was advanced to E-5. I wasn't the only one shocked. Several of my classmates were fleet returnees, or crewmembers from submarines who were receiving extra training, and they were also duly flabbergasted. However, this didn't stop them from congratulating me.

So it happened that when I reported to SSN-760, I was wearing the ET2 insignia. For the next several weeks, as I met each of my new shipmates, I heard the question, "Were you a nuke?" from almost all of them. This demands that I explain to you what they meant by this question.

The term "nuke" (or as some spell it, "nuc") refers to an enlisted sailor who has received specialized training in nuclear engineering. These sailors would be the ones operating the reactor, the propulsion plant, and the electric plant. Part of Rickover's legacy meant that these sailors were regarded as the "cream of the crop" (if you know a nuke, just ask him, he'll tell you!). Since nuclear training was always in high demand, the sailors in nuclear training programs received large enlistment bonuses and early advancement to higher paygrades. Some of these students, however, weren't prepared for the rigors of nuclear training, despite what their ASVAB scores indicated, and were dropped from the program. Most of them went on to SAEF schools to become ESM, Sonar, Fire Control, or Radio technicians, and reported to these schools with the E-5 "crow" already on their arms. They also reported to their submarines with higher ranks than their contemporaries from BERT and C School, and it was widely known that any nub who reported to the boat as an E-5 was either a "nuke drop" or a former surface-fleet sailor.

Thus it was that I received the question from nearly everyone I met on board, and had to explain that no, I'm not a "nuke drop" or a "skimmer" (the "affectionate" term used by submariners to refer to surface fleet sailors), but a normal, everyday coner (a crewmember whose watchstation is forward of the reactor compartment, the sub's nose cone) who happened to get lucky. This seemed to satisfy most of them, and surely served to increase the expectations of some of the salty dogs on board.

"Dive, dive!"

And so, after narrowly avoiding galley duty and being pulled from my extra tech school before I could complete it, I was off on my first deployment on my new boat. It was a cold November day when we set the Maneuvering Watch and transited downriver to Long Island Sound on our way out to the Atlantic Ocean. Prior to reaching the dive point, the ship was verified rigged for dive, and we slipped under the waves.

It didn't really feel any different from being tied to the pier, with the exception of brief sensations of movement and the occasional angle of the deck. I remember looking around at hatches I used to enter the boat while we were in port and think, "Wow, there's really several hundred feet of water above this hatch!" and try to determine if it worried me or made me feel different. Aside from a sense of awe that I was finally in a submerged submarine for the first time on a real operational mission, it felt "natural" to be there.

Compared with the hustle and bustle of stationing the Maneuvering Watch, during which every crewmember has a specific duty (except, at the time, me, so I sat in the radio/ESM room studying), and the early part of the surface transit, or the "piloting party," the normal underway watch routine was positively calm. My shipmates who weren't on watch, or due to soon relieve the watch, went to the rack (their bunks), to "sleep defensively," a practice recommended by officers and senior enlisted men alike that hedges against long periods of activity without sleep, often caused by casualty drills, field days, or extended watches.

Non-quals like me, however, dared not jump in the rack immediately after the dive, especially when there were watchstations to qualify, and dolphins to be earned. The dolphins, which are the insignia of one who is "Qualified in Submarines," aren't just a mark of prestige. Your dolphins tell other submariners and sailors that you have the skills and knowledge it will take to help save your ship and your shipmates' lives in the event of an emergency. They also show that you know the ship and its systems from bow to stern, which not all sailors can claim. Surface fleet sailors, for example, weren't required to earn the ESWS pin (Enlisted Surface Warfare Specialist), but all enlisted submariners are required to qualify in submarines within a year of reporting to the boat. This adherence to high standards of qualification is part of what makes submariners what they are, whether fast-attack (SSN) or boomer (SSBN) sailors.

This is also why non-quals are generally given quite a hard time when they head for the rack.

When it was finally time for me to sneak into the berthing area, it was to a sleeping space not much larger than a coffin. At a little over six feet long, two feet deep, and two feet high, a bunk on board a submarine is large enough for the bare necessities only: you, your bedding, your linens, blanket and pillow, and a small fluorescent light. Underneath your mattress, which sits inside a hinged lid, is your rack pan, where all of your personal belongings are stowed. The sailor's clothes, toiletries, and a few small personal items are supposed to fit in a space as long and wide as the bunk itself, but only two or three inches deep.

This is, of course, possible, but it takes the nub a bit of practice to become efficient at submarine stowage. This, and several shared lockers in each berthing area, are the only "official" location on the boat in which he is supposed to stow personal gear, but after a while, the nub starts to become familiar with his divisional spaces, and can find other unofficial stowage locations. As long as they didn't create a "sound short" (a path through which sound can propagate through the sub's hull into the surrounding water), some other hazard, or become known to those who wouldn't approve, our belongings were generally safe in these spaces. Items as large as guitars have made it on board many submarines, including SSN-760, to the delight — or chagrin — of the crew.

For some sailors, sleeping in their bunks can be as much of a challenge as stowage, especially for those of larger-than-average stature. I have seen guys that had to get out of their bunks just to roll over. Since I am only a hair over 5'10", I didn't have this problem, but I did have another one. Most people move quite a bit in their sleep over the course of a night. During duty nights and underway periods on the boat, I must have been moving around in my sleep when I encountered (bumped into) the side, ends, and top of my "cozy" submarine rack. This sensory information was fed into my dreams, which featured me trapped in a bilge or some other confined space on the boat, and panicking to get free. I awoke from these dreams lashing out at nothing, and feeling relief that the only thing "trapping" me in my rack was a privacy curtain. Thankfully, the dreams didn't last very long, and I became almost as comfortable in my rack as I was in my own bed at home.

Our "northern run," as it was called, was short by comparison with a similar run of normal length (three to four months), and we made it home for the Christmas and New Year holidays, but not before enduring a surface transit that made even old salts queasy. The ashen faces of some of my seasick and suffering shipmates was caused by a winter storm sliding up the Eastern Seaboard, resulting in the ship taking seventeen-degree rolls to port and starboard. While the conditions completely debilitated some of my shipmates, others of them, while still suffering, made light of the situation.

Our navigator, for example, had a clear plastic trash bag tied to his belt (as did several others in the crew who were on watch at the time), in case the nausea became too much for him, which, at some point, it did. However, having no choice but to stay in the control room during that period of the transit, he continued his duties, and distracted himself from the conditions by proudly displaying the contents of his makeshift seasickness bag to crewmembers that were quite clearly on the verge of losing their own dinners to the deep. It wasn't funny at the time (I was rather sick myself, but never made it to the cookie-tossing stage), but it made for good humor the following day, after we had endured the storm.

Crew leave was staggered into two groups, one of which would take leave during the first half of the holiday period, which included Christmas. The second group, in which I always tried to position myself, took leave during the New Year holiday. I always requested that period because I didn't like to miss New Year's Eve parties! Crewmembers not on leave during one of the two periods stood extra watches to make up for their shipmates who were on leave.

Since I had just completed the qualifications for the in-port topside watch, and was still a nub, there was a day on which one of my watches was scheduled for one of the least desirable times (at least for some), the midwatch, which started at midnight. The weather on that day happened to be rather miserable; as I recall, it was raining sideways, and it was a freezing rain to boot. Although topside watchstanders had plenty of protective gear to wear in such conditions, sheets of frozen rain on your "pumpkin suit" made for a miserable time. My belowdecks watch that night was a fire control technician (FT2/SS) named Jon, who later became one of my best buddies on the boat. On one of his trips topside, he brought me hot cocoa and some advice: "Get qualified, and then qualify belowdecks. Then you won't have to stand watches like this."

Once a nub starts qualifying watchstations and standing real watches, he starts to feel like he is making a real contribution. During the month-long northern run we had just finished, I had qualified ESM, Duty ET, and topside watch, which kept me out of the galley for sure, but didn't end my responsibility for continuing to make progress with my ship's quals. And so it came to pass that, when our holiday leave period ended, *Annapolis* was once again tasked to brave the icy north.

Underway periods are the best time for a nub to study qualifications. Not only is every crewmember on board so you can get help learning the ship's systems if you need it, but (ideally) no maintenance is being performed on any major systems, which means it can be studied in its operational state, and many times in action, performing the tasks for which it is assigned. This, along with the fact that I was standing port-and-starboard on-call ESM watches, contributed greatly to my relatively speedy qualification time. By the end of our approximately two-month deployment, I was wearing my fish.

That last paragraph probably deserves an explanation. ESM (Electronic Surveillance Measures) is a lot like passive sonar, in a way, in that both the ESM and Sonar operators receive signals, analyze them, and classify them, based on various characteristics of the signal. The similarity ends there. Sonar watches are stationed twenty-four hours a day, while ESM watches are stationed only when the boat is at periscope depth (PD). During most operations, except certain ops to be described later, excursions to PD occurred on a regular schedule based on the need to copy the submarine satellite broadcast, or SSIXS (we pronounced it "sissicks," and we copied it as often as our operations orders required). Once we had copied the entire SSIXS broadcast, the excursion to PD was complete, unless we happened to be ventilating the boat, in which case we would stay up a little longer before "going deep." The ESM operator's job for the short time we were at PD was to detect and report possible collision threats (by signal strength) and detection threats (by signal type).

Since we only had two qualified ESM operators at the time, we stood port-and-starboard on-call ESM watches for twelve hours each. Furthermore — and without going into too much detail — our PD schedule was such that I really only stood a few hours of ESM watch during any given day. The rest of the day was devoted to my study of ship's qualifications, and this I did aggressively. I wanted to rid myself of that "nub" tag as quickly as possible.

"Get qualified, nub!"

Ah, the familiar refrain of the qualified submariner to the hapless nub.

The qualification process starts as soon as the new non-qual reports to the boat. That day, he receives his "qual card," or qualification card, which serves as a record of the nub's progress as he gains more and more knowledge of the ship's systems and procedures. A qual card is divided into "blocks," or topics of study, which are further divided into individual knowledge requirements. The nub must get an authorized petty officer's or chief petty officer's signature indicating that he has mastered each of the topics contained in the block. Blocks consisted of such areas of study as damage control, sonar, fire control, Radio/ESM, navigation equipment, steering and diving equipment, hydraulic systems, air systems, trim and drain systems, propulsion and electrical systems, and environmental systems.

Once the non-qual felt he had mastered the material contained in a particular sub-topic, he would go to the petty officer or chief petty officer who was authorized to give a signature for that topic and ask for the signature. It was at this point that the nub would be quizzed by the PO/CPO. Often, the nub left his first request for the signature with one or more "lookups." A lookup consisted of information that the PO/CPO had asked about the topic that wasn't covered in the study material, or was buried deep in it or in some other obscure place, which the PO/CPO felt was necessary for complete knowledge of the topic. The nub would have to go find the information, either by doing a thorough search of the study material (often a Ship Systems Manual, or SSM), or by asking a qualified crewmember who was familiar with the system or procedure. In some cases, the lookups were reasonable, "hey you missed this part, go look it up" pieces of information that were quite important to the knowledge of the topic. In other cases, the lookups were picayune and arcane bits of information about a procedure or system that only a qualified watchstander or technician on the system would be required to know. Guys who assigned such lookups usually had a bit of an over-inflated sense of importance, but this was rare.

A nub was assigned a specific schedule in which he had to accumulate a certain number of qualification points. If he was behind schedule in accumulating these points, he became "delinquent" in his qualifications, which was a very undesirable position. The delinquent nub was commonly referred to as *dink*, and everyone knew who was dink and who wasn't by a progress chart in the middle level passageway. Dinks were given extra motivation and "encouragement" by their qualified shipmates, and weren't permitted to watch movies or partake in any recreational activities in the crew's mess or elsewhere. They were also given an extra-hard time when found to be heading for the rack. The admonition "You can sleep when you're qualified!" seems to ring a bell, though I don't remember ever finding myself on the dink list.

Once the nub had obtained all of his block signatures, he was ready for his qualification board interview. This was an oral examination by a board consisting of a senior petty officer, a chief petty officer, and an officer. The interview covered all of the blocks on the qual card, and the nub was tasked to draw diagrams of critical ship's systems in order to prove his knowledge of them. Even at this advanced stage of the qualification process, the non-qual could come away with lookups, but if this happened, it was nearly assured that he would pass the board if he found the correct answers to the lookups. Once this was done, all that remained was the qualification ceremony, in which the CO pinned the newly qualified submariner's dolphins onto the left breast of his uniform.

My ceremony occurred the day before we were scheduled to return from our northern run, which meant that I would be wearing my dolphins when we returned to port. It also meant that there would be precious little time for the tradition of *tacking the dolphins*, a symbolic and forceful gesture meant to assure that they never fell off. This was something that all newly qualified submariners had to endure stoically, and completed the sailor's entry into the ranks of qualified submariners. It also left quite a painful bruise. However, as I mentioned, we were due to return to port on the day after my dolphin ceremony. There was a flurry of activity, so many of my shipmates were too busy to notice my shiny new fish. As a result, I only got tacked by about twenty of my shipmates; there was an unofficial twenty-four hour limit to the "tacking," after this period, a handshake and congratulations would have to suffice. I didn't complain a bit.

Pride and Tradition meet Political Correctness

I served in the U.S. Navy during the 1990s, a period in which American culture began to be afflicted by the ridiculous, yet pervasive, notion of Political Correctness, or PC. You might think that the military would be immune to such societal shifts that would undermine tradition, yet this wasn't so. Suddenly, it became inappropriate to offend someone, to degrade someone's self-esteem by even the slightest of means, or to ask them to participate in a traditional activity that might be considered humiliating by some. Thus it was that many time-honored traditions started to go by the wayside, or, if still practiced, were done so "underground," without the knowledge of the senior enlisted staff or the wardroom.

This creeping PC began on my boat with the Bluenose ceremony. This typically raucous and somewhat humiliating (but still fun) tradition was celebrated for sailors, junior and senior alike, who were making their first trip across the Arctic Circle. Some very cold, very foul, and sometimes embarrassing things were done to each Bluenose candidate, none of which I will recount here (for the sake of brevity). After the ceremony, the crew, through a sharing of endured "hardships," became a bit more bonded, and a good time was usually had by all.

When it came time, during the aforementioned second northern run, for me and many of my shipmates to participate in the Bluenose ceremony, we were told by our CO that participation was strictly voluntary, which came as a surprise to many of us. We were told that this was because no crewmember could be forced or coerced to participate in something that might be perceived as a form of hazing, or to be "humiliated" or "degraded" against his will. Few, if any, took advantage of this pronouncement, and the cold, messy ceremony proceeded as planned.

Events that occurred during my time on *Annapolis*, such as the Order of the Rock ceremony for crossing the Strait of Gibraltar during our Mediterranean deployment, were devoid of any such ceremonies, and the certificates of recognition thereof were handed out without fanfare or celebration.

The next tradition to fall victim to PC was the tacking of dolphins. While I had mixed feelings about this tradition myself — mostly due to a few overzealous sailors who were truly out to inflict pain on the newly qualified, simply because some nasty old salt had done it to them in the past, I still wanted to participate. It seemed right that I should honor a tradition practiced throughout much of the history of submarining, and because my more manly instincts told me that I should endure the pain to prove my worth as a submariner. However, dolphin tacking had become a forbidden activity during my qualification process, and if it was discovered that a tacking had taken place, the guilty parties could be taken to Captain's Mast for disciplinary action. A sailor who objected to the ceremony could report a tacking, but would generally be regarded as a whiner if he did. Even so, the rules were the rules, and all tackings, including mine, now took place out of sight of the CPOs and wardroom.

Even the qualification process wasn't immune to the PC movement. Such terms as "nub," "dink," and "non-qual" were to be forbidden in the "new Navy," because they might cause embarrassment, humiliation, or a loss of self-esteem to the persons at which they were directed. In fact, the use of these terms was often intended as a good-natured motivator when used by most well-meaning shipmates. They were replaced by their more official names "non-qualified sailor" and "delinquent." Non-quals could no longer be harassed (with, of course, the best of intentions) by their qualified shipmates if they sat down to enjoy a movie in the crew's mess, and those who were dink weren't to be labeled with the unflattering term.

At the time of this writing, I have been out of the Navy for nine years, and I am unaware whether or not the PC madness has been stopped and proud old traditions restored. I am sure some of them will never return; PC will leave scars all over society, and the Navy, being a microcosm of society, will be no exception.

The Daily Grind

Daily life on board a submarine differed starkly depending on whether the ship was in port or out at sea. While this may seem obvious to some, I was asked many times if I had to sleep on the submarine while I was in port.

The in-port routine was much like a normal civilian job, with notable exceptions. Although there was a ship's schedule, or "Plan of the Day," which defined the times for morning muster and afternoon liberty, any sailor could count on staying well past the official liberty time at the discretion of his division chief, depending on how much work needed to be done on a given day. There were certain evolutions that required, if not all hands, a large number of the crew, or an entire department, to work from very early in the morning to late at night; these included weapons loads and reactor startups. Other, shorter evolutions that required all hands included stores loads (during which food and other supplies are loaded on to the ship by the sheer backbreaking manual labor of the crew), and field days.

Also like a civilian job, we got to go home — the barracks for some, Navy housing for others, and civilian housing for those who chose it, almost every night. Notice I said "almost every night." Every person on board, with a few obvious exceptions, was part of a duty rotation. This rotation was typically four days, but could be reduced to three if necessitated by the ship's schedule. The duty section slept onboard ship, to provide a safety and security watch complement.

When a submarine is at sea, the schedule is entirely different. Although we observed a normal twenty-four-hour clock like everyone else, the watch rotations observed an eighteen-hour cycle of six hours on-watch and twelve off-watch. As you might expect, a sailor spends his six hours of on-watch time at his primary watchstation. Sonar, maneuvering, helm were among the many stations to be manned. The twelve hours off-watch, however, wasn't entirely "free" time. Off-watch periods were often occupied by training, maintenance, administrative duties, and the two activities that sleepy sailors dread: field day and drills.

"Field day," for the sleep-needy, wasn't always the happy activity that the name implies. Field day, the ship's weekly cleaning, is an all-hands evolution, meaning that no sailor could be in his bunk (unless authorized by the Doc or the COB). Every internal space on the boat was scrubbed, dusted, vacuumed, and polished. This activity, although reviled by some, is absolutely necessary to the health of the crew and the proper functioning of ship's equipment. Believe it or not, one-hundred-plus sailors living in the same confined space generate a lot of dust and dirt, which, even with proper filtering, can cause problems if allowed to accumulate. Despite its reputation as a sleep-thief, field day even contributes to morale. It is one of the few times during underway periods that all hands are awake, and can socialize to some degree while working toward a common goal.

Drills come in different varieties. Casualty drills are simulated accidents or equipment failures that require an immediate crew response. These include fire, flooding, toxic gas (from the battery, torpedo fuel, or refrigeration plants), reactor accidents, and other problems. Tactical drills were a bit different, and included simulated collisions, man-overboard, battle stations, torpedo evasions, and other simulated combat actions. Most drills were scheduled, to give the crew advance warning so that they could sleep defensively during the rest of their off-watch time. A few drills were unscheduled, but not many.

The drill schedule depended on what the ship's long-term schedule would be. For example, if we were about to deploy with a carrier battlegroup, we would run numerous tactical drills while we trained with the various other ships in the group. This was also the case during the TRE, or Tactical Readiness Evaluation, a grueling period in which the crew's combat skills and response to casualties are tested. If, on the other hand, the ship was preparing for an ORSE, or Operational Reactor Safeguards Examination, the drill schedule was almost entirely composed of casualty drills. Additionally, since the ORSE team was on board to inspect the material condition (maintenance and cleanliness) of the engineering spaces, all-hands field days of the engine room were thrown into the schedule. This, as you might imagine, made for a stressful time for everyone.

Evenings and weekends, for the most part, could be expected to be free of field days, drills, training, or other work. Each sailor's off-watch time could be spent pursuing his own interests or personal necessities, such as sleep (a personal favorite of mine), movies, music (listening or playing) card games, reading, qualification studies or correspondence courses, and exercise. I participated in almost all of the activities listed above; any sailor who doesn't is asking to go nutty in a hurry. It pays to keep busy, one way or the other.

The Hunt For....

I have been asked many times about what we did out there in the big blue ocean. The official answer, and the truth, is training.

In the 1990s, the Cold War was largely considered to be over, so the Soviet threat was diminished significantly. However, it wasn't completely gone, and there were other potential threats to United States' interests and national security that we had to keep an eye on, such as the Balkan conflict (while not, strictly speaking, a security threat to the U.S., could have eventually threatened our allies in NATO).

Such vigilance required continuous training, sometimes alone, sometimes in conjunction with other U.S. naval units, and sometimes with naval units of other nations. This training really kept us sharp, and honed the skills of our junior crewmembers as well as our "old salts." Much of this training involved ASW, or Anti-Submarine Warfare, and we were frequently the "OPFOR" (opposing force) submarine that had to be found and prosecuted by the "good guys" in surface ships, ASW aircraft such as P-3 Orions and S-3 Vikings, and other submarines. I fondly remember being asked by multiple ASW platforms if we could broach our sail out of the water a bit so that their lookouts or radar operators could actually detect us! We also played the "good guys" at times, searching for other friendly submarines that were simulating potentially hostile platforms, conducting submerged approaches on surface targets, or practicing our reconnaissance skills on land-based or littoral (coastal) facilities.

It was during such training exercises, as well as the TRE previously described, that I really earned my keep as an ESM operator. During routine operations, as mentioned before, I was on-call for twelve hours a day, and probably spent no more than three hours sitting at the ESM stack. However, during battle-group or ASW training ops, as well as our workup period for the TRE, I sat many consecutive hours, sometimes my entire on-call shift of twelve hours, at the stack. My only breaks during these times were when I was relieved for a head call or to wolf down a quick meal.

At times, I had only one or two contacts to track. There were other times, however, when the receivers were saturated with contacts; when I went to the rack after watches like this, the beeps and screeches of the demodulated signals that I tracked followed me into my dreams. Although operations like these were stressful, they were also the most rewarding.

I may have made it sound like all we ever did was conduct training. I'm not saying that this isn't true. I will say this: our training stood us in good stead on multiple occasions.

Submariners are experts at cross-training. Although my primary watchstation was ESM, I was also qualified to operate the ship's radar, periscope, and navigation equipment, as well as HF, UHF and satellite radio gear. Like many of my crewmates, I could also monitor a battery charge, pump bilges, and change the ship's ventilation lineup. My most important and most frequent watches, other than ESM and in-port watches, were the radar watch and piloting party. I became an expert at tracking surface contacts with radar during our long and eerie nighttime surface transits back to Groton. During our approaches to Groton and other ports, I plotted our position on the secondary navigational chart as a backup to the primary plot, using data from visual fixes, radar, and GPS (which, at the time, was merely a backup, but was to become our primary means of obtaining navigational fixes).

Some of my fondest memories of my time on the boat were of our midwatch surface transits, when the control room was darkened, but still quietly abuzz with activity at the radar, chart station, and attack center, where the fire control watch (FTOW) tracked contacts using data passed to him by sonar, radar, and the Contact Coordinator, who passed visual information from the periscope to the FTOW. At these times, we fought sleep by quietly joking around, and telling each other about our in-port plans. During one particular transit, there was a quiet parade of crewmembers to the periscope stand to observe a comet. The middle of the ocean proved to be the perfect place to admire the phenomenon, since there was no light pollution from any population centers.

Operational units of the Navy, as well as those of the other armed services, don't take a break for the major holidays. I have spent several major holidays underwater, including Thanksgiving, Christmas, and New Year's. We were, however, prepared to celebrate these holidays, submerged or not.

I should have expected that, considering the Submarine Force's reputation for having the best chow in the Navy, we would be well stocked for the holidays. Even so, I was surprised by the spread that greeted me for my first Thanksgiving underway, during our Med run in 1994 — nothing from the traditional holiday dinner was missing. Turkey with gravy, mashed potatoes, vegetables, stuffing, rolls, and various other goodies awaited me when I got off watch for dinner. Nobody from the crew missed it, and we stuffed ourselves with abandon. The only thing missing was family, but we were a family of sorts, and made the best of being away from home for the holiday.

A similar spread greeted us for that year's Christmas dinner, but that holiday wouldn't have been complete without a visit from old Saint Nick. As the clock struck midnight on Christmas Eve, I sat at the ESM stack and reported the detection of air search and navigation radars from Santa's sleigh. Sonar consequently reported a "fly-by" of the sleigh, whose detection signature reported by sonar included "reindeer farts." I also seem to remember someone requesting that the Chief of the Watch raise the snorkel mast so that the fat man in the red suit could come down the "chimney" to bring us our gifts.

Join the Navy, See the World

One thing that some civilians don't expect to hear from submariners is that we get to visit foreign ports, as well as domestic ones, during our deployments. Just like any ship's crew, we need time to de-stress, as well as time to make repairs and take on new supplies.

My first liberty port was Bergen, Norway. We tied up to the pier during the month of February; our proximity to the Arctic Circle guaranteed a frosty reception, but only from Mother Nature. Sailors from the Norwegian Navy greeted us, and we exchanged pins, hats, patches, and other uniform items, as well as tours of each others' ships. Additionally, most Norwegians speak at least some English along with their native tongue, so there was very little problem understanding each other, either at the Bergen naval base, or out in town.

This lack of a language barrier was to prove fortunate in another way when I met a local woman named Margrethe, who took me on a tour of the city and its surroundings. Quite aside from my tour of the town, the stop in Bergen was to be instructional to me in how not to overdo it in a foreign port. My friend Shawn, who was the belowdecks watch following our first night in port, heard me groaning in the rack at four p.m., and brought me a small handful of aspirin and a cold glass of water. It felt like he saved my life.

During the *Annapolis'* Med run (Mediterranean deployment, typically lasting six months) of 1994-1995, we pulled into several ports, each with its own interesting (depending on how you looked at it) characteristics. La Maddalena, Italy, was a small, and I mean small, NATO base staffed by the Italians and the semi-permanent home port of a U.S. submarine tender. Actually, the base was just a little rock sticking out of the ocean between Corsica and Sardinia. La Maddalena was a larger island to the north with a small Italian population that catered to tourists during the spring and summer, and tolerated American sailors for the rest of the year, while welcoming our money. Our Med run happened to be during the off-season, so the beaches and resort areas at most of our liberty ports were deserted.

Gibraltar, a small possession of the United Kingdom on the southern coast of Spain, is a microcosm of the UK itself, while also being a major tourist trap. Being sailors, of course, we were very encouraged by the advance word we received about the amount of pubs per square mile in Gibraltar, and several crewmates participated in the Gibraltar Mile. This insane competition featured a three-legged race in which the participants drank a pint (of beer or ale) and a shot of booze at each designated pub along a specific route. All that the poor saps received for their efforts was a lousy T-shirt, which was usually covered with regurgitated ale at the end of the race. Gibraltar does have some non-alcoholic attractions, such as the

Rock of Gibraltar, which many of us visited for its historic significance and the beautiful view of the surrounding area.

Toulon, France boasts a base of the French Navy, and, in my mind, not much else. There were some historical sights to be seen, and it is located on the southern coast of France, which puts it on or near the French Riviera. However, as I mentioned before, it was decidedly not the tourist season. If we needed any proof, the raw and cold rain that attacked us sideways served that purpose. Some of my crewmates did make themselves useful by traveling inland to perform charity work at an abbey, while others traveled by train to Paris. Without a passport, I wasn't sure whether or not I could travel safely inland, so I stayed in town, somewhat disappointed, but happy for the break from the underway routine.

Limassol, Cyprus was another tease, because its weather seemed perfect. However, the tourists were still not present, so many of the resorts and beaches were empty. This made the "tourist season" feel like an arbitrary and unfair designation in this beautiful place. Despite the fact that we were tied up at perhaps one of the busiest commercial ports in Europe, the water was a bright, stunning blue, instead of murky and sheened with oil. Our friends, the British, had military personnel stationed nearby, so there were quite a few pubs in town in which we could feel relaxed and take the edge off. In fact, some of our crew played softball with the British during our stay, and some of the crew made trips to Greek historical sites.

At the time of our deployment, the Middle East was relatively quiet, so our stop in Haifa, Israel was, thankfully, uneventful. Still, the backfiring of a car engine outside the walls of old Jerusalem rattled those of us who took a bus tour to the Holy Land. Our tour guide was unfazed; earlier in the day, while we were riding the bus en route to our tour area, he explained to us that while terrorism was an unfortunate fact of life in the Middle East, Israelis didn't let it stop them from living their lives. The stark contrast between normal people performing everyday activities such as shopping and dining in outdoor cafés and IDF troops patrolling the streets with large-caliber weapons underscored our tour guide's words.

The trip from Haifa to Jerusalem was a bit of an eye-opener for me. I expected to see mostly desert, but there were many well-irrigated farms and gardens along our route. Because I'm not a religious person, I also didn't expect to be moved by the sights and sounds of Jerusalem and Bethlehem. However, I could feel the millennia of history in and around the places we visited. I expect that even the most devout atheist would come away from these places with at least an appreciation of the history of the area. Along with the obvious places to visit, there was something of an unexpected "holy" place along the route to our destination: a roadside rest in the form of an Elvis museum. It was unmistakably a shrine to the King.

Our domestic port calls were usually not noteworthy, except that, like our foreign stops, they gave the married sailors an excuse to go out and party with the younger single sailors out in town. We would usually pull into domestic ports if we were picking up or dropping off "riders" (personnel not part of the normal crew that either served a special purpose or none at all) or were needed for emergency maintenance.

It was a circumstance of the latter variety, combined with a series of other required system overhauls planned for a future time, that found *Annapolis* — unexpectedly, for most of the crew — in a floating drydock at the naval base at Kings Bay, Georgia. On one fine evening at a local nightclub, my crewmate Monty and I joined a two-piece cover band on stage and regaled some of our other crewmates with a rendition of the Rolling Stones' *Sympathy for the Devil*. The rest of that month wasn't so eventful, but the boat got fixed, and we were on our way back out to sea.

Shipmate – Warrior

Our stay in Georgia took place during the final year of my enlistment. For reasons irrelevant to the telling of this story, I decided to end my naval career.

Crewmembers leaving USS *Annapolis*, no matter the reason, were bid farewell with a short ceremony in which their contributions to the ships operations were fondly, but concisely recounted. The departing crewmember received a plaque describing him as "Shipmate – Warrior" (the phrase often used to designate the plaque itself), and went on his way to continue with his paperwork processing him to his next command, or into the civilian world. Shipmates said goodbye and continued with the day-to-day operation of the ship, pausing only briefly to remember their departing comrade before returning to the daily grind.

Even knowing this, I walked off the pier a bit disappointed that there wasn't more acknowledgement of my leaving by some of my other shipmates. Looking back, I know there was, and still is, a job to do in the Submarine Force that no single crewmember's departure can delay or prohibit. This is a fact for which I will always be grateful, because I have been out there; I know what the life is like and what the job entails. I sleep well at night knowing that there are men who are willing to do that job to keep U.S. submarines patrolling the world's oceans.

Submarines as Time Machines
The Story of *USS Cod's* Restoration
Paul Farace, *Cod* curator

The power of a historic place or object to transport a person back in time is well known to veterans revisiting battlefields or anyone who has discovered a long-lost personal treasure in the back of their attic. The power of a well-restored warship to take a visitor back in time is infinitely greater, for both veterans who served aboard it, or one like it, as well as civilians with little or no connection to the Navy or the time period the ship represents. After more than thirty years of working aboard The *USS Cod* Submarine Memorial, a *Gato* class fleet submarine docked in Cleveland, I can well appreciate the effect this historic submarine has on her visitors, who often regard the memorial in its present state of restoration as a time machine.

Stepping aboard the sub, veterans rediscover a thousand tiny details lost to fading memory. The civilian embarks on a journey to a strange world locked in time and previously only accessible in movies and books.

Recently we hosted a visit by the *Cod's* chief radioman, a plank owner who served aboard *Cod* for all seven of her war patrols. He had not seen *Cod* since 1945 and had intended never to see her again. Only after repeated prodding by his family did he agree to allow them to take him to Cleveland to see where he had served his country in WWII. As he climbed down the *Cod's* vertical ladder into the forward torpedo room, he was no longer an eighty-two-year-old man in Cleveland, but a teenager in Perth, Australia in wartime. And as if using a time machine, he was able to relive history with the added insight of all he had learned in the years that had passed. In addition to a handful of happy memories of shipmates, the overall experience was distilled in one painful lesson he passed to his family and the *Cod's* present day crew: "War is a terrible thing ... war is kids killing kids."

Cod would have proved to be an effective time machine to her former crewman whether she was fully restored or a neglected wreck. Without significant restoration our veteran would have sought out those details that remained to reconnect with his past. His family would have had to work harder to see the boat as he would have remembered it. But *Cod* is not typical of most fleet submarine memorial vessels. Thanks to a wonderful combination of luck and wise leadership, she survives today as the sole unmodified example of a WWII fleet submarine, complete in virtually every visible respect as possible. Out veteran did not have to waste time explaining to his family how things were different now than how he remembered them, or to explain what was missing. All his effort could be spent on passing on that one lesson he felt was most important to teach them.

As he walked through the submarine, our radioman saw all of the crews' bunks, complete with Navy blankets, flashcovers on mattresses, and pillows with Navy pillowcases. Bunkbags stuffed with underwear and other personal items hang from bunks. The Navy's fouled anchor dishes and silver were set for coffee in the officer's wardroom; a rumba record is cued on the turntable of the Navy record player that sits on the shelf behind the captain's chair. The radio room where he spent countless hours listening to coded dispatches was ready for his next watch, complete with telegraph keys, earphone headsets, radiomen's typewriters, and stacks of paper to be jammed into the carriages when the FoxSked contained a message intended for *Cod*. The galley is ready for her cooks to don their aprons and fire up the electric ranges to feed eighty-plus hungry submariners.

The radioman's family was present with him in 1945, thanks in part to the *Cod's* level of restoration. Few modern day elements interfered with their journey back in time. Each item present aboard the boat spoke of a time when their grandfather walked the decks as a young man. For his family, the fully restored crew spaces filled in a thousand details he did not have to mention. They could easily relate to the aspects of the submarine that were common to their experience: where grandpa ate, slept, showered, etc. Grandpa had only to translate those aspects that were alien to them, and pass on his sad but important lesson.

How *Cod* Got to Be an Effective Time Machine

There are currently sixteen fleet submarine memorials in the U.S. Although they all started out as instruments of war bent on the destruction of the Imperial Japanese war machine, today each has a different mission dictated by the realities of their individual situations. Like *Cod*, most of the subs on display ended their naval service as dockside-training vessels towed to American ports to maintain the valuable skills of Navy submarine reservists. The subs selected for this duty were often thin-skinned *Gato* class subs unworthy of extensive modernization because of their limited diving capabilities in comparison to later types of fleet subs, or were thick-skinned fleet boats not needed for foreign military transfers. When a fleet sub became a reserve training vessel, the modifications to the boat were limited to converting the main crew berthing space to a classroom (necessitating the destruction of the priceless bunks and support structures in that compartment), and modifying the diving station by replacing the shallow water depth gages with replicas that could be set by instructors to indicate various depths while the boat remained safely at the pier.

Unlike surface ships that underwent extensive mothballing procedures for long-term storage prior to becoming memorials, former reserve training subs were often transferred to their civilian guardians with little or no preparation for their new role as memorial ships since systems used for diving and propulsion had already been "mothballed" prior to their duty as dockside trainers. In the case of *Cod*, this meant that water-cooling jackets on engines left operational for reservists were not drained before the arrival of freezing winter temperatures. Battle lanterns mounted in each compartment were soon being eaten from within by caustic chemicals leaking from their batteries. *Cod's* memorial crew also had to deal with leaks from bursting water pipes and sanitary tanks that still held a mysterious brew of whatever was put down *Cod's* drains by reservists. In time these "conservation time bombs" were discovered and corrected, even if too late to be defused.

Lastly, many important items aboard the boat not welded to the hull walked away as souvenirs with Navy reservists convinced that *Cod* would end up in the scrap yard. Unsupervised civilian visitors removed others. Although the donation agreement stipulated that clocks and other valuable items be secured by the Navy reserve center until transfer to her civilian owners, not a single clock remained aboard *Cod*. Thankfully, those who removed the clocks left their telltale shockmounts in place on bulkheads, so that replacements could be substituted when they became available. For some unknown reason, the waterglass holders and towel racks in the officer's staterooms all vanished, along with the metal wastepaper holders. It was rumored that *Cod's* coffee cups, dishes, knives, forks, spoons and 125-lb Hobart galley mixer ended up in several local restaurants and hash houses. Blankets, pillows, door curtains, and sheets also vanished over her career as a dockside trainer. In short, we had a submarine, but there was little aboard *Cod* that would tell the visitor that this was a home to more than ninety-five men while at sea. And it would remain that way for many years.

Cod began her memorial career in January 1976, with an important advantage that most of her sisters lacked, a free berthing space. Since the submarine reserve program had ended in 1971, no new sub would be arriving in Cleveland to replace *Cod*. There would be no towing expense. She could stay at the pier she had occupied since her arrival in Cleveland in 1959, when she replaced *USS Gar* (SS 206) as the reserve training boat (*Gar* was then towed to Ashtabula, Ohio, for scrapping). Following the *Cod's* turnover ceremony, the Navy handed the keys to the fence surrounding *Cod's* Lake Erie berth to her civilian guardians and then transferred the monthly electric bill to them. When spring arrived, along with her first paying visitors, all of her admission income, the sole source of support, could be spent on the care and maintenance of the sub. Money not spent on the small staff and expenses was saved for future needs.

Like most other fleet sub memorials, the early leadership of the *Cod* memorial considered cutting access holes through her pressure hull to accommodate visitors. In addition to the hull access, much of the unique features of the forward and after torpedo rooms would have had to be destroyed to accommodate staircases. Plans were even considered to beach the vessel on lakefront park property to create a dryberth display. The limited income generated during the early years of the *Cod's* memorial status and other, more pressing matters prevented these modifications from being carried out. Luckily, by the time money became available for these modifications, *Cod's* caretakers realized she was the only sub on display at that time to remain unmodified (*USS Razorback* on display in North Little Rock, Ark since 2004, also uses her original hatches and ladders). After several years of operating the boat as a memorial, it was discovered that only a few visitors were unwilling to use the ship's original vertical ladders and hatches. Casual comparisons with fleet sub memorials using staircase modifications indicated they had a similar percentage of visitors unwilling to venture below decks!

While *Cod* and her civilian guardians were blessed with these unique advantages, they would also encounter problems common to many other ship memorials, lack of curatorial management. The small group of Cleveland business leaders and retired Navy veterans who made up *Cod's* board of directors wanted to save *Cod* from the scrap yard, but did not necessarily want to run a memorial ship. When several local historical foundations declined to manage the boat, the group was left to operate the venture on its own. There soon developed a system by which a group of

WWII submarine veterans would handle the daily operation of *Cod*, freeing the board of trustees to handle strategic issues. While it worked fine for the overall survival of the boat, *Cod's* early organizational structure did not include effective curatorial oversight. Beyond a professional tourguide brochure prepared for *Cod's* visitors by her board, much of the interpretation and restoration of *Cod* was left to her shipkeepers.

While submarine veterans are highly valuable to any submarine memorial program, they lack the training and skills necessary to successfully operate the submarine in her new role, that of historic ship. Without curatorial oversight, *Cod's* well-intentioned shipkeepers damaged compartments in the process of creating display cases for undocumented artifacts. Original painted surfaces were covered in new colors because someone had donated spare house paint after a home renovation project. Rusting deck structures were cut off and discarded with no provision for repair or replacement. Handmade signs soon covered much of the interior of the boat, often carrying inaccurate information.

In the fall of 1987 *Cod's* board joined the Historic Naval Ships Association, an international body consisting of those who manage historic warships. I attended the 1988 conference in Erie, Pa, held there to commemorate the launching of the rebuilt brig *USS Niagara*. Through interaction with individuals with similar vessels, especially Russ Booth, manager of the *USS Pampanito* (SS 383) project in San Francisco, I experienced an epiphany: I suddenly had something to compare *Cod* against, I learned how to organize and begin a cohesive restoration program, and I discovered sources for missing parts! Russ Booth provided a wealth of invaluable information on working with submarines, interpretation methods, as well as friendship. Our good-natured rivalry spurred both organizations to new successes in restoration. Slowly, and sometimes painfully, I began assuming the duties of curator for *Cod*. The shipkeepers, sometimes reluctantly, began to understand that there was a right and wrong way to do things aboard *Cod* and that the curator had to be consulted when working aboard the boat.

Typical of the way our friendly sub-to-sub rivalry helped to improve *Cod* was the issue of the submarine's periscopes. *Cod's* scopes had been pulled up and blocked in an elevated position in the late 1970s with the use of a crane. They did not rotate and if the blocks propping them up were to ever fail, a ton of high-precision optical gear would come crashing down the periscope wells, destroying these prized instruments. Russ Booth was very proud of the fact that they had isolated the hydraulic circuit for the

Pampanito's periscopes and installed a hydraulic pump that allowed the scopes to be raised and lowered at will.

The prospect of enabling *Cod's* scopes to raise and lower made my mouth water. To his credit, just after enjoying a moment or two of victory over *Cod*, Russ mentioned that two hydraulic motors had been donated to his sub for the periscope project and that he would gladly ship on to *Cod* if we needed it. At the next meeting of *Cod's* advisory committee we discussed the *Pampanito's* achievement and their kind offer of a hydraulic motor.

Several of our crew believed that the ship's original hydraulic (IMO) pump could be returned to service safely and without need of an external pump. They set about making their ideas work. Within a few weeks I was mailing Russ a video of *Cod's* scopes smoothly raising and lowering under their original hydraulic power. I waited a week for the full impact of our achievement to register with Russ before I called him for his reaction. When I did finally call him I was greeted with a nonchalant reply: "Oh, yes, we saw your video ... good work. And by the way, we just got our IMO pump operational." That good-spirited rivalry and assistance between two submarine memorials continued even after Russ' untimely death in 1997.

One of the most important lessons Russ taught me was to understand what is really important to visitors. Later called the Russ Booth Rule, it is simply this: the average visitor can best appreciate those things aboard your vessel that are common to their lives: where did the crew eat, sleep, and go to the bathroom? All of the technical wizardry of a torpedo data computer or ballast control panel is lost on the average visitor. Guided by the Russ Booth principle I began working on *Cod's* crew spaces.

Thankfully, we discovered a wealth of photos of *Cod* both in the National Archives and through the wonderful donation of the widow of our last wartime skipper. Mrs. Edwin Westbrook donated her husband's footlocker, filled with hundreds of candid photos of *Cod's* last war patrol. These, along with photos obtained from attendees of *Cod* crew reunions, provided a wealth of information to guide us in restoring compartments throughout the boat.

As we began to outfit the many spaces and compartments aboard *Cod*, we were constantly warned by our friends in the historic ship community to visitor-proof our work, keep the public out of compartments with Plexiglas barriers, and to otherwise safeguard everything or it would walk away! At first our projects involved large items, like replica mattresses, pillows, and curtains. Since these things were not usually on top of souvenir hunter's wish lists, security barriers were limited to keeping the small chains that had been strung across doorways years earlier to keep honest visitors out of various compartments.

WWII-vintage Navy fouled anchor pattern china was provided by the Naval Historical Center, another great benefit of our association with HNSA. But fearing for the safety of these Navy artifacts, I kept the china under lock and key and would only display the modern era Navy teacups I was able to purchase commercially on the wardroom table, along with look-alike galley and pantry items.

The first restoration projects aboard *Cod* had a dramatic effect on both the boat and her visitors. Compartments under restoration began to look alive. And our visitors reacted to the work as well. Previously, when visitors stepped through the watertight door between the forward torpedo room and the officers' country, they would often turn their heads and yell back to those behind them that there was a "shower and kitchen (wardroom pantry)…" Now, with staterooms outfitted with mattresses and blankets on the bunks and curtains on the doorways, they began talking in hushed tones, as if not to disturb anyone asleep!

I also noticed that the table items in the wardroom and pantry were not disappearing. Sure, sometimes a fouled anchor pattern teacup was moved from one end of the table to the other, but nothing vanished. Our friends aboard other ships either dismissed us as blind fools or believed Clevelanders to be much more law-abiding and honest than Americans in their neck of the woods.

Then came eBay and with it access to the attics of millions of Americans. The full set of Navy china and silver provided by the Navy Curator Branch remains under lock and key since I amassed a duplicate set of Navy wardroom silver via the Internet. Emboldened by the fact that my look-alikes did not walk away, I began to substitute the real thing. Today, our visitors often sit in the captain's chair in the wardroom and pick up his coffee cup. The King Neptune pattern silverware set on the table is sometimes misaligned by careless fingers of visitors, only to be realigned by later visitors.

The behavior of *Cod* visitors is in stark contrast to that expected by our friends in the ship memorial community. Those who consider the topic either scratch their heads in disbelief or quickly grow agitated in seeing something they cannot understand.

Cod's restoration quickly extended beyond the realm of her crew accommodations to include her missing deck guns. While *Gar* had gone to the scrap yard with a typical mid-war outfitting of a 4-in. 50-cal main deck gun, and both a 40 mm and 20 mm gun on her conning tower, *Cod* arrived in Cleveland with nothing offensive beyond her ten torpedo tubes. For years a 3-in. deck gun from a minesweeper stood in for the 5-in. wet mount she carried on her last two war patrols and during her Cold War commissioning. Only the mounts and gun-barrel travel lock indicated that she once carried two 40-mm guns on her forward and after conning tower gun decks.

The Navy was unable to locate replacements for the missing guns when we began our restoration efforts. Over several years we were able to locate pieces of two gun mounts. They looked like Thanksgiving turkeys two days after the meal. Except for the pedestal mounts, receivers, and barrels, everything else was missing, including seats, sights, and traversing equipment. We were glad to have something and they went up on the conning tower in the hopes that someday we would either be able to fabricate the missing parts or find complete guns.

Ultimately the end of the Cold War and the resulting base closure program helped us. As military bases began closing around the nation in the early 1990s, the many ornamental guns displayed there became available to us. A dedicated volunteer with a metal fabrication business often had more gratis *Cod* work in for form of old rusty deck guns on his shop floor than income-generating work from customers. Dedication like that is priceless. With time, we were able to install complete and fully articulating 40-mm guns on *Cod's* conning tower.

The missing 5-in. wet mount proved to be far more difficult. While countless thousands of 40-mm guns were manufactured for both the Army and Navy, only a few hundred of the 5-in. wet-mount deck guns carried by our fleet boats were ever built. Most went to scrap with their boats or were left aboard when the sub was expended as a target ship. The best advice we were given was to search for them in the front yards of American Legion Halls and VFW Posts in towns across America. And we searched. It is how *Pampanito* found the 4-in. deck gun used by USS *Tautog*, the boat credited with the most enemy ship sinkings of WWII. I was told the USS *Lionfish* found her wet mount in this way.

Salvation for *Cod* came one day during a visit to the Washington Navy Yard to pick up various artifacts for *Cod*. During a brief chat with Navy Curator Henry Vadnais, Jr, in his office, I asked him if it would be possible to get the 5-in. wet mount on display next door in the Navy Museum. It was a question I had asked him every time I visited him. But this time after telling me the Navy Museum would never part with their prized gun, he added that he had located one for *Cod*! I remember asking quite seriously if he was kidding me. Several months later, just in time for a crew reunion, we installed the seven-ton gun on its mount on *Cod's* main deck. Today, that gun fires salutes to visiting ships and commemorates important events with a very loud bang!

As time passed more and more of the items on our wish list were crossed off. Many things remain, and possibly will never be found, like 252 Sargo Exide batteries. I take comfort knowing *USS Ling* has a full set, as well as *USS Becuna*. Code machines are another matter. Dedicated volunteers aboard *USS Pampanito* obtained a non-functional WWII-era cipher machine and completely restored it before proudly displaying it aboard their boat in the radio room. Sadly, their restoration efforts were too good, because within a short time the curators of the National Security Agency, who had placed the cipher machine on long-term loan to the *Pampanito*, terminated the loan agreement in order to obtain the fully functioning machine for their own use. *Cod* was able to obtain a highly accurate replica of this code machine from a film studio in England, following the release of the film *Below*. Although the film's shoreline left much to be desired, the props and sets built for the movie were first rate. The code machine donation was in recognition of our assistance in helping them recreate Momson Lungs for the film. As much as I would love to have the real thing, I can be satisfied with a highly detailed replica of the code machine and avoid the heartache the crew of the *Pampanito* endured.

We were very happy to realize just a few years ago that we had finally obtained and installed the last major piece of missing deck equipment with the installation of a twenty-seven-foot-long whip VHF antenna. *Cod* is now identical to the way she looked when she left the Boston Navy Yard after being recommissioned in 1952. This is our focus date for restoration. *Cod* is the last completely intact WWII fleet submarine afloat. The few minor modifications that were made during her 1952 refit simply brought her up to the March, 1945 fleet submarine standards.

Restoration of *Cod* continues. Thankfully, missing parts now arrive in small boxes, not in giant crates or aboard heavy lift trucks. Our eBay auction purchases continue, but at a lower tempo. I was once personally responsible for driving the price of Navy blankets from $10 to more than $30, before I took a break to allow their prices to settle back down. Our goal is to equip *Cod* with virtually everything she would have had aboard during her operational service, from ironing board to MK 14 torpedoes. The narcotics and period surgical supplies for the pharmacist mate's cabinet are a challenge best put off for the future. Like most goals, they are good for providing direction, even if achieving them is a distant event.

Pieces of the boat continue to return to us through a variety of channels. Each week I make sweeps through Cleveland thrift stores. On several occasions I have been able to retrieve *Cod* galley equipment that left the ship decades earlier and was donated to charity. Thrift stores are also a source of inexpensive Navy blankets, uniforms, and other valuable vintage equipment to outfit the sub. And when brought down the hatches and stowed in their proper places below deck, the effect of these items is to turn back the hands of time.

Speaking of time, the future holds important challenges for *Cod*. The base closure program that provided us with our missing deck guns now calls for the closing of the Naval Reserve Training Center that has provided us with a secure berthing space for the first thirty years of *Cod's* mission as a memorial. Who will be *Cod's* future landlord and will we receive the same break on dockage fee?

The *Cod* was last dry-docked by the Navy in 1963. Lake Erie's fresh water has been kind to her hull, but time and weather take their toll. She is due for a drydocking, an evolution that will cost at least $1 million. A campaign has begun to enlist the aid of all Americans to ensure the future of this National Historic Landmark submarine. Cleveland is no longer a center of heavy industry. The Fortune 500 corporate headquarters that used to fund our world-class orchestra and museums have either been merged out of existence or moved away. Sadly, the Cleveland that built *Cod's* diesels, milled the steel in her hull, and built much of the equipment inside her, has vanished.

The WWII veterans, who once rallied behind *Cod* at the mere appearance of a threat to the boat, are dying off at heartbreaking rates. In just a few years military veterans of any service will be a tiny minority of our population, and grouped predominantly below the Mason-Dixon Line.

If *Cod's* visitors think she saw her last action in 1945, they're wrong.

I first found Subsim sometime in the late '90s while looking for help with Fast Attack (great game but unfortunately it was buggier than a straw mattress in a boardinghouse). At the time I was a major naval sim nut, playing USS Ticonderoga, Silent Hunter, and even Wolfpack on an ancient DOS machine.

Adam "Bort" Aldrich

Ourselves Alone

The Lost Patrols of *U-49*
Clifford J. Hurgin, Jr.

Day 7: I had a strange dream that I am playing a sub game and I'm not really a U-boat captain....

On Friday, Apr 8, 2005, a Subsim skipper named Kapt. Wrratt made the first of many posts in a forum thread titled *Ahoy-no Time Compression-Expansion Pack-May 1940*. His goal was to replicate a real-life WWII war patrol by playing the recently released U-boat subsim Silent Hunter III virtually around the clock, without using time compression (which speeds up the game). He started the game, and let it run, only stopping to save when he had to go to work as a rural carrier. The rest of the time, for days, Wrratt (real name, Clifford Hurgin, Jr.) was manning his station as the captain of the *U-49*, scourge of Allied merchant shipping and home to Onkel Willie's mouth-watering apple strudel. He ate his meals at the computer, slept on a cot next to the computer, and was awakened whenever his virtual crew sighted enemy ships or needed permission to snorkel.

At first, even longtime Subsim vets were stunned. Playing 24/7? Initial comments were, "Truly hardcore stuff", "Yep, running SHIII on a ZX Spectrum does have its disadvantages!", "I suppose you haven't bathed in a couple of weeks, either, and have bananas and sausages hanging from your living room ceiling?", and "He's so hardcore, if his sub goes down with all hands, I bet he drowns himself in his bathtub!"

But disbelief turned into delight as thousands of Subsim players were mesmerized by his witty, engrossing tale of a skipper, U-boat, and crew melded for a single purpose — to test the endurance of a subsim skipper and unveil one hell of a story. Wrratt won fame as a true subsim pioneer and his adventures were even written up in *PC Gamer* magazine.

What follows is Wrratt's captain's log, a sensational combination of humorous in-game fantasy and clever role-playing, interlaced with features of the game and his real-life struggle to keep his sub and crew alive.

⊕⊕⊕⊕⊕⊕⊕⊕

I did my first career patrol with no TC (time compression), and it was a truly grueling affair. A six-day patrol took me eight days in real time. It took me two days to reach Grid AN81, leaving from Wilhelmshaven. I would let the boat run all day and night, but the crew were total slackers. They couldn't do a four-hour shift with eight hours rest. So during my sleep I would be awakened to a German telling me "not enough crew in diesel compartment" and I would have to assign fresh crewmen. On the third afternoon we sank our first ship at Grid AN81, a small merchant.

On our fourth night we started heading SSE and had finished our 24 hours of patrol. Late that evening I was awakened to find a C2 cargo ship. We moved into position and fired tubes one and two at about 650m to his port side. Both were set for surface run impact only. Two hits but not sinking, he was dead in the water. We pulled away 1700m to surface and watch to see if it sank.

I went back to bed and checked two hours later, the crew was all-aflutter, a bunch of land-lovers. Anyway, we fired a third at 700m and sank it.

And so it went like that for the rest of the week… Now if you are still reading this you might be asking, well, what is this guy's point? I work as a rural carrier, go in to work at 6 a.m., home by 1 p.m. or so. I am single and own my own home and I can use my time as I like, so I might as well sail a U-boat for fun. Plus our *U-49* has a top-secret shower, keep it hush-hush though.

Well, on to my log…

U-49 Patrol Log – War Patrol V

My VIIB U-boot (German for "boat") is *U-49*, not *U-47*, that's Gunther Prien's boot, the "Bull of Scapa Flow". My boot was built 1938-1939, by Krupp Germania Werft, Keil. It is one of 24 units, *U-45* through *U-55* and *U-99* through *U-102*, (except *U-69* and *U-72*) Builder: Bremer

Vulkan-Vegesacker Werft, 1938-1940 made *U-83* through *U-87*.

Log. Update, Jan 14, 1940, 7:30: weather report, clouds clear, precipitation nil, fog nil, wind speed 5 knots direction 0 degrees.

Log update 21:16, 550 km to Grid AN47, gave order ahead full. Planning on running till 75% fuel level or reaching our patrol grid.

Log update Jan 15, 1940, at 1:50, which is 24 hours since leaving our pen at Wilhelmshaven, we have traveled 553 km as the crow flies from our home port. We have 432 km to reach Grid AN47, estimated time of arrival is approximately 15 hours. This first patrol (5th overall) in my new VIIB is vastly better than my old Type II (which I used in my past four patrols). The improved speed and fuel load has made for a fast 24 hours' sailing. I shall endeavor to make the most of my extra tubes and "eels" once nearing enemy territory.

Thank goodness for the vitamin mod! My crew is now able to work instead of sleeping all day.[102] I went to sleep last night around 10 p.m. EST. Woke up around 1:15 or so and none of the crew were tired yet, but rotated crew anyway and went to back sleep. I woke up at 06:00, shut down the game for a few hours. Restarted sailing around 2 p.m. EST, we will begin our patrol and hunt.

I gave our first-time crewmen the ol' "how merchant ships do more damage than a man-o-war ship to our armies because the supplies the merchants have are the lifeblood of our enemy, our blockade will work, and Britain will lose this battle in the Atlantic" speech. Needless to say they are very enthusiastic for our mission, we are all volunteers, the officers are all career naval men and the mates have been working most vigorously the past 24 hours.

Today's sailing was calm and uneventful. Let's hope our next 24 hours are fruitful and soon some tonnage will be unusable. We shall try not to dwell on the merchantmen who will die, by explosion, drowning, caught in raging fires or trapped below deck as the ship goes under. Back at their homes in England or Ireland soon the new widows and orphans will be grieving!

⊕⊕⊕⊕⊕⊕⊕⊕

---

[102] Editor's note: The "Vitamin mod" is a small program that modifies the Silent Hunter III game to keep the virtual crew at their stations longer, before they register as "fatigued", needing the player to rotate them with rested crew. Essentially, this means Wrratt could sleep uninterrupted for six hours at a time.

Before I start my log today, I will respond to a few of my fellow skippers' comments. Wonderful sense of humor, a bit crazy and odd has always been my personality. Why change now? I'm over forty, at ease with how I live and earn my living. I have been employed as a rural carrier since 1984, a job that has served me well. Would I enjoy a new lover right about now? You bet I would! Got an older sister in her late thirties/early forties who looks nice, is a vegetarian like myself, and lives in Danbury? I will be in port soon and looking for a good time, nudge, nudge.

My home is a small Cape style, so hearing the Watch Officer call out during my sleep isn't a problem — having some excited German calling out in the middle of the night tends to wake a person.

Log Jan. 15 1940; Right now game clock is 13:00. We sailed all through my sleep with no contacts. *U-49* is 108 km from Grid AN47, estimated time of arrival is 3 hours forty minutes. Has anyone been to the Firth of Forth? My *U-49* isn't cowardly, it's just we try to avoid all contact with man-o-wars unless ordered to do so. I did my first four patrols starting 1939, we have yet to be sighted or pinged, closest the enemy came was when we were 25 km from Southend. We sank a C2 cargo ship after we had spent two days sailing past V&W destroyers, a light cruiser and some Elco torpedo boats, and it seemed they all started closing in on our position. I had the chief engineer move our boot about 80 meters off the port side of the sunken C2. The water was very shallow so we didn't budge for several hours. Finally, as the two V&Ws swept past, we slipped out ahead slow, at first through the sound backwash of their screws. Then it was just a matter of slowly building speed. We surfaced 20 minutes later to get fresh air and to get more speed and to recharge the batteries before daybreak came. We slipped away totally safe and undetected.

Gottfried Zesller is the senior officer on the boot, but he respects my role as skipper. He told me he can work as Chief Engineer. He was a machinist by trade and has been working with motors since his youth as an apprentice in a shipyard. He told me after my last patrol, that I should upgrade my batteries to the newer quicker recharging ones, and to upgrade *U-49*'s diesels. Which I did. I told him that the engines were his to work on during our patrol. He is free to tell me how best to make sure the motors don't fail us at sea. I would hate to have to try my English on the Brits... "Halloo, mein comrade, I mean, dickey ole chap. I sardine fishman. Have you und spare can opener?"

Log Jan. 15 1940 1:50; Forty-eight hours since leaving port. We have been patrolling our grid for over eight hours. All is quiet. Perhaps BdU[103] should try contacting some Vertrauensleute (spies who send information on convoys). We shall try to keep our spirits up, after all, it has only been two days sailing and only eight hours active hunting (two *real* days, mein comrades). There was one somewhat tragic event this afternoon. Onkel Willie, our cook, dropped a heavy skillet on a case of eggs, crushing most of them. It looks like we will be feasting on spretzel for supper and dinner tonight and having ham without eggs for breakfast for the next few days.

⊕⊕⊕⊕⊕⊕⊕⊕

What follows is an excerpt of Udo Hartenstein's (my Navigator) dairy:

Die Überspringvorrichtung hat die meisten des Tages an seinem cubbyhole und an Schreibtisch aufgewendet und Diagramme, Schreiben und das Hoffen für ein Nachrichte

von BdU oder vom Hydrophon überprüft. Er schien nicht unten hearted an unserem Mangel an Kontakten. Der Koch (wer alle wir Onkel Willie anrufen), fiel ein wundervolles Abendessen aus! Wie er, kochen Sie für rüber fünfzig von uns in solch einem kleinem sperren? Sein Sonntag an der Mitternachtüberspringvorrichtung bestellte vollen Anschlag für halbe Stunde. Er ließ Fritz das Grammophon spielen. Uns Ingruppen von 6 zugestehen, zu gehen oben auf die Brücke für eine Zigarette oder ein Rohr. Einige gingen oben gerade für etwas frische Nachtluft. Die Nacht ist so noch und ruhig, ist der Mond fast Hälfte-voll und Einwachsen. Die Musik frequentiert ein wenig, während sie unsere Aufladung füllt und den Schlag des Diesel und der Schraube ersetzt. Wie kann solch ein ruhiges Gefühl unter Krieg- und Bluthalle soviel bestehen? Ein anderes gutes Merkmal unserer Überspringvorrichtung ist, wenn er auf der Brücke ist, die er jeder unter informiertem hält von, was los oben ist. Eins der neuen crewmembers hat die Idee Ingrid, welches die Krankenschwester zurück an unserer Unterseite am Dockside Nude ist. Woher erhalten sie solche Ideen irgendwie?

Translation…

The skipper has spent most of the day at his cubbyhole and desk, checking charts, writing and hoping for an update from BdU or the hydrophone. He did not seem downhearted at our lack of contacts. The cook (who we all call Onkel Willie) turned out a wonderful dinner! How does he cook for over fifty of us in such a small space? Being Sunday at midnight, the skipper ordered full stop for half an hour. He had Fritz play

---

[103] BdU: Befehlshaber der Unterseeboote, Commander of Submarines

the gramophone, allowing us in groups of six to go up on the bridge for a cigarette or a pipe. Some went up just for some fresh night air. The night is so still and peaceful, the moon is nearly half-full and waxing. The music is somewhat haunting as it fills our boot and replaces the beat of the diesel and screw. How can such a peaceful feeling exist amongst so much war and bloodshed? Another good trait of our skipper is when he is on the bridge he keeps everyone below informed of what is going on above. One of the new crewmembers has the idea Ingrid the nurse back at our base will be at dockside nude. Where do they get such ideas anyway?

Log Update Jan 17 1940; *U-49* is sad to report Grid AN47 is devoid of any shipping. We patrolled our grid for over 28 hours and all was clear. How delightful the cruise has been, it's now over three full days sailing, and not one sighting. We are heading south towards the Channel, and hopefully I can still recall how to use the periscope and TDC[104].

> ....well I have to say running the game without time compression is like having an interactive "screensaver" running ... go outside in the garden for a hour or so, check the game ...eat dinner and shower, check the game ... I think it is safe to say this mission will be going on for more than one week, unless we are sunk or come across 14 ships in the next four days!

⊕⊕⊕⊕⊕⊕⊕

Log: Jan.17, 1940; The Watch Officer called out at 6:21, "Kapitän! Es gibt ein Schiff! Lager 000" I calmly put down my coffee (actually I hurried, I haven't seen a ship in days!) and climbed up on the bridge...still dark, not yet twilight. I see the ship through the UZO[105].

A "Schlepperboot", dead ahead. Okay, not the best target to see after three days. But at least it was not a "Kreuzerboot" bearing down on us. I told the Helmsmen to " Stoppen Sie langsam das Boot", und "fünf Grad Steuerbord".

Next, I had the Watch Officer assemble his gun crew on deck. I told him that this is a low threat target of opportunity, that only five high-explosive shells could be used (please keep it out of your record that I

---

[104] TDC: Torpedo Data Computer, a real-time target tracking system that computed the firing solution.

[105] UZO: U-Boot-Ziel-Optik, aiming binoculars mounted on the bridge.

manned the gun myself). The first shell fell short and was fired at 2000 meters, the next two found their mark. I couldn't tell if the fourth went over or in front of the now burning Schlepperboot. The fifth looks like it hit the boiler, because there was a catastrophic secondary explosion, moments later it slipped under the sea. 6: 29 Grid AN58, sank Schlepperboot (unknown name British), 1139t.

PS: Karl (Radio Operator) was listening to the BBC since the BdU doesn't transmit much, and heard a joke: "Was ist braun und klingt wie eine Glocke?" …"Dung !"

Log Jan 17, 1940; Today was a good day. We headed due east after sinking the tug. I went to bed and was abruptly awoken after three hours sleeping.

12:17 Watch Officer had spotted a Tribal Destroyer, long range 233 degrees, closing fast. We went to periscope depth. We might get a shot at her, the crew scrambles into the forward torpedo room. The attack periscope is up, range is 3900 … too far away. Flank speed and draw down the scope … a few moments later up scope and ahead slow. Dead ahead, but no time for a shot. All stop, and down scope … slowly turn 90 degrees away and ahead slow, silent running for the next 10 minutes…

14:07; Watch reports warship closing fast and over the next hour, we see two armed trawlers and two Tribal Destroyers doing a patrol in Grid AN72. They aren't actively looking for us so we keep on our course.

15:18; I can hear faint sounds of screws at 330 degrees. We stay at periscope depth, still too much daylight for surface running. 15:26; SO[106] reports warship closing fast … all stop … in very shallow water. I check the battery, we've used first fifth of power, diesel still about 75%. We let another Tribal go by as we safely move on.

16:20; SO reports merchant, 8 degrees, moving slow.

17:10; the helmsmen has moved us to within 2200 meters. Ahead slow and we begin to move to intercept her course. Hopefully, while we are still moving we can shoot and move before any of the nearby ships (not close at all, not even on the horizon) can respond. Double-check that the first tube fired is number three, an electric set on slow, surface impact only. Then number one tube, steam — fast, surface impact only.

17:16; We slow down and I push the scope up a bit higher. There she is, closing slow. This will be a nice shot, dead ahead at about 575 meters,

---

[106] SO: Sound Operator

but we have to creep along at 1 knot as to not overrun the 500-meter mark. Time slows to a crawl… tick, tick….

17:20; Seems like two hours … approximately 575 meters … has time frozen?

17:21; Tube 3 is fired at her stack, tube one, rear mast. I order ahead full, 10 degrees to port … 25 seconds later the electric strikes her starboard side, followed quickly by number two and the ship sinks.

17:23; Small Merchant, 2404 tons, in the shallows of Grid AN72. We put her smoking grave 180 degrees behind us and her between us and the last known whereabouts of the warships south of the wreck. We ran submerged till 19:50.

21:00; The WO[107] called out, ship spotted, bearing 313. He has good eyes because I couldn't see it, when I came up on the bridge. She is moving at medium speed and we are sailing ahead full, it will be a close try. We are already on a tack to cross her bow, so it will be our first surface attack at flank speed, (the diesel willing). The watch calls out the bearing as the men in the conning tower set their TDC data. I call out, "Fire two" at 1900 meters. We fired tube one seconds later. We came to a stop to wait. The first fish found its mark and we destroyed an enemy unit, we don't know what happened to the second fish. I could have fired too late and it passed the aft of the merchant ship.

We radio our report to the BdU; three ships sunk, 10 torpedoes remain. No word back from them yet. We have had a glorious day, the crew is topnotch and in fine spirits. Neptune has given us seas as smooth as glass, perhaps Mars will grant us bigger game to hunt.

Jan 17, 1940; At 23:38 Fritz reported that the battery was fully recharged. We plot a course due east across southern edge of Grid AN59, ahead slow and I go to my cubbyhole desk and make a few notes before going to sleep.

Jan 18, 1940 00:31; Offizier der Uhr shouted, "Schiff stationierte aufwachen oben, wachen Sie auf! das Schiff kommt nahe!"

I rush through the command room and climb up the ladder into the cold night air, on the bridge. Leutnant zur See Arend Akermann, my most valuable officer, reports that its bearing is 066. I am sleepy and can't see anything out there. I can't make up my mind as to what we should do.

---

[107] WO: Watch Officer

00:32; All stop.

00:32; Ahead slow.

00:32; All stop.

00:32; Ahead slow.

00.33; Dive to 12 meters, ahead slow. I check the hydrophone myself, a warship … all stop and I go to the command room and raise up the observation scope. I can't see much.

00:34; A Hunt destroyer passed us heading east to west parallel to our course, our port side. I order full stop, down scope. We will sit still for a while, and after the Hunt is well past, I return to sleep.

Jan 18, 1940 4:40; I awake and order ahead slow and surface the boot, the crew is rotated. We check the navigation charts and adjust our course slightly.

(Notes not to include in the ship's log) I went up on the deck around 05:35 to clear my ears of the hydrophone coma (very much like the famous Bob Ross coma) I was in and smoke a pipe on the deck casing. Looking out over the slim bow cutting effortlessly through the darkened sea, an image came to mind of the Old Norse Vikings and their seafaring ways. Our boot would look good with a sea serpent's head made of steel and horns as sharp as dragon's teeth. Ready to rip asunder any ship that would dare cross our bow. Of course, 40 knots and 50,000 more tons of mass would be needed to give it the old Cap'n Nemo treatment. Later I wonder why our surface fleet doesn't board and capture the ships and confiscate the goods?

⊕⊕⊕⊕⊕⊕⊕⊕

Jan 18, 1940; A very slow day, only sound contact was at 08:34. I could hear a faint sound at 310 degrees. We plot a course towards the sound, ahead full, we want to see it before the sun comes over the horizon.

08:43; We can see the smoke of a ship's boiler topping the horizon. Crew is assembled in the forward torpedo room. WO reports ship at 350° and closing. Looks like a merchant through my binoculars. The UZO shows me it's a small merchant, but when I check our navigation map I see a green ship[108], so it's a no go on an attack. We turn 180 degrees to show a

---

[108] In the Silent Hunter III game, the player can reference a map and choose to have the computer identify the sighted vessels as friendly (blue), hostile (red), or neutral (green).

slim profile in the water and we slowly move ahead. Twenty minutes later we return to our previous plotted course. Beginning a fruitless day of sailing and sound checks in Grid AN81. I feel *der Gott des Krieges lacht an uns dieser Tag* (the god of war laughs at us this day).

Jan 18, 1940; I wake after an uninterrupted night's rest.

At 19:54; I get a solid sound contact at bearing 003, we check on it and it is a small merchant, a red icon on my map. We get into an intercept course.

20:14; WE[109] fires tubes two and one.

20:15; Forty-five seconds later the WE reports impact. I watch it blow up and sink, but the WE doesn't report unit destroyed and there is no Log entry and no sunken ship icon? Now I have to figure, is this just a one-time deal, or has this happened before and I am unaware of this bug? My last save was before I went to sleep last night.

> Before I went to work this morning, I decided to save the patrol as it is. If there is no Time Compression then there should be no "Dr. Who-type Reverse Time Travel". So, I will have to wait and see what happens with the next ship. The game has run great up till the ship that's sunk but no kill. SH3 ran nonstop for over 36 hours (last weekend) and it has never crashed in the three weeks I have played it. So having this one glitch is a "Enttäus-chung", me and the crew must keep on keeping on, and I will try not to forget what "blinder Haß" stole from us.
>
> And to give credit where it's due, the "what's brown..." joke was said by Eric Idle (MPFC) as Arthur Name, "name by name, but not by choice".
>
> And why is it that nobody remembers the name of the wundervoller Deutscher, Johann Gambolputty de von Ausfern-schplenden-schlitter-crass-crenbon-fried-digger-dangle-dongle-dungle-burstein-von-knacker-thrasher-apple-banger-horowitz-ticolensic-grander-knotty-spelltinkle-grandlich-grumble-meyer-spelterwasser-kurstlich-himbleen-bahnwagen-luber-hundsfut-gumber-aber-shonedanker-kalbsfleisch-mittler-aucher von Hautkopft of Ulm?

⊕⊕⊕⊕⊕⊕⊕⊕

---

[109] WE: Weapons Engineer

Log. Jan 18, 1940 2250; We are patrolling Grid AN84. Have been actively using hydrophone, all is still. Where have all the ships gone? Long time passing.

(Yeah? Six days? My last patrol on 1x TC was over a week and a half. I know now that this patrol will be many more days, I still have 8 fish and oddles of cannon ammo … My office is small enough to be a sub, and I would like to have a better logo on the side of my delivery truck, an eagle is nice, but I would like a black skull with flames…)

Log Jan 19, 1940 13:48; *U-49* has been actively patrolling, I have been standing watch myself and stopping every twenty minutes so I can man the hydrophone. We spent most of yesterday in Grid AN84. Last night and today we are actively patrolling Grid AN87, all is calm. We are heading SSW towards der Kanal and plan on patrolling Grid BF33, 32, and 31. We have about 70% diesel fuel remaining. While we are patrolling we are sailing slow, or 1/3, to make our fuel load last as long as possible. We hope to sail past the Dover area at night, giving us the cover of darkness, as protection from any British planes. I am reading the more experienced Kapitanleutants' logs, after action reports, and learning a great deal. I wish I could write as well as they do, and use the proper sailing lingo and jargon. Your adventures with DDs, sinking tonnage is always dramatic. Whereas this journey has been very atypical of the slow pace of the below-average patrol. The whole crew is working day and night for the Vaterland. But alas, our results so far are even worse than we did in our Type II.

Log. Jan 20, 1940; Having some morning coffee, after helping Onkel Willie do the laundry, being Sunday, it's wash day.

Log supplemental for Jan. 19, 1940; At 16:06 during an all stop, I am manning the hydrophone and have a faint sound contact, 328 degrees.

16:24; 312 degrees and at 16:47; 298 degrees

17:07; 356 degrees and at 17:11 faint smoke, dead ahead.

17:14; We plot a course parallel to the port side of the unknown ship.

17:45; We begin slipping in behind the ship.

17:48; The watch and I identity it as a coastal merchant and we increase speed to close the gap slightly.

18:00; 5560 meters ahead. WO assembles his crew, Adol Bahn, Kurt Marks, Gotz Abel. Will allow them to fire when we reach about 4500 meters. We slow and then stop to give them as stable a platform as possible; the sea is as smooth as can be.

18:19; The order to open fire is given, the first few shells misses, but they soon find their mark and the ship is ablaze in deck fires. After several high explosive shells the gun crew switched to armor-piercing shells.

18:23; We hold our fire and watch as she burns and begins to take on water. Their power must be off because they no longer seem to be pumping water on the fires, or perhaps the few sailors, the captain and mate are already dead from asphyxiation or explosion.

18:35; Enemy unit sunk. Grid AN87, 1996 tons. The pursuit and sinking took us away from the plotted course, so we turn 180 degrees and head back towards the channel.

Log Jan 20, 1940; During my sleep at about 04:00 the WO called, "Wake up, sir, there is a ship." I rush to the bridge and he has identified it as a warship. We dive and I order ahead slow. The sound operator reports three more warships around. I check the navigation map and see we are just outside the channel. I make sure we are plotting a safe course through the two-armed trawlers and the two-elco boats. There are two other sounds reported, but we can't see them through the observation scope. After an hour's slow cruise we plot a course and remain submerged there the rest of the night and I return to quarters.

Log Jan 20 6:55; CE[110] orders ahead full and we surface the boat and begin recharging the batteries. It looks like another clear day with calm seas. We begin our daily patrol ritual: sail, stop, sound check and sail.

Log Jan 20, 1940 07:46; I pick up a screw sound at 098 degrees. We move ahead standard.

Log Jan 20, 1940 07:50; We see smoke at 310 degrees. We plot an interception course.

07:55; We dive to 12 meters and assemble the crew in the forward torpedo room.

07:55; Up goes the attack scope and I see a coastal merchant moving slowly along.

08:09; All Stop. We prepare to fire tubes two then one at 600 meters, 90 degrees.

08:17; WE fires tube two, then seconds later tube one. Forty-two

---

[110] CE: Chief Engineer

seconds to impact. Nothing from the first fish, must have had a bad angle because it was set impact only, the second one finds its mark and hits the bow. We are doing some damage, but their crew are working the pumps and they are keeping up.

We surface and the gun crew is made ready. They put on lifejackets and their helmets. I give them full confidence to do the job themselves, we are pointblank range, 500 meters so the WE has his crew fire armor piercing shells right through her hull. Five shells later, it begins to sink lower in the water.

08:23; We have destroyed an enemy ship, tonnage 1997, in Grid BF33. (Someone in our gun crew will be given a medal if we earn any and, knock on wood, we safely return to port) We move away ahead full, and thirty minutes later we resume our patrolling.

Log Jan 20, 1940 at 10:24; We see smoke at 290 degrees. Moments later I use the UZO and see a Hunt 1 DD heading NNE and closing towards us. We calmly begin tacking away from his course.

10:28; The DD is still closing.

10:34; Our tacking begins to separate our boot from the dastardly British warship, we are now 7300 meters and separating quickly to resume our patrol of der Kanaal.

(Personal notes) Jan 20, 1940 16:42; We have been running submerged for a few hours at ahead slow and will remain under till the twilight hours. After the noontime mess, some of the crew are reading, sleeping, some doing routine maintenance on the diesel while it is shut down, others are quietly playing cards in the rear crew area.

Albrect Eppen, old friend from my youth, Udo Hartenstein, a very scholarly gentleman and thinker, and myself were sitting around my cubbyhole and bunk. We were enjoying some interesting discussion about oaths; as officers and servicemen we take our oath as an almost physical bond to what the oath decrees. Udo's father, as well as my own, served in the German Imperial Navy. The oath they took was to Germany, whereas the one we took was to the Führer. Udo rightly points out that the events of 1933 now seem to take on an almost unreal or detached feeling as time has passed. Udo was witness to a mass book burning, and he had to hide himself away because he wept at the loss of knowledge. He says he is always reminded of what Heinrich Heine (1797-1856) said, "Dort, wo man Bücher verbrennt, verbrennt man auch am Ende Menschen." Translation: "Wherever they burn books, they will also, in the end, burn people." May God have mercy on our souls and the souls of our children, if such a crime should come to pass.

What Udo was trying to get Alberct and myself to understand was how very smart and deceptive the National Socialist Party is. They know our character and our best traits and use them against us. They use our pride and love of our nation to their own nefarious ends. The Party exploits the basic human trait of pride. He told of us how the work and trade unions always wanted May Day off, the Führer gives it to them and then the next day bans the unions! He said the Party makes us intentionally afraid of enemies, while undermining our rights and making us dependent on the Party, instead of using our own judgment.

Udo has me thinking that life on the sea is looking better all the time. I wonder what Rio is like?

⊕⊕⊕⊕⊕⊕⊕⊕

Log Jan, 21, 1940; Grid BF 32, We sailed under the surface most of the afternoon and had no solid contacts.

17:46; The SO reported two warships at 113 degrees and closing fast. We plot a new course and hope to keep a safe distance from the unknown ships.

18:35; We no longer have a sound contact with one of the ships, but one is still about 80 degrees and has been for a while not in view, just out of range. I don't like having him so close for this long.

19:10; Surfaced towards the end of twilight and began recharging our battery. Can see the smoke of the warship who is now at 40 degrees and on a course to cross our plotted course. We plot a possible interception course at 19:12.

19:15; Darkness is falling, but we have marked his estimated course. We shall proceed with extreme caution, heeding the veteran skipper's advice. Better to let the man-o-war go than jeopardize my boot and crew.

I started allowing the crew up on deck around 23:30, so the men could have more time for a smoke or some fresh night air. I told those going up on deck to smoke down in the conning tower not on deck. We don't want any enemy eyes to see the glowing ember of a cigarette.

I feel the unease of the proximity to England. Unlike last Sunday's stop we are at an extreme state of readiness! We played the gramophone for a while, to help ease our minds. Then we had Karl Creutz tune in the BBC radio. Edwald Albrect, who knows English very well, translated the late news broadcast for us. The football scores brought on some talk. It seems South League 'A': Arsenal had a friendly against South League 'C': Tottenham Hotspur and Arsenal won three, nil. (Normal league competition was suspended with the Declaration of War in 1939 and smaller regional leagues were set up).

The highlight for me was that President Roosevelt was determined to help England even if the USA Congress was less then enthusiastic for his foreign adventures. It seems the Amerikans are looking to the Far East as a possible war ground for Amerika, not the Continent.

Log Jan. 21, 1940, time 10:48; We have been patrolling the northern quarter of BF36, all is still. Last night at 01:05 we had a sound contact during one of our dives. And to make a long story short, we plotted our course to intercept. It took over two hours to catch up, we were doing 10 knots. Finally, we can see an enemy tug moving at 8 knots. We assembled the gun crew and after a very slow approach they opened fire at 1200 meters. The tug was ablaze with in seconds, the crew fired several more high explosive shells, and at 03:48 we had sunk a tug 1141 tons, Grid BF 32.

After sinking it we plotted a course due south. Although it was only a tug I think we learned a bit about how to close in without being detected. The darkness and staying to the aft of the ship worked on a merchant. I am thinking it would work on a naval ship, too.

Log Jan 21, 1940; We have spent the day patrolling BF31, it was a fruitless search, but those on watch weren't too downhearted, for they knew they tried their best.

Log Jan 21, 1940; At 20:14 during a dive and sound check the Sonar Operator reported merchant bearing 023. The navigator and I plotted a course and ahead full to try to close in.

21:19; We are about 3500 meters behind a C-2 cargo ship. We are plotting a course to follow it and then move into a flanking position and then pass it to set up for a shot.

21:40; Watch reports ship spotted. I am up on the bridge, but I don't see anything. Again the WO reports a ship. I'm scanning where he's pointing and I think he's still showing me the C-2, then I see it, coming in the opposite direction and almost on a crash course with the C-2 is a V&W destroyer.

21:47; We dive to 12 meters and all stop, as we rig for silent running, and watch as the DD goes so slowly past our position, his closest point was 1300 (so maybe it wasn't too close after all).

22:24; Ahead standard and secure from silent running as we surface the boot. We begin our pursuit of the C-2.

22:47; Suddenly a small fishing boat comes trolling between our boot and our prey the C-2. The hands are on deck rigging the nets for the day's fishing so we dive our boot under the sea as a precaution.

22:50; Finally the fishermen move on and we surface *U-49* and once again resume stalking the C-2.

23:54; We have passed the C-2 to our starboard side at about 4000 meters wide of its position.

Log Jan 22, 1940 00:08; Began our final positioning to try for a 90-degree shot at 500 meters submerged. At 00:19 all stop and the tension builds as the final moments turn to hours. Tick … Tick … Tick … I think my chronometer is broken. I should have brought a Swiss watch, instead.

The time has arrived and at 00:22 I give the order to fire tube one then tube two. We begin moving forward slowly, seconds later the WE reports torpedo impact! 00:22 WE reports enemy unit destroyed, even before it finishes sinking. He's good. C-2 cargo, 6395 tons, sunk western edge of Grid BF32.

There is a sigh of relief. We don't know what happened to torpedo number two. I believe the first one split the ship right apart. It was set four meters, impact, fast. Carefully we moved past the stricken ship and at 00:23 we rig our boot for silent running and then stop the boot.

00:26 SO reports warship moving fast and closing in, bearing 210.

00:27 SO reports warship fast and closing 270 degrees. We begin the almost unbearable wait to see if the dreaded warships will come near. Onkel Willie breaks the tension with a surprise treat of a wonderful apple strudel. When he brings it out of his oven it fills the boot with such a delectable aroma! We eat in quiet and we all seem to be more confident that our ruse of waiting for the warships to move on will work before beginning our journey out of Der Kanaal. I tell the crew that I am proud of how they worked as a team in sinking the C-2. That our patience paid off and our slow, safe approach allowed us to sink the ship with seemingly little danger to our boot.

Log Jan 22, 1940; *U-49* began our departure at 01:51 ahead slow. At 02:05 all stop for a sound check, all is still, except for the mournful groan of the sunken ship. 02:08 we have plotted our course SSE towards French waters of Grid BF33. We plan to lay low during the daylight hours and exit the Channel in the dark.

Log Jan, 22, 1940; I gave the order for the radio operator to send in our patrol report. 02:55 *U-49* Grid BF32. Torpedoes left, 6 (I thought I was down to four). I have no more reloads for tubes 1 and 2. Number four tube still has an electric from when we left port. Seven ships, 17,477 tons. We are still over 50% fuel. I must be extra vigilant tonight. I don't want to stumble across any British warships. Perhaps after we are clear of the channel we will try for one of the man-o-wars that are patrolling the Dover to French Coast choke point, or save the fish for softer targets. *Viel Glueck and gute Jagd!*

Log Jan. 22, 1940; We have been sailing along and I have retired for the night. I am trying to sleep as lightly as possible, and had the volume turned up loud. At 03:20 the watch officer calls down through the conning tower to have Fritz wake me up. They see a large ship heading due west. I climb up the ladder through the hatch and into the cold dark night. The crisp air is exhilarating and I am looking for the ship, I see it, but it is very far off. We stop and wait as the ship passes by our port side at 3000+ meters. We watch as a C class DD passes by. Twenty minutes later we resume our course, slowly at first, and after a half-hour, I return to sleep.

Log Jan 22, 1940 05:21; WO spots a warship approaching. 05:23 I am up on the bridge and order all stop. It is still very dark so we will wait under the cover of darkness as a V&W DD passes by our starboard side heading due west at 2700 meters closest point to our boot. After the DD is past, we resume our course. At 05:40 WO reports warship in sight, again we stop and wait, and again a V&W DD comes slowly past us heading SSW. Dash it all, why don't these bloody Brits go sail on the Thames or something and let me get some sleep.

⊕⊕⊕⊕⊕⊕⊕⊕

Log Jan 22, 1940 15:05; It has been a very good sail this afternoon. We have had no warship contacts and have been steadily making our way. *U-49* has been running on the surface most of the day, about 27 kilometers from the French coast. We sailed in Grid BF36 and angled through the small bit of water in BF37. We are currently in Grid BF33 and approximately 80 kilometers from where we anticipate the British patrols that we encountered on our way into Der Kanal, and 160 kilometers from the Nord Sea. If all goes as planned we will be free from the Channel in between 8 and 10 hours from now. Of course, that is without having to stop and wait, or if we have to slow down to slip our way past any pesky patrol ships we see or hear.

Log Jan 22, 1940 20:37; The watch reports ship sighted. I get out of my bunk and investigate. At 20:40 we go to 16 meters and rig for silent running. And so begins a few hours of waiting for ships to move along.

20:52; SO, warship, closing fast, 343 degrees.

21:05; SO, warship, distance constant, 016 degrees.

21:18; SO, warship, closing, bearing 018 and closing.

2123; SO, warship, 010, constant distance.

21:52; SO, warship, bearing 349, moving away, finally a break.

22:38; Warship, constant distance.

23:39; We surface and resume our sailing out of Der Kanal. *U-49* is about 90 minutes from the Nord Sea. We are in seagull range of the French coast, and the water is shallow. This has been a seemingly safe night and we will be greatly relived to be free of these waters.

> I earned a wound badge! Not on the U-boot, but in my own house, before work on Friday morning. I whacked my right foot into a metal bookcase, and really did a number on my little toe. Went to work anyway, but when I came home and took off my work boots my foot was swollen and my toe was as purple as a plum. I kept off my feet and kept ice on the "wound". I was in so much pain yesterday that I only sailed *U-49* for a few hours. Today, Saturday, I am icing my foot and will resume sailing shortly. I just came home from work and my foot is hurting, but it's raining and a bit chilly today so I will be playing Skipper all day and night. I am going to tell BdU that I slipped climbing the conning tower ladder and would like them to have daily "safety talks" so things like this could be avoided. I hope they give me my badge, or at least have Ingrid the naughty nurse give me a sponge bath or something.

⊕⊕⊕⊕⊕⊕⊕⊕

Log Jan 23, 1940; At 06:13 the WO, who is getting better at seeing ships, called out in a rather alarming fashion, "Kriegsschiff, schnell, schnell!" I hobbled up through the command room, gimped up the ladder and limped up on to the bridge (much like a Fred Sanford shuffle). A Hunt 1 DD was about 4000 meters passing west to east.

06:20; I have the CE halt the ship and we will watch the DD go harmlessly by.

06:26; DD turns 90 degrees and starts heading straight towards us. It's closing at 25 knots, bearing 239 degrees. I hurry everyone off the bridge and we rig for silent running, and we go ahead 1 knot to help get the boot under, we set our depth at 15 meters. I do not know if it's coming to kill us or not. I check the hydrophone and the sound of the onrushing ship is loud. I walk out into the command area and look as calm as I can pretend to be. I raise up the observation scope up a few inches so I can see the keel of the DD when it comes near.

Oh, mein Gott … at 06:38 I can see the hull splitting the water, and I can hear the hum of its engines…Oh, mein Gott, this is too close for comfort. I am ready to go flank ahead if I see any drums hitting the water. Luckily the DD just kept on sailing. That is the closest a ship has come to our boot, I am glad I didn't panic and alert the DD that I was there.

07:05; We surface *U-49* and begin a new course away from where we saw the DD sail towards us. At 07:21 during a sound check dive the SO reports that he could hear a merchant ship bearing 337 degrees.

Jan. 23, 1940 08:07; *U-49* is carefully patrolling Grid AN79. The Sonar Operator, while using the hydrophone, picked up a merchant that was moving along slowly, whose distance was constant and bearing 029. I order the boot surfaced at 08:08 and we plot our interception course. After a very short sail we can see a small merchant. I decide to use the deck gun since it is still dark and the seas are like glass. The WO and myself feel confident we can put the ship under quickly if we load HE shells.

Log Jan 23, 1940 08:27; We open fire from 2500 meters At this distance I felt it best to do the shooting myself and I was lucky to hit it with just about each of the seven shells I fired. The deck was burning and the ship began turning erratically.

Log Jan 23, 1940 08:29; We sink a small merchant, 2409 tons of British "tea", sitting in the shallows of Grid AN79. We dive the boot to 12 meters for a sound check and at 08:31 SO reports a merchant moving away slowly at bearing 335. We plot a course to investigate and surface the boot. I put our best watch on deck. After sailing for over 45 minutes, still no ship in sight.

At 09:19 we go to 12 meters for a sound check, and SO reports merchant slow at 335. 09:20 we bring our boot back to the surface and resuming pursuit of the lone merchant. 09:29 WO reports ship in view. We will close up from its aft and open fire again using our deck gun. The WO assembles the crew and gives them their orders. They will use HE and we plan to open fire at 3200 meters because we no longer have the cover of darkness.

09:57; We open fire on the helpless merchant. Our first four shells hit, but the next three miss. Arend yelled down at his gun crew to lower the gun and aim for the hull with armor piercing shells. They find the mark and moments later the ship begins to slip down into the cold embrace of the sea. 10:00 enemy unit sunk, a small merchant, 2408 tons of "scones" to go with the "tea" from earlier this morning.

Log Jan 23, 1940; At 10:30 the SO reports a warship bearing 260 degrees, moving away. That's a good direction for a warship to be heading.

Log Jan 23, 1940; Our SO reports at 10:58 a merchant ship moving slowly along, we plot a new course and at 12:10 we can see smoke on the water, but no fire in the sky. We will investigate. At 12:34 we can see and identify a small merchant heading towards the Southend area. We plan to pass to our starboard side very wide of the ship. 12:49 we pass at 90 degrees; at 6630 meters we are doing 15 knots, and will be able to move into an ambush point easily.

Log Jan 23, 1940 13:15; We are 4 km from the ambush area. 13:18, Arend readies his torpedo crew in the aft room. Our plans are to dive and try to fire an aft shot at 500 meters to the merchant, who is traveling at 5 knots. We are sailing in very shallow water. We are in the process of waiting for the merchant. I can see it at about 4000 meters and slowly closing. 13:59 I go to the hydrophone and do a scan, I hear something I don't like. A very loud warship about 149 degrees. I tell everyone of the new situation as I walk back to the attack scope and push it up through the surface. Oh, mein Gott, he is only 870 meters and closing fast. We have no time to do a shot at the DD. I pass the word to everyone on our boot that the attack is off, but everyone is to remain at his station. We rig for silent running. The Hunt 1 DD passes us at 14:08. This was one lucky merchant. He will sail into port safe and sound. The warship was in sight as the merchant passed our predicted course. At 14:19 we plot a new course and leave slowly while in the sound backwash of the warship, ahead slow.

He who hunts while hiding away. Finds when he might become the prey. Is best to flee and live to hunt another day.

Log Jan 23, 1940; We moved away after half an hour, slowly as first, then we raised the boot and resumed our patrolling of Grid AN79. We were hoping to have our first three-ship day. We had warship contacts at 15:24 and 16:02 and 21:06. Finally, we pick up a merchant sound at 21:47. There are also two warships within our hydrophone area. We will track the merchant to safer waters and try to sink it with the deck gun.

After a long pursuit, at 23:45 we are 1855 meters behind our target. The gun crew is told to load HE shells and we plan to open fire at 1200 meters. They are very proud to be trying for another ship and are ready for action after the long day's hunt. The WO, Arend, has his crew open up at 00:04, 1200 meters. The first comes up short sending a spray of water on the now alerted crew looking over the aft of the ship. That is the wrong place to be standing because the next shell blows them 20 meters in the air. I guess that's what the Brits call a little "TNT time". Arend's crew fired several more high explosive shells onto the deck and bridge of the fire-engulfed boat. The Captain must have died at the wheel and went down very gallantly with his ship. He must have been a very good skipper; he tried to maneuver even though his ship was ablaze and his life in peril. He and his ship went under at 00:07. Enemy Unit Destroyed, Coastal Merchant, 1994 tons, Grid AN79…This was the most fun I have had on this patrol so far. I know war is hell, but today it was a blast.

Log Jan. 24, 1940; WO reports a merchant and a warship at 8:07. By 8:16 we have identified a Tribal DD and a C2. We discuss our options and we decide to attack the DD since it is just about at 90 degrees now and 4000 meters away.

We ready the crew and make our calculations. We will fire tubes 3 and 4, number 3 fish set impact, fast, two meters deep. At 0823, 1900 meters away from the DD, WE fires tube 3 then 4 (electric, slow, impact, one meter deep). 08:25 torpedo impact. It didn't hit where I was aiming, but it hit the aft part of the ship and seemed to ignite the depth charges on its deck. There was a huge secondary explosion that sent the British sailors flying into the air. The ship slipped under the sea quickly. The second fish must have missed. 08:25 ship sunk, Grid AN76, Tribal DD, 1850 tons. The crew is very pleased at our first warship sunk .We were confident that we would hit it and that it was worth it because now we could attack the C2 with the deck gun.

Arend has his deck gun ready by the time we break the surface and begin pursuit of the C2. He taps each one on the helmet as they come up onto the bridge; they look more sure of themselves each time they climb out onto the gun deck. At 8:33 the order to open fire is given at 1100 meters. We go to ahead slow to give the gunners a better platform to shoot from. The cannon shells are causing damage to the front deck and bridge area, and the ship begins to become engulfed in fire. I calculate they fired 23 HE shells to sink the C2. Ship sunk, Grid AN76, C2, 6400 tons. The long dull hours now seem like a distant memory after such an action-packed morning.

Log Jan. 24, 1940 9:07; The SO reports a merchant closing at 303 degrees, we plot a new course to investigate. A short time later we can see a C2 coming towards us. We will use our cannon again. Arend goes to the stern quarters to get his gun crew and they can hardly believe their luck when he tells them to come on deck ready for action. The target is 1200 meters and closing, we open fire at 1000 meters. HE shells come raining down on the doomed ship. Cargo and crew are being obliterated with each hit from the deck gun. Soon the ship goes dead in the water, suddenly the boiler seems to explode, cracking its hull in two. It sinks in very shallow water leaving its burning decks above the water line. This smoldering beacon is soon put behind us as we go due North, ahead flank speed for twenty minutes, then resume our patrolling. Ship sunk 9:49, Grid AN84, C2, 6401tons. The crew and I feel we are making the most of these last few days of our patrol.

Log Jan. 24, 1940; I check our torpedo info (key I) and I see we still have an external torpedo for the stern room so we have begun winching it below deck. It will take us about 35 minutes so we have a flack gunner on deck to give us some cover if any planes come near. After they had the fish loaded, Onkel Willie serves them a hearty lunch of ham, potato soup, pumpernickel bread, and cheddar cheese We are down to four torpedoes, but I am glad to have a fish in tube one as well as the three that are for the aft room. We have 23 HE shells and 2 AP, 199 AA, and the star shells. So we will carry on the patrol, our fuel is about 50 percent and our food supply will last a good while longer. Everyone on the boot is in good spirits. *U-49* is running on the surface this afternoon and we are keeping an eye out for any planes or ships.

Log Jan. 24, 1940 19:08; SO has a contact closing, bearing 279. We plot a new course and will see what we can see on the sea. 19:30 we can see what looks to me bigger then a C2. 20:00 we have identified a C3 large freighter heading towards us. Our plan is to move across its bow and set our ship at 90 degrees to the C3's path.

Log Jan. 24, 1940 20:35; *U-49* is submerged at 12 meters and is 500 meters from the C3's course. Arend has his crew in the stern torpedo room have the fish set to swim 8 meters deep, impact only and fast. We hope to stop the ship DITW and finish it with the deck gun.

WE fired on my command at 20:40. Thirty-five seconds later we could hear the impact and we surfaced our boot. We rushed up on deck as fast as we could. I could see the C3 stopped dead in the water, but not much damage otherwise. I took over the gun crew for Arend, who was still reloading the next torpedo. I put on a helmet and climbed onto the gun deck to help load shells. We were so close to the C3 we could hear men yelling in English. We fired our cannon pointblank; they had to elevate the gun as if the target was 3000 meters away. The first HE shell we fired blew apart the their port side lifeboat. The next shell hit the bridge as did the next several. We were now sailing very slowly past a quickly dying ship. Their crew stopped fighting the fires and manning the pumps as the ship began to list to port. I ordered them to switch shells and we fired our remaining two AP shells into the boiler area.

Log Jan 24, 1940 20:45; Ship sunk, Grid AN73, C3 freighter, 7907 tons. Being this close to the action was a great experience. I better understand what the deckhands are facing when out manning the deck gun; it seems rather dangerous work. It actually felt like combat. Although nobody on the C3 fired back, I have a feeling this won't always be the case. I hope England capitulates sooner rather than prolong the suffering.

20:50; We have plotted a course NNE. At 20:53 we began moving our last external aft torpedo, we think it will take about 40 minutes. We are down to 17 HE shells and two aft torpedoes. We will resume patrolling after the fish is secured.

Log Jan 24, 1940; Today I played a new game one of the seamen made up. He calls it "Fighting Ships". It is played by two people; one is the U-boot commander the other Fleet Commander of three ships — a battleship, a DD, and an armed trawler. The three ships are placed by the player on a pegboard that has been numbered 1 through 32 vertically and "A" through "Z" horizontally. Each boat has holes to represent how many hits it will take to sink it. The Fleet Commander can put his ships anywhere on the pegboard, but can't move them once the battle has begun. The U-boot commander has 30 "pings" he gets to call out, such as "W-17". If he misses the ship, the Fleet Commander says "miss" and the U-boot commander puts a white peg in his peg board that is numbered the same as the Fleet Commander's. If he hits, he will put a red peg representing hitting a ship. The Fleet Commander wins if the battleship is not sunk. It is a fun game and now the catchphrase on our boot is, "Oh, mein Gott, you sunk my battleship!"

⊕⊕⊕⊕⊕⊕⊕

Log Jan. 25, 1940; While I am working the hydrophone at 01:45, I can hear an unknown ship at 355 degrees. We set our speed ahead slow and I go up on the bridge to be ready when the ship comes into sight. 02:03 WO calls my attention to an armed trawler moving SE and another going NNW. WO reports a merchant moving ESE (this was the ship I could hear), we plot a course ESE to stay in front of the merchant. The armed trawlers are moving away from our expected ambush point some 6 km ESE of our current position.

WO and I identify a C2 at 02:50, we plot a course to set up for an aft shot at 500 meters. We have to move further down the line to have enough time to get our solution worked out, and to be in the proper place for the aft shot.

03:55; The C2 is 1190 meters away and closing at 5 knots. Everything is being set and our anticipation grows. 04:01 I call to the WE to fire. The whoosh of the compressed air tells us that our fish is on its way. There is a loud ... KAAABOOOOM, as the fish rips the unfortunate merchant in half. I believe it goes down with all hands. Soon the hagfish will be entering those poor unfortunate scurvy sea dogs.

Log Jan. 25, 1940 04:02 Navigator reports the demise of the C2, 6403 tons, in the shallow water of Grid AN73.

We plot a course NNE and at 04:09 our final aft torpedo is loaded. The boot is filled with the smell of frying ham as Onkel Willie serves the first round of breakfast. I go out to the command room and give them a speech. I say, "We got our motor running...We sailed along the byway... We were looking for adventure or whatever would come our way... Yeah, sailors we are going to make it happen, grip the world with a war embrace, fire all our guns at once and explode into space...Like a true nature's child, we were born, born to be wild...." or something like that....

Log Jan. 25, 1940; During our 07:50 dive and sound check the SO reports a merchant, slowly closing and bearing 045. We dive *U-49* again at 07:50, merchant is now at 045, so we adjust our course.

Log Jan. 25, 1940 8:22; We have identified a C2, 08:29 we dive to 12 meters and begin our ambush run.

Log Jan. 25, 1940; Everyone knows his job and does it efficiently and without talking; everyone seems to get "deep" within himself before an attack. I wonder what each man thinks of. My first thoughts are, "Is the solution right? Did they make sure to set impact only? How fast is the merchant really going? Then I think about the sailors I am about to kill on board the target ship, but they started this war so I figure it's them or me.

Then I think about having a pipe of fine tobacco to calm my nerves ... then it's back to the business at hand and the death-dealing blow from our torpedo.

08:46 WE fires our last fish ... swim little fishy, swim if you can, and it swam and it swam all over the dam, British ship ... going BOOM, BOOM-BATTA, BAM....

Log Jan. 25, 1940 8:47; NA[111] confirms enemy unit destroyed, C2 Cargo, 6443tons, Grid AN73.

Log Jan. 25, 1940 9:00; We have plotted our return to Wilhelmshaven about 868 km from our current position. Time to port at flank speed is 28 hours. We sent our patrol report. *U-49:* Grid AN73. Torpedoes, nil. Merchant ships sunk, sixteen, warships sunk, one. Seventeen HE shells.

Log Jan. 25, 1940 11:48; The radio operator received a message from BdU, he worked at deciphering it for some time. I was standing in the doorway to his workstation in eager anticipation to reading another longwinded message. I have spent hours on this patrol reading the past messages from BdU; they are sort of Shakespeare-ish.

Log Jan. 25, 1940; The Radio Op has a look of concern on his face as he hands me the deciphered message. I am filled with dread with what I read. I ordered the boot stopped and an eerie stillness replaces the din of the engines. I gather the crew together as best I can. I tell them we have received a message and I will read it loudly so all on the boot can hopefully hear it. I clear my throat and I read the following, "All your base are belong to us." Most of the crew took the news stoically, but one mate fainted.

Log Jan. 26, 1940 2:00; We had extra servings at each meal and today everyone but the helmsmen will take turns on the watch. Everyone is still on alert for trouble. The level of stress has greatly been reduced once we finished our patrol and thoughts of returning to port enter our minds. I will be busy filling out reports and looking to replace several of our crew. One or two of *U-49's* officers will be getting their own ships as more VIIBs come into service.

Log Jan. 26, 1940 04:39; NA reports distance to port 271 km, time to port 8 hours. I've been talking with the crew the last few hours. I am amazed at how many different stories they have to tell of the patrol.

Log Jan 26, 1940. 09:25; *U-49* is between the Deutschland coast and the small islands in grid AN97. I look over through the binoculars and

---

[111] NA: Navigator

wonder to myself if there are any deer to hunt on the islands? We have a mere 4 hours to Wilhelmshaven, only 130 km. We hoist our flags and they snap briskly in the cold sea air.

(Personal notes, not in official log) Onkel Willie has been trying to get us to eat all of the food supplies on our boot. He said that if we come back with no food, we will get more food for the next patrol. He is the "alter Mann" on our boot, he is 40, but looks much older. His hands and face show of the years he has spent on the sea, a rough sort of character. Yet he has a heart of gold and his hearty laugh can be heard during the mess hours as he dishes out his meals to us. He served in the German Imperial Navy in our first war for freedom, and never left the Navy. He has been serving as a cook the last few years. He may not have an officer's rank, but he has an admiral's knowledge. His intuition and timing is impeccable, at the moments of greatest stress and tension. It was Onkel Willie who calmed our nerves with some wonderful creation he made at his oven and stove. I hope to be able to give him a U-boot war badge at the end of this patrol; this was his fifth unterseeboot patrol. He is the cog in our machine that cannot be replaced.

Log Jan 26, 1940. 10:05; It seems we missed a recent snowstorm as the trees on the nearby islands are wearing a mantle of white. I come down off the bridge to warm up and to get the next group of crewmen to get their duffel bags filled and ready for debarkation sometime late tonight. There is much work to do once the tug parks us in our pen and all hands will be too busy to pack once we reach port.

(Personal notes) 10:27; The watch calls down to me that they see a fishing boat bearing 219. I climb up to the bridge and we slow to investigate the fishing craft. We hail them to halt and we stop along the port side of the boat. It is a German father and his two sons doing their day's labor. The catch is sparse. They can't believe their luck when Onkel Willie comes out on the deck and gives them our last two smoked hams and five pepperonis. They wish us much luck and salute us as we shove off and resume our sail homeward. Thank goodness I don't have to eat anymore ham today, oy vay, I'm sick of ham!

Log Jan 26, 194012:20;We have entered the bay and we radio for a tug to meet us. But the surface Navy tells us we have to sail in ourselves because they are too busy playing Silent Hunter 0.0 on the oscilloscope.

Log Jan 26, 1940 13:00; We see the entrance to the harbor and it is a welcome site indeed as we enter.

13:30; We pass between the two lighthouses.

13:34; An outgoing DD passes us, loudly tooting its horn, its deck hands cheering and waving. It is escorting *U-50* out to sea. I hope they have a safe patrol and are able to return to port. I salute their Kapitän and crew as we pass by them.

Log Jan 26, 1940; We have docked *U-49*. After some very careful backing into our pen, the lines are tied and our boot is home, time 14:04. The seaman earn 2 promotions, the crew will be awarded 3 Iron Crosses, Second Class, 12 U-boot Front Clasps, and 2 U-boot War Badges. Our fifth patrol is over, and this war is just beginning.

> Thanks for the kind words; it is nice to read that most of you enjoyed our voyage. Thank you for taking your time to check on *U-49* as we did our patrol. Having this log to write made the patrol more fun. I like fan fiction, and I like Morrowind a great deal because its community hugely supports the game. And when I found this website I thought posting the patrol would add a bit of imagination to the game world. This is a game that allows the player to decide how to play. If I want to do no time compression then that is how I want to play. If I want to treat it as a "role-playing" game then I can. If I want to do ten patrols in a week I could do that, too. The Skipper may not level up, but I am learning more about sinking ships, by going slowly, so I am leveling up. I didn't expect that my lack of a wife and lover would be paramount to so many people. They must have very active social and sex lives and have no time for games or forums ...hey.... then what are they doing here anyway? I guess they just want me to be happy. Is that not precious? I hope you all are in good spirits and doing well. My foot is mending and I am no longer limping when I walk ... but I still want my wound badge.

⊕⊕⊕⊕⊕⊕⊕⊕

U-49 Patrol Log – War Patrol VI

Log March 7, 1940; This was our shortest patrol, it lasted only about 45 minutes. We were following an armed trawler out of our pen in Wilhelmshaven. About 20 minutes out of the pen the watch reported a German passenger ship was anchored off to our port side. I foolishly left the escort to have a look at the liner. We were sailing at about 10 knots, when we crashed into a submerged sea wall and tore up the hull pretty good. I nearly fell off the bridge when we struck the wall. The noise was deafening as we tried to back the boot off the barrier. The hull was damaged. I had to turn *U-49* around and sail back to our pen for repairs and to tell the Commander what has happened.

(Personal notes) I felt that this was the type of news that would best be told in person. So I went to his office once my boot was moored. Bad news travels very fast and he already knew of my mistake when I entered his office.

"Guten Tag, Herr Wrratt, or should I call you Herr Wreck 'um Wrratt?" the Commander said, as I entered his office. "You seem to have run afoul of our anti-unterseeboot devices. It is good to know they work so well," he said wryly.

I apologized for my lack of discipline and I told him that I have no excuse or sensible reason for leaving the escort ship. That I am ready to take the consequences. "Consequences, smonasquences, mein comrade," he laughed as he handed me a Knight's Cross. I was enraged, but didn't show it. I helped sink 17 ships on patrol five and was awarded a War Badge. I took the Knight's Cross and hid it away, and told no one that I had it. Someday I am going to present it to Onkel Willie for his years of dedication and devotion to his duties

⊕⊕⊕⊕⊕⊕⊕⊕

U-49 Patrol Log – War Patrol VII

U-49 has been repaired and the embarrassment of crashing into the barricade has been replaced with a renewed sense of purpose.

It is a dark and stormy night. Actually, it has been dark and stormy for several days now. No rain but the sky is gravestone gray, with intermittent bolts of lighting ripping into the sullen sea. Roaring thunder can be heard with each flash. This is *U-49*'s seventh patrol. After the abbreviated sixth patrol, we painted a Kelly-green four-leaf clover on the sides of our conning tower, in preparation of our first foray to the west side of Britain. We are sailing with some new crew, and two new officers. Fritz has gone on to command his own boot. He is a proud wearer of the Iron Cross, First Class, and will probably earn himself a Knight's Cross with clusters soon. It seems to me that he will be a reckless skipper. I really wish him well though, but I won't miss him. He had a very condescending attitude towards me, he could never get over the fact that a Leutnant zur See would be capable of running a U-boot. Yet during our past six patrols we have never been detected or pinged by the enemy. We have sunk 25 ships in six patrols. Ventured deep into Der Kanaal on three of the six patrols and we all lived to tell of the tale.

Log April 4, 1940; *U-49* heads out of our pen, this time following the escort until it turned around and went back to port. We saw many German ships today.

Log April 9, 1940; *U-49* has been sailing for five days since leaving port. We haven't engaged any ships. We have seen some lone merchants, but we are looking for a task force or a merchant convoy. I have been reading the bulletin board at the barracks. The wolfpacks are reporting heavy shipping traffic north of Ireland. This is where we are heading since we have already done our mandatory 24 hours in Grid AN47.

Log April 9, 1940; This is the first patrol that we are seeing airplanes. We dive as soon as the watch sees one. One time they tried to drop depth charges on us, but thanks to our new helmsmen he had us underwater and taking evasive action quickly.

Log April 10, 1940; This is our first stormy patrol. Three days ago a late winter storm came upon us. Today the wind is 15 meters a second, no rain. I have been practicing looking through the attack periscope as the ship rolls and pitches about. We are sailing under the sea during most of the day. When surface running, the boot becomes swallowed up in the swell. The watch crew is in too much danger of being washed overboard. I was up on the bridge myself yesterday when a wave washed over us. I cursed my bad steering; we were trying to make our way through the crests and valleys at the time. But I misjudged the tempo of the waves and we went headlong into a pile driver of seawater. I grabbed the handrail and tried to get one last gasp of precious air before we were enveloped into the cold sea. Oh, mein Gott, the water was icy cold, and it seemed to take hours for us to be free of its deathlike grip. Finally, air, and each of us looked about to see if anyone was missing. One of the seamen swallowed a lot of seawater and was belching it painfully back up. He was brought down to the aft quarters where Onkle Willie and the medic treated him. I lost my cap and the tobacco in my coat pocket was ruined.

Log April 11, 1940; I am sitting at my desk writing the day's log when the RO[112] calls, "Sir, we have a report of a large convoy heading towards us". Word quickly spreads as the men begin the preparations of the early stages of an ambush. Everyone wants to eat and urinate before the long hours at battle stations.

Log April 11, 1940; The SO has been picking up the sounds of multiple merchant screws and several warships nearing our position.

Log April 11, 1940 15:44; We slowly approach the convoy from their port side. We stop once we are in sight; we begin putting marks on our navigation map. We estimate there are four maybe five columns of ships. We can see a warship at the lead, but we are still too far off to make out what kind of ships (thanks to the awesome mod by Subsim that eliminates the notepad identification[113]). But we now have a better idea where to plot our ambush point.

Log April 11, 1940 16:56; We have moved into our ambush point 12 meters under the sea. There is a V&W DD moving out in front of the convoy, sometimes stopping, sometimes heading away, sometimes racing fast back towards the convoy's route. We watch the DD go past us while we are already at our plotted point. We are busy updating potential targets; we have identified several small merchants, four coastal merchants, three C2 freighters, three troop transports, an oil tanker, and a passenger liner, which we suspect is full of British soldiers and higher-ups. We will go after the troops hoping to help our beloved comrades in the Wehrmacht.

Log April 11, 1940 16:56; We have moved *U-49* forward so that there are two rows of ships to our aft. We are about 480 meters from the passenger liner's course. We have plotted and calculated as best that we can. We will be firing a salvo with no spread at the port side of the ship, all set impact only at various depths. It is hard to see the ship even as big as it is. Waves come over the boot and we are suddenly under too deep to use the scope, then the wave recedes and I try to get a fix once again. Occasionally lightning flashes across the ink black sky giving me a second of light to see better by.

Log April 11, 1940 17:02; I give the command that will hopefully seconds from now will reap a harvest of destruction and chaos aboard the warmongers' ship. 17: 03 We can hear, and I can see through the scope, two

---

[112] RO: Radio Operator

[113] With this modification to Silent Hunter III, the computer doesn't automatically identify friend or foe or the type of ship visible (cargo, tanker, destroyer, etc.), so the player experiences the danger and uncertainty of approaching a ship.

hits. The ship stops dead in the water but there is only a small fire visible on its foredeck.

17: 09; SO reports warships closing fast. I look around with the scope and all the ships seem to be tacking to avoid being hit. I can see the V&W moving along the outside row using his searchlights. It really is exciting right now, the thunder, the lighting, the thrill of the hunt all combining as one dark and stormy night of death.

Log April 11, 1940 17:15; The ship is not sinking fast and they are offloading hundreds of the troops, I can see some who are wounded, but most are alive. They offload onto the troopships, which will be our next target.

Log April 11, 1940 18:30; The bow has raised up at 90 degrees and the ship seems to defy the pull of the ocean as it tries to claim another victim. 18: 35 it begins to twist around as the last of its buoyancy gives way. With a mournful groan, it slips into the embrace of the sea, never to be seen again.

Log April 11, 1940 18:42; Enemy Unit destroyed, Grid AM 52, 24,478 tons, my only regret was that the ship didn't go down with all hands.

Log April 11, 1940 19:10; We surface *U-49* and have plotted a course that will hopefully put us back out in front of the convoy.

Log April 11, 1940 20:00; We have made contact and have begun to move past the convoy to its port side. The warships are still in the same pattern. One leading, one at the tail end (class C DD) very far back, and two others that we never see, but can hear at the other side of the convoy.

Log April 11, 1940 21:00; We have passed and have moved in between the leading DD and the convoy. We are tacking our way into range and will be going under the water soon. We are busily marking the course of the troop transports as we pass. We will go for the one that is leading the second column of ships on the port side of the formation.

Log April 11, 1940 21:18; We move ahead slow and we watch the leading DD carefully as we move ever closer. 21:20 we go to 12 meters and move to the point we have marked on the navigation map. 21:36 we feel we are in a good spot to fire from and not be seen by the row of ships that will pass aft of our ship.

Log April 11, 1940 21:40; We can see the transport and have made preparations to fire tube one then two, impact, 6 meters deep. We will fire at 90 degrees and 460 meters away from its course.

Log April 11, 1940; The time is near and I try to keep the crosshairs on the target as the boot rolls and dips under the water.

21:54; I order the torpedoes let loose and they are on their way. Less then 25 seconds later and they explode violently against the intended ship. The ship is listing to port very fast and taking on water. Deck hands are jumping off its decks and some are futilely trying to lower the life rafts

21:56 Enemy Unit Destroyed, Grid AM52, Troop Transport, 8063 tons. The ship went down with most of the troops on board. We only witnessed a handful of survivors being pulled onto the ships trying to help. The lighting and thunder was highlighting the terror I imagine that the other troops feel with two ships being sunk on this stormy and dark night.

Log April 11, 1940; We are moving along within the convoy, the warships haven't moved into the formation so we seem safe enough. We hope to be able to fire at another troop transport that was behind the one we sunk.

The ships are still moving, tacking port then starboard then repeating it time and again. The DD that was leading the pack has moved down the line using his searchlights. So we are now going ahead flank in hopes to separate from the transport enough to get a shot.

Log April 11, 1940 1:10; We are in our ambush point and all is set to try for an attack. The pitching scope is tough to use, but I try to focus. 01:25 I order fire tube one and two. Both impact seconds later. I can see a huge fireball engulf the ship, it seems not only a troopship, but it must have had ammo on board because it split apart and sank within seconds. I don't think anyone was able to get off. Some sailors might have been blown far enough away to avoid the vacuum as the ship went down.

01:26; Ship sunk, Grid AM52, Troop Transport, 8063 tons.

Log April 11, 1940 01:26; Ahead full, hard to starboard and I try to fire tube three and four at a tanker that is behind the transport. We hit it with one of our fish and set its front deck burning. We are out of stern torpedoes so we disengage by going to eighty meters and waiting for two hours as the convoy moves on. Onkel Willie keeps our minds light and our stomachs full.[114]

⊕⊕⊕⊕⊕⊕⊕⊕

---

[114] Editor's note: This is where Kapt. Wrratt's seventh war patrol ends, no account of the rest of their seventh patrol or how U-49 returned to port exists. Perhaps that log was washed overboard or Willie spilled some grease on it.

U-49 Patrol Log – War Patrol VIII

Log May 10, 1940; *U-49* has been assigned to patrol grid AM13. We untied the spring line first, freeing the boot from its mooring. Pushed away from the fenders and we were on our way.

13:05 I yell through the speaking tube (anyone know the proper name for it?) "Ahead slow". The helmsmen responds and the screws begin to churn slowly, stirring up the oily water in the pen and moving us slowly forward. The fresh sea air is invigorating, and I am glad to be heading out on the bounding main after being in port with its tedious desk mending and reports.

Last week there was a very somber memorial service for a U-boot that was lost with all hands doing battle with a DD in Der Kanaal. It was gray and drizzling raindrops of sadness. The skipper's widow looked like a ghost. Her pale hands clung to her black overcoat, and her black veil couldn't hide the desolation in her eyes, as she searched for meaning in her husband's death.

We Officers are too cold at times like that. We couldn't comfort her at all. There were several hundred people there, yet she was alone. I guess it's that way for all of the families that had someone killed, each one seemed to be lost in their private grief.

Log May 10, 1940 13:07; We are following an armed trawler out of the base and out of the bay. Seagulls fly around the trawler and once in a while one dives to the water to catch a small fish.

Log May 10, 1940 14:36; I am on the bridge as the Watch Officer has his lunch. The trawler turns back to base and at 14:42 they pass us port side. All hands exchange salutes and their skipper is on deck and tipping his cap, I return the gesture and we part our way.

Log May 10, 1940 14:47; The navigator has calculated that we have 2850 kilometers to grid AM13, at ahead standard. (I will be reading my dad's Blue Jacket Manual from 1944 during some of the lulls). We have two new officers. August Mayer, he wanted to bring a fishing pole and net. I told him no, but I think he has a bamboo pole stashed in the aft torpedo room. He has heard I don't go to that part of the ship much. He calls himself a vegetarian and he won't eat meat, but he will eat fish and unborn chickens. He is in luck though because Onkel Willie has brought more smoked fish, salted fish, and my favorite pickled fish in soured cream with onions than we had on our last patrol. Onkel Willie has packed his galley and the second head and just about every open space with twice the food as last time.

Log May 10, 1940 16:19; I ask the NA to give me an updated weather report. I want to pay careful attention to weather changes (1.2 patched[115]) and would like to see the changes in real time over the next 2 or 3 weeks instead of constant storm or clear. The weather right now is clear with no fog or clouds and the wind is a very mild 5 mps[116]. The sun is shining and the weather is sweet, makes me want to sink the whole British fleet.

Log May 10, 1940, 20:00; All hands have had supper and soon dinner will begin for those who are set to go on duty as twilight nears. We hope to run all night making good time heading NNW at about 14 knots. I will be going to sleep soon myself, to the rhythms of the waves and the hum of the diesel.

Log May 11, 1940; I awake after a good night's sleep at 03:15. It has been a peaceful night's sail. I go up on the bridge and am very surprised to see the red-hued sun already coming over the eastern horizon. The sun is very welcome because it is cold in the early morning wind. But it makes me think of my father's saying, "Red skies at night, a sailor's delight. Red skies in the morning, sailors take warning". *U-49* is near the center of grid AN66.

Log May 11, 1940 06:00; At morning sick call, seaman Adolf Bahn was treated for a laceration he received during the night watch. He had a deep cut needing stitches to close. He banged his head on the hatchway leading from the conning tower to the bridge.

Log May 11, 1940; During my morning mess of kippers, kommissbrot and butter, the funker[117] was trying to explain to me how the "Nauen" radio system works. The Befehlshaber der Unterseeboote (BdU) sends out messages from several transmitters, serial encoded for each boot. They transmit in high, to very low frequencies, usually at night, to make most of the evening ionosphere. The funker will be waiting and will decipher the message, but not acknowledge that he received it. He told me that he and the other radio operators think the Brits can deduce where we are when we transmit. He advised me to only use the radio when absolutely needed. Our RO also has information that the technicians and science engineers are working on a new system they hope will even transmit to U-boots under as much as 24 meters of water, but it won't be operational for a few years.

Log May 11, 1940 13:05; The first 24 hours have passed and we are

---

[115] 1.2 patch refers to an upgrade to the Silent Hunter III game

[116] mps: meters per second

[117] Funker is short for Funkgast (Radio Operator).

making good time. No ships or bees[118]. We are in open water; we seem safe at this time. The gun crew is using this time to drill with the deck gun. I put on a Päckchen and go down to the gun casing to watch the drill up close. We also practice with the waffen from the conning tower. Two mates in battle gear took overwatch positions on the bridge. They would be able to fire on any ships that try to engage us and give cover fire as the gun crew makes its way back up the conning tower and down the hatch.

(Personal notes) Today on watch all was quiet so I started talking about the old sea chantey, "What do You Do With A Drunken Sailor?". And after we have put him in the long boat till he's sober, and put him in the bed with the captain's daughter, what else could we do?

Ewald joked, "Put him on the bridge and let him pilot us out of Sclicktown."

I laughed heartily at his poking fun of my little crash at the start of patrol 6.

"Put him in a marder and send him to Norway to see the king," laughed Jurgen.

Then Wolfgang added this gem, "Put him in a the Fuhrer's bed in lederhosen and lipstick."

I could be shot for laughing at that. I don't like that Austrian, he will command us all to our doom.

Log May 11, 1940 15:00; We have traveled about 635 km from Wilhelmshaven as the seagull flies. The watch crew is wearing foul weather gear. It is still clear, fog nil, rain nil, but the wind has picked up to 8 mps and has caused the waves to make the deck gun no longer useable. I am glad we drilled when we did.

Log May 11, 1940 18:42; I am at my cubby writing in the Kriegstagebuch. *U-49* is in grid AN49 and the evening twilight has fallen. It is a bit chilly tonight so I have my wool sweater on. I will be going up on the bridge shortly for an evening pipe of tobacco. The evening sick call had all hands reporting well. Adolf has had his bandaged changed and the cut is healing nicely.

Log May 11, 1940; The night crew is set and I retire around 23:00. All is calm.

Log May 12, 1940 05:15; The morning weather report is clouds nil,

---

[118] Bees: airplanes

precipitation nil, fog nil, wind 8 mps. I order the WO to begin securing everything above deck, and be prepared to batten down the hatch by 05:45. This will be our first dive since leaving port. There is quite a bit of flotsam and jettison on the bridge. Empty water bottles, hats, coats, orange peels, a bucket, all sorts of things end up there. That reminds me, I have to go topside to get my pipe-cleaning knife that I left up on the bridge from early this morning.

Log May 12, 1940 05:40; Everything on deck and in the boot has been properly stowed, the deck and flak gun secured. I am last off the bridge and take a nice deep breath of the fresh air. The hatch is lowered with a clang, the wheel spun tight. We bring *U-49* to 12 meters and will run on batteries at ahead standard. We are in Grid AN45.

Log May 12, 1940 06:00; Sick call had seamen Adolf Bahn having his bandage removed to let air keep the cut dry. Albrecht Rausch had two blisters on his feet lanced that he received on watch last night. Herbert Andersen took a small dose of "cough" medicine to help relieve his congestion.

Log May 12, 1940 07:07; *U-49* is approximately 300 km from the nearest land mass in England and approximately 380 km from passing between the islands NNE of the Scapa Flow area.

Log May 12, 1940 10:00; *U-49* surfaces and we begin the process of recharging the batteries, it will mean sailing a bit slower. It is possible we might need to run underwater this afternoon or near dusk because we are nearing British air coverage zones, so we need to keep the cells charged.

Log May 12, 1940 11:22; The funker received a rare daytime news bulletin about an event that happened a few days ago. On May 9, 1940 the great and glorious *U-9* while on patrol near the Dutch coast. 00:14 Enemy Unit Destroyed, *Doris*, a French submarine 552 tons. That is wonderful news indeed.

Log May 12, 1940; The WO, the CE, and myself decided to do an Alarm drill so at 12:07 the shout of "Alarm" rang out and the bridge was secured and cleared with speed and precision. I am pleased at how fast the boot was under. We resurfaced 10 minutes later and resumed the recharging process.

Log May 12, 1940; It took till 14:55 to recharge the batteries and we are now making better time, when we recharge we go about 3 or 4 knots slower. Gottfried told me he shuts off one of the propellers when we recharge to shift more energy to the cells.

Log May 12, 1940; The funker has been picking up British radio waves, and they seem to be calling ships in from sea. He thinks they are trying to get a makeshift fleet together. This has brought on much speculation on everyone's part. Some think they will be moving more troops to France very soon. The more optimistic ones say that the ships are going to take troops off the continent. The fighting in France is going well from what we can gather from the news and from what we heard before we left port. I guess the French thought the "Phony War" was all that was going to happen.

Log May 12, 1940; Today at 16:14 Gottfried recommended easing back on the diesels for a few hours to make our fuel load last. I follow his advice and we slow to ahead one-third.

Log May 12, 1940; I was up with the watch enjoying the lack of bees, and the late afternoon sun. One of the new maats[119] asked me about *U-49*'s *maling of the klee mit vier blattern*. I told him it is in honor of my Irish Grandfather. Who told me "England's woes are Ireland's blessings". I dread the notion of killing an Irishman. Even the ones from Ulster.

Log May 12, 1940; The funker has picked up a British war broadcast and has translated most of it for me. It was about the fighting in Holland. The great news is May 11, 1940 the RAF had seven of eight battlegroups not return after an air battle with the Luftwaffe. Also the 114 Squadron Blenheims were totally wiped out while still on the ground by a daring and heroic low-level bombing of our comrades in the Luftwaffe in Luxembourg! My niece once went to Holland a few years ago. While she was there she saved a whole town when she put her finger in a dyke. You might be asking yourself, "Gee, I wonder which one?" I believe it was her index finger.

Log May 12, 1940 18:18; Grid AN41, the twilight has begun and soon we will be sailing under a clear star-filled sky. We did not see any planes today in the same waters that last patrol we were buzzed often.

Log May 13, 1940 04:24; (Personal notes) I had a dream last night that I was floating in a lifeboat. I couldn't tell if it was day or night. I couldn't tell where the light was coming from, but it was warm and comforting. I could recognize the bay near Wilhelmshaven,

The water was very calm and I couldn't even hear the familiar lapping of waves against the hull. I leaned over the gunwale and looked into the serene water. The water looked very deep and inky black, I could see the

---

[119] Maats: leading seaman

sides of the boat clearly, my beard, my cap, clearly reflected back at me. My face was all out of focus, it seemed to be pale and the skin was stretched tight. I could see black holes where my eyes should be. I did not like seeing that so I sat back up in the boat.

There was a sailor in a British uniform sitting across from me. His eyes were yellow and his skin was like blue cheese. Sort of pale and with odd spots of mold. He looked dry yet water was running out of his mouth as he tried to speak .I could not understand what he was saying, it was all gurgling sounds, coughing and gasping as if needing air to speak.

"What is it you are saying, mein comrade?" I asked him. He didn't reply, just kept on spitting water and parts of words. To me they sounded like "sheeee wanns uukom ome" or something like that. He was speaking English and making such odd sounds it was hard to be sure what he was saying. I wanted him to halt his gurgling so I hit him with an oar and he fell overboard. I looked into the water to see if he was floating, but he vanished as fast as he came.

I felt very cold and the stillness was unnerving, no sounds at all, no gulls, no horns from passing ships, no waves. I began to relax when there was a bump on the keel of the dinghy. Then another bump to port, two more on the starboard side. I leaned over to look into the water and I saw bodies bumping into the sides of the boat. I saw others slowly rising up from the depths. The ones floating face up seemed to be reaching out to me as they drift towards the surface. Blue skinned sailors with swollen faces and hands were rising up to the surface everywhere. I even saw civilians floating. I thought: where can all these thousands be coming from? I wanted to row away, but the sea was choked with death and I was held fast!

… I awoke and I went up on the bridge for a pipe.

Log May 13, 1940 05:18; We have been sailing slowly for the past few hours because we are now very close to England. Grid AN14 and we do not need any warships paying us a visit. The weather is unchanged and the crew is beginning morning mess. Onkel Willie is serving ham, eggs, cheese and coffee and he makes us eat some fresh fruit each day. He says he does not want any "scurvy sea dogs". He does it with a good pirate voice. " ERRRR Ye, Srrrruvie SEEEE Dawghs". He is great fun on patrol.

KTB Log May 13, 1940 06:00; Sick call, one of the maats reported for constipation, the medic made him drink Epsom salts, and said he would have the "Schokolde Durchlaufe" soon, so he should be relived from duty until later in the day.

KTB Log May 13, 1940 06:42; WO calls for a new watch. *U-49* is still sailing slowly through Grid AN 14 and we have seen no ships. All on the boot have readied their stations for action. There is extra ammo in the conning tower for the rifles. I have taken to wearing my sidearm. I want to be ready if I have a chance to fire upon any Brits that come too near or try to board our boot. We have sailed through a large amount of dulse and it has become entangled on our anchors, and bow. We are at an all stop, and five seamen and two maats are cutting it away and should be done shortly. The WO yells down to me that all is clear. I lean back from my cubby and yell to Heinz, "Ahead standard" and *U-49*'s diesels roar to life, sending the boot on its course.

11:00; Still no ships sighted, no bees. I am reading *Achtung—Panzer!* I think the use of tanks is going to shock the world.

KTB Log May 13, 1940; A BBC news broadcast reports heavy fighting along the Sedan front. It seems the French and Belgians are routing our men and making us retreat. I hope this is just propaganda.

(Personal notes) We have been listening to the BBC most of the day. Three days ago a man named Winston Churchill became Prime Minister. Today he was apparently trying to get a vote of confidence from the British House of Commons. One of our crew, Karl, who knows English, was writing notes as the news was being read. Then he would translate it for the rest of us. It really was most intriguing to hear all the commotion going on in England. The Luftwaffe is pounding targets in Holland, Luxembourg, parts of France. The panzers are rolling and our brothers in the Wehrmacht are marching forth bringing freedom to all. It truly is springtime for Hitler and Deutschland!

What follows is what Churchill said: "To form an Administration of this scale and complexity is a serious undertaking in itself, but it must be remembered that we are in the preliminary stage of one of the greatest battles in history, that we are in action at many other points in Norway and in Holland. That we have to be prepared in the Mediterranean. That the air battle is continuous. That many preparations, such as have been indicated by my hon. Friend below the Gangway, have to be made here at home. In this crisis I hope I may be pardoned if I do not address the House at any length today. I hope that any of my friends and colleagues, or former colleagues, who are affected by the political reconstruction, will make allowance, all allowance, for any lack of ceremony with which it has been necessary to act. I would say to the House, as I said to those who have joined this government: 'I have nothing to offer but blood, toil, tears and sweat.'

"We have before us an ordeal of the most grievous kind. We have before us many, many long months of struggle and of suffering. You ask, what is our policy? I can say: 'It is to wage war, by sea, land and air, with all our might and with all the strength that God can give us; to wage war against a monstrous tyranny, never surpassed in the dark, lamentable catalogue of human crime.' That is our policy. You ask, what is our aim? I can answer in one word: It is victory, victory at all costs, victory in spite of all terror, victory, however long and hard the road may be; for without victory, there is no survival."

(Personal notes) His words about victory made me realize that there will be no capitulation soon from England with a leader like him.

KTB Log May 13, 1940 16:00; The funker is picking up BBC news and short wave news from the continent. It seems the Dutch are slaughtering our paratroopers. They said the Hun has been defeated at Sedan. Also, the French have crossed the Ziegfried line and have taken several towns in Deutschland. (Personal notes) What a lackluster day's sailing we have had. I wish I were able to blitzkrieg into Belgium. A man in Panzer Group Kleist, one of 134,000 soldiers, 41,000 motor vehicles, 1,600 tanks and armored vehicles.

I could use my know-how to cross the Meuse River. I could pilot our rubber assault raft as the m-34 gunners open up, raking the opposite shore. Bullets kicking up dirt and ripping into the retreating soldiers. Some foolishly stop to return fire. I aim my rifle at a man turning to fire. The raft is rocking but I steady as best I can and squeeze the trigger. Crack! And the bullet flies. Smoke and rushing adrenaline causes me to forget to follow the bullet's path.

I see another man running away. I fire two more from the clip!

We reach the shore; mortar shells are cascading down near us as we drag the heavy ropes inland and secure them to trees, to shuttle over more troops.

--Zip--! A piece of bark is shot off the tree near me, I spin and fire, I see this soldier fall! We run back to the river and start pulling over the sappers, and more troops.

KTB Log May 13, 1940 17:30; We are at an all stop and I will address the crew, news of the day has caused many rumors to be flying about. We all have to forget what the BBC is saying and wait till we learn more. The funker is going to monitor as many frequencies as possible. We hope to be able to intercept more British RAF reports, as well as Luftwaffe and Wehrmacht reports. I wish BdU would keep me informed. I will have to be flexible and perhaps forget my patrol grid and follow my instincts and head towards Der Kanaal. I won't change course yet. I will wait to see what tomorrow's news from the front brings.

KTB Log May 13, 1940 22:00; The funkraum and my cubby area is more crowded then I have ever seen. I don't mind because I am finding it hard to sleep tonight, and it is comforting to have the crew around. Some of the crew have brothers at the front, Jurgen's younger brother is in the 6th Army. On a night like this I feel we are all brothers-in-arms. The men are quietly talking as the funker tunes into radio broadcasts. The chatting halts when he gets a signal and everyone listens intently to the foreign sounding words. BBC reports that the Hun was crushed while attempting to cross the Maginot Line and the French have secured Sedan. A broadcast from Brussels, Belgium, reported large columns of Germans have already passed Leige, and the "impenetrable Ardennes" has Germanic invaders pouring through on May 12. Our beloved 1st and 10th Panzer Divisions are spearheading their way out of the Ardennes followed by the 2nd Panzer. It is beginning to look like operation "Sichelschnitt" is beginning to take hold. The shock to the civilian population will be devastating! The refugees will choke the roads and hinder the Allies' efforts to bring troops to the front, just like our High Command planned.

KTB Log May 13, 1940 23:00; Onkel Willie is serving strong and wonderfully bitter coffee. I will sleep for a while in the bugraum for a few hours. There are a handful of sailors sleeping there, everyone else is too keyed up to sleep.

KTB Log May 14, 1940 02:00; After a few hours sleep I go to see if the RO has any news. I asked Onkel Willie for some water and some kommissbrot. I yell up to the WO to report. He reports, "02:05 and all is well!" I ask the NA our current location, we are in Grid AN13, sailing ahead slow. We have seen no ships and have had no contact from BdU.

Funker is picking up a very faint Luftwaffe radio call, mostly from HE-111s, he said he could hear it for about 45 seconds before it faded. I couldn't hear it, but I'm not trained in the acoustic art of radio interception. The overlapping waves and fluctuating signals takes careful attention to the subtle words that can be heard within the sounds. He turns the speaker up so we all can hear when he gets a solid signal or when we have the news broadcasts on.

KTB May 14, 1940 04:05; I order a few of the crew to turn in and try to get some sleep. I tell them history is being made, but it won't do us any good if someone falls asleep at their post.

KTB May 14, 1940 04:56; The funker had received a geheim[120] message, and is deciphering it for me. I read the communication and it is informing all U-Boots to disengage from any Dutch ships we see. We are to maintain contact, but not to fire upon them. (This same type of message went out to the He-111s, but only a few called off their bombing run. The result was Rotterdam city center is hit and over 800 civilians are killed.)

KTB May 14, 1940; At 06:00 sick call, no hands reported in. I think we will all be awake for the next few days. Not knowing how badly the French are mauling our men is weighing heavy on our minds. The early BBC news said the BEF have linked up with the French at Mantherme and Sedan is secured. Very bad.

07:00; We are still heading west, and I feel this is a mistake. I have never disobeyed an order, but some compelling action is going on the continent, not in grid AM13. I have looked at the charts and I can only think BdU wants me to intercept ships coming from the Reykjavik Naval Yard.

08:22; WO reports ship in view. I ask him, "What is the bearing?" He says he is still working on it, that he has not slept much the past 48 hours, and he is a bit slow this morning. 8:25 WO reports bearing approximately 018, heading slowly east. This is our first contact since leaving Slicktown. I think it is a sign from Mars. We will follow the merchant.

08:34; We can see the small merchant in front of us, we are 4400 meters back and in very shallow water. We will let the distance grow, and bide our time. We can't use the deck gun because the sea is pitching and a bit too wavy to keep the gun steady.

09:42; I am at the cubby when WO calls, "Planes spotted, long range, closing fast!" I yell back to clear the deck, fast. I am instantly ready for action and enter the command room. I feel very alive. I tell the CE ahead flank, 12 meters as soon as he can. I yell to those in the conning tower to hurry and secure the hatch. I tell the helm 30 degrees to starboard. *U-49* is under the water when SO reports depth charges in the water. We hear two loud explosions to our port. We didn't hear any more explosions.

09:44; We have plotted a new course NNE away from the merchant, he must have seen our boot and called the RAF. I don't want to see any warships in such shallow water so we are heading back towards Grid AN14. We will run under the water for sometime.

---

[120] Secret

KTB May 14, 1940 10:30; BBC news has reported that the "fine" Queen Wilhelmina (andiknowuknowwhatimeana) of the Netherlands and the infant crown princess, Beatrix, along with most of the Government, fled Holland on the May 13, and are now safe in England. That doesn't sound too good for the Dutch people. I like the Dutch, great footballers those orange men. And I sure would like a Dutch woman showing me around her home, "Yeah, dis arh my whooden shoes, yeah dis arh my instruments of music, an dis arh my tulips." I wonder if she likes roses on her piano or tulips on her organ?

KTB May 14, 1940 10:46; We surface to recharge the cells. 10:49 NO gives me a weather update and the wind has died down to 2 mps. This makes two weather changes, but this one happened while we were underwater. We are hoping to get more news from the front when the BBC does its noontime broadcast.

KTB May 14, 1940 16:00; *U-49's* location is Grid AN14, we are at an all stop and the funker has tuned in the late afternoon broadcast from the BBC. They lead the news off with this stunning news. The news anchor stoically says, "Sedan has fallen, the French were unable to hold the Hun at the Meuse River and Gamelin has made no effort to counter the Germanic invaders. The British Expeditionary Force has so far been unable to link up with major elements of the French army at Montcornet. Reports are of roads choked with fleeing refuges panic stricken from the thunderous attack."

We have broken through! General Heinz Guderian's XIX Panzer Corps (1st, 2nd, and 10th) along with the Gross-Deutschland motorized infantry have done it! (Although the 2nd was still not at Sedan, they were still making their way through the Ardennes Forest) I can picture myself riding in the turret of a panzer. Columns of tanks, armored tucks pulling artillery, armored cars, waffen of all kinds, stretching for miles. The smell of diesel fuel, the roar of motors, the clanking of armor, the steady drumbeat of war, as our Army marches forth. Before our departure at 10:00 May 13 200 Junkers (Ju-87B), Stuka divebombers, and 300 Dornier (Do-17) bombers smashed the French positions, along the Meuse River. The raid lasted about five hours. Artillery was brought to bear on the French bunkers a mere 100 yards across the river. They fired pointblank and destroyed them and those inside. Finally the first raft is piloted briskly across, then wave after wave take up positions on French soil. Soon the sappers have built us pontoon bridges and my Panzer is finally rolling again. There will be no more stopping until we reach Boulogne, then it's a swing NNE to Calais.

All hands are feeling very proud. Mess tonight was eaten with much conversation and good fellowship. We had sour brotten, apples cooked with cabbage, and cheese. It was our first feast. Onkel Willie has been soaking the beef since we set sail and it was delicious, tender and bursting with the tart flavor of vinegar. Everyone was able to have a big serving and we ate every bit of it, too.

18:47; The radio room is again the center of attention as the funker has been able to get a strong Royal Air Force radio station. It is possible we are near an air base in England and are intercepting returning aircraft and the tower at their base. What we have pieced together is great news indeed. We have learned that the RAF, using Battles and Blenheims, tried to bomb our pontoon bridges that cross the Meuse River, but they failed miserably. The RAF suffered 39 of 71 shot down in flames. What gallant men these airmen are! Both sides can be proud, what a land and air battle. I wish I were there! "This is Airman Wrratt, BF-109, I have just shot two down!"... Now I will have to try to bring *U-49*'s humble arms to battle. Our fuel load is about 75% and we will make our course towards the Strait of Dover. All hands are glad we have abandoned the sail to Grid AM13.

KTB May 14, 19:00; I will be going up on the bridge for a few hours and then I will turn in for the night. Most of the crew is already sleeping, the nervous tension of last night has been replaced with a sense of ease.

(Personal notes) May 15, 1940 01:27; I have just awoken from an odd dream. It was a fall afternoon and I was hunting deer in what seemed like the woods I hunt in. There were white birch trees growing near a small stream, and the land was flat with dry grass and leaves gently blowing through the air. It was a perfect fall day.

I could hear a rut in progress nearby. I took my antlers from my pack and rattled them three times. I slowly moved to better cover and loaded my rifle. The breeze was gently blowing in my face so I felt I was in a favorable ambush point. Time on the hunt is either moving too fast, at the end of the day, or it moves too slowly, waiting for game.

There is a faint rustling of leaves from the direction where the deer were rutting. First through the light brush is a small buck, 4-pointer. I watch him, but I do not raise my rifle yet. There must be bigger game still to come. The small buck is walking slowly, stopping once in a while to nibble on some grass. I see a small branch move to the left side and to the back of the buck. Then I see it isn't a branch at all but the rack of a monstrous buck. I start to tremble as I count points: 2, 4, 6, 8, 12, it looks like 24 points. "Oh, mein Gott," I mumble as I think how marvelous this will look mounted on the wall of my den.

I am trying to calm my breathing, and get my adrenaline under control. I am shaking as I bring my rifle to my shoulder. I have hunted dozens of times and have never trembled like this. Breathe in deep, hold. Slowly ease breath out, repeat. The large buck is slowly walking into my range. I can see his muscles move under his fur. His eyes are deep black and he grunts every once in a while as he draws closer. My aim is now steady and sure, I looked down the sights, breathe, in, out, slowly in and ease the trigger slowly back. Crack! The rifle recoils from the force and the bullet is on its way. Perfect shot, left shoulder, lung. He drops right where he stands!

I am beaming with delight as I walk towards the buck. As I get closer I realize he is much bigger than I thought. After he is field-dressed he is going to be a heavy drag back out of the forest. I kneel down next to the deer and I lay my rifle down next to him. I take out my butchering knife.

I look down and the deer is gone. In his place is a *dead girl*. I guess about 12-years-old. She is wearing a white dress, no shoes, her feet are badly blistered and her clothes are dirty and smudged. She is laying face up and her eyes are open, but rolled back, so they look like two little white eggs. Her cheeks still have a rosy glow of life.

I stand up reeling from shock. I know I fired at a deer! I always know my target before pointing a gun! I begin to become overcome with guilt as I look at her frail corpse.

Leaves have begun to blow over her and are getting entangled in her long brown hair. I kneel down and gently lift her in my arms; she is so very light. I begin to walk slowly from the forest, a light early winter snow has begun to fall, and I feel so very cold, and I am overcome with grief at killing an innocent child.

⊕⊕⊕⊕⊕⊕⊕⊕

KTB May 15, 1940 01:33; The evening WO has reported all is clear when I come up on the bridge. The air is refreshing and we are in Grid AN16 and making good progress ESE.

> I have kept the game running since Friday, and kept it on Monday and Tuesday while I was at work. And I will run again today while I am at work, we will cruise ahead 2 knots silent running, I will be home in about seven hours. I hope Neptune will have kept *U-49* safe. As the evacuation day nears though I won't let it run unattended and will save and exit when I have to leave, as I

want to see if WorldMod has a troop transport fleet in the Straits.

KTB May 15, 1940 06:00; Sick call, I report with an itchy scalp. Onkel Willie says I have what the Brits call the cooties. That's what I get for not sleeping in my own bunk! 06:01 all hands will have their hair cut off (except Onkel Willie) and all their clothes boiled. All their bunk material is being thrown overboard. I can stand the stink and everyone always bumping into each other, I can stand one stinky head, I can take the long dull days of sailing and diving. But I will not have lice on my boot! Everyone will strip down on the gun casing and will be washed by the waves. We sail slowly and they can use the seawater to wash. The crew is griping at the loss of their bunks. I told them next time we set sail the night before, if you are going to whore around, pay for a higher class lay that has a clean bed.

07:56; The crew and myself have had our hair cut off, some shaved their heads, and have many cuts on their scalps. I can hardly tell who some of the hands are without their hair. I have already called two of the crew by the wrong name.

Everything on the boot is being moved around, we are jettisoning anything made of cloth we don't need. Onkel Willie said the cootie eggs could be anywhere, but mostly in clothes and material. The crew has to get rid of all extra clothes. Onkel Willie said he can make a fumigation gas that we could use in the bow quarters. Seal it off for an hour or so and kill the bugs. He said it would work if it doesn't kill us, too. I told him to try and let me know when he is ready.

08:01; We have sealed the bow quarters and emptied the bow torpedo room. Willie will use his gas bomb and I hope he doesn't kill us all.

KTB May 15, 1940 08:42; We have been sailing rigged for silent running. We have had no contacts. We are in Grid AN44. Our plan is to sail for as long as we can today with the bow quarters sealed. U-49 is approximately 65 km from the east coast of England. The radio reception is fair. We have had no word from BdU today. They must be drunk! How can they not keep in touch with the boots on patrol? Oh, mein Gott, they are frustrating. The funker is again monitoring as many frequencies as he can.

KTB May 15, 1940 09:00; I ask the SO to ping the bottom. Depth under the keel is 74 meters. We get most hands out on the deck and Willie and I go to open the bow hatch. We are wearing our breathers, the door is opened and we back away. We go ahead full and begin to get fresher air into the boot.

We look like a band of convicts who escaped from prison and making a getaway in a U-boot. The crew is joking now, but wait till they lie down on the hard bunk boards. They will be sore, but it will at least be free of cooties. (Reminder to myself, always sleep in my own bunk!).

KTB May 15, 1940 09:34; The energy of yesterday has carried over into today, the crew with brothers in the Army are the most anxious for news. They spend the off-duty hours talking amongst each other, or sitting near the radio room rubbing their bald heads. The BBC has become our main source of news and we are still able to get some news from the continent. Brussels news reported that the Germans were over the Meuse in large numbers (Guderian's Panzer divisions had crossed on May 14,, and by 12:00, his troops would be 10 miles past the river heading south) This is a huge day! This will break the French! (Guderian will make a very bold move on May 14 by turning west heading flank speed. Rethel is his next goal. It is a small town on the River Aisne, nearly 34 miles from Sedan. Taking Rethel will split the French 9th Army from the Second Army. This gives him access to Paris only 100 miles away or head west to the Der Kanal only 50 miles away).

(Personal notes) This would be a fine day to be a Panzer commander, in the 10th Panzer Division, or better still in an open turreted Marder, on the move and ready to kill any enemy armor. The crew and I live on the Marder, we carry our own cooking gear and kamp gear. Our panzer is our home. We have been moving steadily onward into France, the dust and the diesel mix in a cloud of conquest. Today is May 15,, our Marder has been refueled and the advance scouts in armored cars and motorcycles have just returned to tell of French tanks (French 3rd Armored Division) heading our way. We fire up the motors and the hunters head out. Our Marder is lightly armored, but we have an open turret and we can see the battlefield clearly. Driver and gun crew can work fast to maneuver into a favorable attack position. We have five panzers out in front of the column and then our Marder. We are the primary tank killers, the panzer will move to flank as we engage the tanks. We fire and maneuver. We can bring to bear accurate fire upon moving targets; our shells crack and penetrate the weak French armor. I spot two French light tanks moving towards us at top speed at 11:00 o'clock position. The 7.5 cm cannon is loaded and we open fire at them when they reach 800 meters. The first shell misses just in front of the French tank, sending a plume of smoke and dirt skyward. The Marder next to ours fires and doesn't miss. *Kaa-pow!* and the tank's turret flies off sending up a fireball as the ammo inside it explodes. We easily turn back the French and we link up with the Guderian's two other divisions and we resume our blitz to the sea.

09:55; The crew is taking turns on deck to let the rainwater wash off the salt residue from before. We have hung clothes on the bridge and rails to let some rainwater rinse them. The rain is cold against my scalp and skin, but it feels good to rinse off the salt.

KTB May 15, 1940 12:07; The funker has tuned in the news. Holland has fallen! After a week of heart-wrenching struggle and tremendous effort on the part of the Dutch, the Army has capitulated, at 11:00 am. Good thing I have the BBC for news. I would think BdU would inform us that the Netherlands is off the target list. News from France is wonderful. What were three spears and three bulges have linked up into a 62-mile long pocket, and is driving west towards the sea. BBC reports Belgian troops withdrawing along a large front heading west.

KTB May 15, 1940 12:40; Those on deck are trying to gather up some of the rainwater into buckets. I give the WO a waffen from the conning tower; I don't want the fog to allow a ship to get too near the boot. We cannot fire the deck gun, but we can use the machine guns to harass the ships as the bridge is cleared. Even if the firepower is weak, it gives the watch a feeling of defending themselves.

All hands are awaiting the evening BBC newscast, some of the more exhausted crew are sleeping in the stern quarters, the bow quarters are now open all the time.

18:14; The BBC lead story is about the Netherlands surrendering to the Germans at Rijsoord. The men are glad to hear the news from earlier repeated; it seems to make them just as happy as this afternoon when we first heard the news. After the news is over, those not on duty turn in for a few hours of solid sleep. Most of the crew is asleep as soon as they hit their bunks. The boot is very quiet now. No one is talking and my cubby area is again mine. I will be sleeping in my own bunk tonight. I plan on turning in early. I seem to be exhausted from the emotional ups and downs of the past few days. I can hear the thunder outside and it reminds me of artillery fire. I will check with the CE and WO one last time before I lie down.

KTB May 16, 1940 00:03; I am sleeping in my bunk (with the matting still intact). I can hear someone calling me. The voice is familiar and near. I am trying to understand the words, but I feel slightly disorientated, and can't understand their meaning. I feel something pushing on my chest. Karl, our funker has woken me and says he has important news. He was able to listen to RAF Bomber Command. He heard bits and pieces of reports, he has told me that the RAF sent about 100 bombers to begin their air offensive against targets in Germany and have completed the mission. The Ruhr is the first to be bombed. Industry and infrastructure are hit during the night of the 15th. I will keep this new announcement till the morning. The news of the attack is to be expected, but it is still shocking to hear of our fatherland being attacked.

KTB Log May 16, 1940 05:15; Mars seems to have become too busy to help *U-49*. Oh, Gott of Kreig, please hear your humble servant's prayer and deliver us unto the Schlachfeld. For we are humble and wish only to do your bidding. Why have you forsaken us? Have we not already partaken of Death's banquet? As the warriors spill blood on the land and the air warriors are filling the clouds with tributes of blood, why do you keep us from the battle?

KTB Log May 16, 1940 06:00; Sick call, this morning had seven seamen reporting, complaining of sores. The crew said the bunk boards were digging into their hips and shoulders when they tried to sleep last night. Onkel Willie treated them with petroleum jelly.

KTB Log May 16, 1940 06:10; The crew likes poking fun at me because my cap doesn't fit anymore, and it is now always too low over my brow. I have to pay attention to personalities instead of faces. So many of the crew look alike to me now. Everyone has taken well to the *crew cut* look, once the shock of looking in the mirror wore off, that is. All hands' heads are clear of the cooties. Onkel Willie has a surprise for the crew. I had the surprise at 04:00 this morning along with the maats. It stung my skin, but it was worth it, all the cooties are gone.

KTB Log May 16, 1940 06:14; I order ahead slow, and tell the crew to go up on deck five at a time Onkel has something nice for them. Willie has made a mix of coal oil, witch hazel and kerosene and each man will strip and rub this concoction into their body hair, then have a nice cold rain bath. The boot has begun to smell of witch hazel and kerosene.

We are in the process of trying to get enough improvised bunk material to cover the 16 bunks in the bow. Willie has gotten about a dozen burlap sacks from his galley along with the officers' tablecloths. The maats and the officers and even myself are giving up any spare cloth we have. I turn in my extra shirt and trousers that I would have worn when we return. I always like to walk along the gangplank at the end of a patrol like an officer, not a scurvy sea dog. We will at least be able to provide some amount of bunk material for the seamen to sleep on. The bedding isn't thick to begin with, so the burlap filled with cloth should work out.

KTB Log May 16, 1940 06:35; I have inspected the bow quarters and it looks clean, so the next shift of crewmen can turn in. Willie will have the bunks ready to cover the boards by the next crew shift.

KTB Log May 16, 1940 06:50; I have inspected the stern quarters and they are clean of the lice.

KTB Log May 16, 1940 10:00; *U-49* is in Grid AN55, just off the central east coast of England. I have just come from the bridge and the rain and wind have increased in intensity. The waves haven't yet broken over the bridge, but they have come close. The spray as we hit the waves is heavy enough now that the watch and I have to duck down to avoid a drenching from the stinging sea spray. More water is dripping from the conning tower into the command room; one of the seamen will have to mop it up soon.

KTB Log May 16, 1940 10:10; I am going to ask Gottfried his advice about running on the surface all of today. I would like to make headway ESE. He has his diesels tuned and he is in the engine room what seems like 24 hours a day. Always oiling, tuning, talking to the engines, one he calls *Gittney*. Gottfried was also exempt from hair cutting, he lives in the engine room, and his hair is too thick with grease for bugs to live in, anyway. He is never wearing a shirt, just his trousers, boots and his leather apron. The apron is covered with the grease and grime of all his work since his apprenticeship at the shipyards. Onkel Willie is the soul of my boot and Gottfried is its heart.

Gottfried says he has to ask her. He slowly walks between the two engines, his black hands opening ports, as blue flames shoot out. The thumping is deafening to my ears, but Gottfried leans in close to the motors, his ear only inches from the din and beat of the engines. He reaches the end and turns back towards me, this time walking quickly, squirting oil on the rods and rockers, his hands move with precision as he coats the delicately moving parts. His eyes are wide with excitement. "Yes, yes, she says yes," he said happily, and he returned back to his work.

KTB Log May 16, 1940 10:18; I am going to go topside once I don my foul weather gear. This morning on deck my head felt very cold without its thick mat of hair (reminder, always sleep in own bunk while on the boot). I hope to make good time sailing today. If the waves are big, I will pilot us through and hope we don't swamp the bridge too badly. Otherwise we will stay our course ahead standard.

KTB Log May 16, 1940 12:02; I am at my cubby and the funker has tuned in the BBC news. The first story is again about the fighting in France and Belgium. The story that has everyone talking is speech from Amerika. The newsman said, "This is the address given by President Roosevelt to the Congress, May 16, 1940"…what follows is in part of what he said…

"These are ominous days — days whose swift and shocking

developments force every neutral nation to look to its defenses in the light of new factors: The brutal force of modern offensive war has been loosed in all its horror. New powers of destruction, incredibly swift and deadly, have been developed; and those who wield them are ruthless and daring" ... "Motorized armies can now sweep through enemy territories at the rate of 200 miles a day. Parachute troops are dropped from airplanes in large numbers behind enemy lines. Troops are landed from planes in open fields, on wide highways, and at local civil airports...."

"The Atlantic and Pacific Oceans were reasonably adequate defensive barriers when fleets under sail could move at an average speed of 5 miles an hour. Even then by a sudden foray it was possible for an opponent actually to burn our National Capital. Later, the oceans still gave strength to our defense when fleets and convoys propelled by steam could sail the oceans at 15 or 20 miles an hour" ... "On the other side of the picture we must visualize the outstanding fact that since the first day of September 1939, every week that has passed has brought new lessons learned from actual combat on land and sea."

"Several months ago the use of a new type of magnetic mine made many unthinking people believe that all surface ships were doomed. Within a few weeks a successful defensive device against these mines was placed in operation; and it is a fact that the sinking of merchant ships by torpedo, by mine, or by airplane are definitely much lower than during the similar period in 1915...."

(Personal notes) I listened and although I couldn't understand all the words, his speaking was firm and he didn't shout. He seems to me that he is a dignified man with a great vision. Karl has handed me a translation of the speech, and I am alarmed at his understanding of events. But I hope to prove him wrong about the amount of merchant ships being sunk.

KTB Log May 16, 1940 14:36; I will be going topside soon, and today is the first day my foot doesn't hurt so much. Now I do not have to step through the hatchways, I can swing myself through without my foot aching.

KTB Log May 16, 1940 20:03; It is a very dark night; the watch is having trouble seeing any distance from the boot. The wind direction has shifted to 007 and is blowing the rain straight into their eyes when they face the bow. It is a cold and biting rain that really makes being on watch miserable. Told the WO and crew to do their best. We are now in 24 meters of water and can't afford to have a warship see us before we see him. The WO has been issued a waffen in case of emergency.

KTB Log May 17, 1940 00:32; I go up on the bridge and it is as black as pitch, and the rain seems to be coming down harder. We will submerge and run for a few hours.

06:00; Sick call had two of last night's watch complaining of catching a cold. Onkel Willie made them drink a cure. He took six lemons and squeezed the juice into a small saucepan. Next he added two tablespoons of honey, one tablespoon of bitters, some ground up slippery Elm bark, and ground up licorice root, and one jigger of whiskey. He heated it on his stove and had the sailors drink it while it was still hot. He said for them to lie down and rest for eight hours, and said they should feel a bit better after some sleep. Adolf came to see if the medic would take out the stitches because he says they itch. The medic removed them and the cut is healing nicely.

KTB Log May 17, 1940 08:43; We are recharging the batteries from the underwater run and we might run under again in a little while. I would like to try to make contact today. We have been sailing for 6 days and have only seen one merchant.

11:46; We have remained on the surface this morning and we are getting ready for the BBC 12-o'clock news. I have already eaten an early lunch of smoked trout and lemon, with soured cabbage.

13:02; We monitored the news and it was tremendous. Paris is in a state of panic, people are fleeing and the government has become even more fractured. The BEF is still being driven west.

KTB Log May 17, 1940 13:03; The funker, Karl, was gotten my attention. 01:07 Karl has picked up a British Naval Base, it fades in and out, but he heard the name Base 20 mentioned twice in the past half-hour. He is continuing his monitoring. I and the NO and CE will look at some of our charts and see how far we are from Naval base 20, *U-49* is currently in Grid AN81.

KTB Log May 17, 1940 13:42; The crew saw me with the charts and were wondering if BdU sent us a contact. I laughed, and said they still haven't sent word about Holland, let alone sending a contact report. I tell them that Karl is picking up British communications occasionally and we might do a little searching around a naval base. Some crewmen smile at the chance to have a look around a British base, others fear the worst. I tell them the storm will give us cover, and if the water is very shallow, we will sail slowly and see how tight the defenses around the base are. I don't know if our boot can cut its way through a net … I am guessing no. I also lack any first-hand experience with mines. We will have to be very cautious.

KTB Log May 17, 1940 13:43; *U-49* is approximately 260 km from

Naval Base 20, about 10 to 15 hours sailing.

(Personal notes) May 17, 1940; I tell Karl to play my 1930s music collection from Amerkia. My Onkel lives in New York and used to send me records. Like my favorite *Beyond the Blue Horizon* and *Who* by Goerge Olsen; the Boswell sisters singing, *Life is Just a Bowl of Cherries* and the flip side *We're in the Money*. Eddie Cantor, Duke Ellington, Hoagy Carmichael ... I love music from the 1920s and 30s. Our boot is full of good decadent music from the States. I don't know what most of the words mean, but the swinging tempo is what appeals to me. I find it strange that such happy music was made during the Great Depression in Amerika. The music begins to fill the boot and all hands are smiling and we are sailing in style now, great tunes, great crew, great food ... now all we need are some great targets!

KTB Log May 17, 1940 20:36; I have just returned from the bridge and the rain has ceased. There is a light fog, but the sky is clear. The storm must have broken up fast, because it was pouring 26 minutes ago. That makes for weather changes in real time and I wasn't on the bridge for even one of them. I told the WO and the CE to take charge if any warships are sighted. I gave each of them a standing order to dive to 15 meters, silent run, and ahead two knots. Have someone wake me, but don't wait. We are in shallow water, 24 meters deep, approximately 8 km from England's coastline. We will pass within 4 km over the next four hours.

Hello, this is where I had to save and exit the game. I had to work this morning, and *U-49* is in too shallow water to leave the boot sailing unattended. But anyhow, I am back in the boot, and we have resumed sailing.

KTB Log May 18, 1940 00:07; I am at my cubby eating a lemon. I think I am catching a cold, but Onkel Willie wouldn't make me a "hot cough medicine", he said no whisky for the skipper. So he has me eating lemons and drinking tea with honey! Now I feel more like a Celt, I wish Grandfather were alive to hear of the troubles England is in now.

(Personal notes) May 18, 1940; Thinking of my Grandfather has made me sad. He was murdered in 1916. He took part in the "Easter Rising". I recall hearing from Grandmother that it was one week of mild street fighting in Dublin. But the heavy-handed heathen British murdered the leaders and others who they suspected of taking part. Grandfather was tried by a kangaroo court, found guilty and shot at sunrise the day after the trial. I will paint under our four-leaf clover "Sinn Fein", which translates to "Ourselves Alone". That is what we are on this boot, "ourselves alone".

KTB Log May 18, 1940 00:34; SO reports a warship moving fast, moving away, bearing 122. That's good. Twilight has already arrived. I will have to go topside and help with the watch. We will stay on the surface for a bit of sailing. We are now very close to shore, about 3 km.

00:50; I am on watch and I am looking with my binoculars through the fog at what I think is a lighthouse on the coast. I am watching it for a few minutes — Oh, mein Gott, that's not a lighthouse light, it's a searchlight on a warship. I yell to the watch, "Schnell, schnell, hinunter die Luke." I yell through the speaking tube as the first of the watch has already descended the ladder, "ahead slow 15 meters, rig for silent running, don't wait for the hatch, take us down fast, but don't crash us into the bottom! They haven't seen us yet, I am on my way down." I don't so much as climb down as jump down. A maat is still on the ladder, off to the side so I can drop in fast and he helps me get the hatch secured, but not before cold sea water starts pouring in on us. I can't help it but I am laughing, and the maat had a big grin on his face. We came seconds from having the warship see us as he broke from the fog bank.

We have two warship sound contacts, the one to aft is fading, the one to our starboard is at 101 and is fading out also. Thank goodness. We will run silent, slow, and shallow for the rest of the daylight hours. We are still a few hours from the area where I suspect nets or mines might be placed. I will have to man the observation scope carefully at that time. I wish I had better charts. I bet that Columbus had better charts when he was sailing for the glorious Queen of Spain, in search of new sea routes to the Far East.

KTB Log May 18, 1940 05:28; We are about 2500 meters from Naval Base 20 area. There are no warships, but there are about 30 to 40 ships (my frame rate is always 100-170, now it is barely 30, ouch). Most of the ships are stationary and already on fire. Dozen of C-3 cargo merchants, one passenger liner, another two dozen or so small merchants. We are sailing slowly through the pack. We will call in a contact report and see if the Luftwaffe comes back and finishes the job, once we are through the pack, that is.

KTB Log May 18, 1940 05:29; SO reports another huge pack of ships. Looks like 30 more, about 2000 meters SSE from our current location. Of all times to have the deck gun unusable.

KTB Log May 18, 1940 05:53; *U-49* is on the surface approximately 2000 meters SSE of the pack of ships. The RO is sending his contact report now. We will then submerge to 12 meters deep, reposition the boot and wait to see if the Luftwaffe comes.

KTB Log May 18, 1940 07:02; We can hear planes, I poke the observation scope through the surface of the sea. The waves are occasionally obscuring the view. But I can see well enough to have sighted 15 to 20 dive-bombers. They have already begun bombing runs and the water is echoing with the percussion of 250kg and 500kg bombs. I can see them score a few hits, but mostly they seem to be missing their targets. It looks like they are skip bombing instead of dive-bombing.

KTB Log May 18, 1940 07:04; Dummies! Goerring's boys nearly got us, too! Mental note to myself: 2000 meters is not far enough away when the Luftwaffe comes to pay a visit.

KTB Log May 18, 1940 08:09; We can no longer hear the explosions and I cannot spot the planes anymore, it seems like the air raid is over. They sunk two ships and have badly damaged seven more. Now most of the ships are either burning from the first raid or have now been hit. This is one fleet that will never get out of base. We plot a new course to investigate the second pack of ships.

KTB Log May 18, 1940 08:58; We are on the surface and we see 30 more ships, a mix of C3s and small merchants. Again, some ships are already burning. We send in a contact report.

KTB Log May 18, 1940 11:17; The Luftwaffe has arrived on the scene. There are 16 Stuka dive-bombers in the formation, and they are aligning for their attack run. We are at the surface, 4500 meters away, so we have a good view of the action. Again the planes seem like they are skip bombing. It doesn't look like they are waiting till the ship is visible through the observation window at the pilot's feet before diving and letting loose their bombs.

KTB Log May 18, 1940 11:21; I can see them scoring hits, sending cargo flying into the air and igniting fires on deck. But mostly I see them missing, sending huge geysers of seawater skyward. Some of the ships that have been struck have begun to list over, and one that was previously burning is going bow up. Black smoke is now coming off at least 2 dozen ships. What a great day this has turned out to be. We helped in sinking ships and we didn't have to fire a shot. This is our first patrol where we saw the enemy engaged by someone other than ourselves.

KTB Log May 18, 1940 18:50; We have been dodging warships for the past five hours. We haven't been pinged, but they have forced us to run at 2 knots rigged for silent running. We all want to get past these patrol ships and be free of this bay. We are only 16 km from the area where the air raid was. We still have 50 km to be out of the bay. About an hour after the raid the WO spotted a ship (V&W DD), we dove and since then it has been SO reports warship contacts, some closing, some fading. He is getting tired as he waves his hand out of his radio room with each contact. We are all talking in hushed tones, and the time is moving very slowly. I wish Onkel Willie would make some rhubarb strudel.

KTB Log May 18, 1940 22:34; It has been very slow sailing. We have finally separated *U-49* from the warships in the bay. We have plotted our course towards grid AN87.

KTB Log May 19, 1940 00:01; Karl has tuned in the midnight BBC news, and we are all sitting in silence as we try to decipher the words. Over the past few days the crew and I have immersed ourselves in English lessons. Trying to conjugate our verbs, and watch our dangling participles. I am afraid it is a hopeless cause though.

KTB Log May 19, 1940 01:00; We have tuned into a French radio station and they are urging calm, but it sounds as if Paris has already been stricken with panic. German radio is reporting that the French government is no longer organizing the army in the field. It seems like what we couldn't do in our first war for freedom, we will this time.

KTB Log May 19, 1940 01:22; We are at a slow sail, 2 knots, some of the crew is topside getting fresh night air. Smoking is allowed only in the conning tower tonight.

(Personal notes) The seamen's quarters is still except for snoring and the muffled sounds of a wireless playing Maurice Chevalier singing, *Paris, Je t' Aime D'Amour*. Its sad melody has me feeling rather melancholy tonight. I think I will go to the conning tower and have an evening pipe of tobacco.

⊕⊕⊕⊕⊕⊕⊕⊕

KTB Log May 20, 1940 01:50; We are in grid AN87 and we are heading SSW along the Netherlands coast. We are in very shallow water and about 800 meters from the shoreline. Sailing on the surface, ahead slow and rigged for silent running.

KTB Log May 20, 1940 02:38; The watch has heard planes flying very high, heading east to west. I went up on the bridge, but I couldn't see them, I could only hear them. (This was the RAF moving squadrons from Belgium and Holland back to England). On a tactical level from bases in England the RAF will deploy Spitfires and Hurricanes to cover the retreating BEF. They also will provide air cover for British ships as they prepare to sail into the Strait of Dover.

KTB Log May 20, 1940 09:47; I am sitting at my cubby reading *Achtung—Panzer!* I can hear the deck hands shouting excitedly. I wonder to myself if it is a fistfight? Although the shouting is loud, it doesn't sound violent. Jurgen sticks his head into the hatch that separates the command room from my cubby area.

"Schnell, schnell, Kaleun! Chef dort ist etwas, der Sie sehen mussen, Portseite!" said Jurgen, with a big toothy grin.

I entered the command room and the ladder leading up through the conning tower had several hands anxiously trying to get topside. But there were already sailors on the ladder and apparently sailors were also crowding the bridge. I looked over at Heintz, our CE, and he said he wanted to go topside, also. I ask myself just what was going on that all hands wanted to get to the bridge?

"Attention!" I shout. All hands below stand ready, but the ones on the ladder have to be told to climb down. There is much moaning at the order. But I must get up to the bridge myself to investigate the disturbance. As I climb up the ladder I can hear the crew on deck is still yelling wildly.

I have my head above the conning hatchway and I can see 10, maybe 12 of the crew all off to the port side, some are yelling and waving their arms like windmills in a squall. Others are peering through binoculars, as still other sailors are trying to take the binoculars away.

"Attention!" I yell. The crew stands ready, but they are pleading with me to let them stand down. Some can't keep their eyes front, and keep looking over their shoulders.

I step to the rail and raise my binoculars to have a look. I am scanning the nearby shoreline, it looks empty enough. I see rolling dunes in the background, and I am now scanning the beach right to left.

"Oh, mein," I murmur when I see what they see. I order, "All eyes to port, as you were." The crew is again cheering and waving.

Granted we have been at sea for only 10 days on this patrol, but there is something very nice on the beach and unfortunately out of reach. I can see very clearly five voluptuous women, frolicking on the shore. They must be jung and free, to be so at ease in their birthday suits. They are running along the shoreline and splashing in the shallow water.

I now know what the old sea captains must have faced when the sirens start to call. The girls on the beach have spotted us and are now blowing kisses and waving to us. The urge to sail to the shore is rather compelling, as I have become fixated on the women. I must resist the hormonal urge. I might strand the boot on a sandbar if we sail into shore.

One in particular has caught my eye. She is tall, with long blonde hair that blows in the sea breeze as she jumps up and down. Oh, mein, she is a healthy looking girl, big hips and what wonderful mammalian glands she has. I would say that in a hat size they are six and seven eighths.

I tell those on deck to go below and send up the CE along with another group of sailors. I can't deprive those still below from this wonderful spectacle. I will stay up on the bridge to make sure no one deserts. Of course, as long as I am here I might as well look some more.

Oh, what wonderful swimmers we all would be.

If naked women swam upon the sea!

⊕⊕⊕⊕⊕⊕⊕⊕

Ahoy, well I had my first CTD[121] early this morning. I guess it was my fault though. I woke up around 03:00, went to my desk and was checking *U-49*'s progress. I must have been aimlessly tapping a key or a function key on the keyboard. That brought up my desktop to show the "sticky keys" option.

Then when I brought back SH3 from the task bar to the desktop, the game became unstable, the graphics looked blurry, and I could not scroll the map in and out with the mouse wheel. About a minute later — CTD. Which wouldn't be too bad if I had been saving after each hour of gameplay. Unfortunately, the game has run days non-stop without any CTD, so I became lethargic about saving.

I lost over 24 hours of progress. My last save was 00:01 May 20, 1940. I will have to sail once more into the channel. I sailed into it yesterday and last night was uneventful. The patrolling warships made us sail at 2 knots submerged, rigged for silent running. I have plotted the same course, as yesterday so we should be safe enough.

There is a somewhat of a silver lining though. A lady where I work gave me a present, *Das Boot* (Director's Cut). She said that I

---

[121] CTD: Crash to desktop, when the game shuts down unexpectedly.

am too lazy to go pick it up myself (which is true). I am really looking forward to watching it later today, to see what I have imagined is right and what is wrong with how I picture sub life.

Plus it will be nice to get the "black spot" removed from my hand that Subsim member "Blind Pew" gave me for not watching it sooner. I have laundry to do then it's film time for a few hours

⊕⊕⊕⊕⊕⊕⊕⊕

I finished *Das Boot*.

What a very compelling film, on so many levels, the acting, the action, the characters all made me feel for them. I liked the hand-held camera and the overall "feel" of the film. The boot is much more crowded than I pictured it to be, and I had the seamen's quarters wrong in my mind. It was much more cluttered. I didn't know they would eat in there also. I didn't think the officers had a mess table and that the crew would ask for permission to pass.

There are many small details I didn't know also, so I will be watching this film again, but looking at the boot's structure and how the crew worked the dive planes, etc. I didn't know about the bolts breaking and all the damage that happens when the pressure on the hull becomes critical. I can understand the hull "crushing in" under the pressure. But I don't understand why the interior pipes burst under the pressure. Well, back to *U-49's* eighth patrol....

⊕⊕⊕⊕⊕⊕⊕⊕

I wake up and the boot is very quiet. I don't hear the familiar sound of diesel and screw. The boot is at an all stop. No faint sounds of the wireless. No one is talking.

I look over where Karl should be but he isn't at his station, and the hydrophone is also not being manned.

I hop out of my bunk and enter the command room; it is empty, too. I ask myself, where is the Chief, why are there no crew at their stations?

This startles me and I become worried that the boot has been boarded by the Brits. I rush back to my cubby and take my Luger from my desk drawer. I check that it is loaded, and I chamber a round.

"Heintz, Karl, Jurgen, where are you?" I call out as I walk back through the command room and towards the seamen's quarters. The hatchway is open and I can see no one is in there. "Heintz, Gottfried, Jurgen, damn it, where are you?" No answer.

I am getting angry and begin to walk at a much faster pace towards the bow torpedo room. The hatch is closed and I can hear noises coming from inside the torpedo room. Thank goodness I found them I think to myself. I am going to yell bloody murder at them for leaving their posts.

I grasp the hatchway latch and open the door. I yell, "All right, you swineherds, how dare…" and I stop in mid-sentence. There is no one to yell at. The room is empty and the noise I hear is a chain swinging and tapping against the torpedo storage rack. "Damn it," I mumble to myself as I walk aft.

The engine room is still, and as I walk between the motors I touch the port side engine and it is cold, as if it has been off for a long time. I am now walking faster and beginning to feel unnerved. I walk all the way aft and no crew!

Walking slowly back towards the command room, I stop in the engine compartment. I take my cap off and run my fingers over my shorn head. Someone has started playing a record on the gramophone, *Beyond the Blue Horizon*. I rush forward but no one is in the radio room. The record is spinning and the boot is filled with soft music.

I step into the command room and look up the conning tower ladder. I slowly began to climb up it, through the hatch and out onto the bridge.

The sky is gray, and there is a soft cool breeze blowing the salty sea air. I am scanning the sea around the boot on the port side when I hear someone yelling, "Herr Kaleun, help!"

I quickly look aft and there is Onkel Willie, Heintz, and the rest of the crew. They are trying to stay afloat by treading water. As they attempt to swim to the boot I can see terror on their faces as the sea tries to tug them under. Their wool sweaters and their leather jackets have become waterlogged and that adds deadly weight to them. Their cries bring on great anguish in my soul.

Some of the crew has managed to get within about 10 meters aft of the boot. I holster my pistol as I rush down the bridge ladder, my feet clanging loudly on the metal rungs. Out to the aft deck and I run to help pull them aboard.

The first one to reach the boot is Karl. He reaches out his hand to me and our fingers touch and I begin to grasp his hand. His eyes are filled with relief.

Suddenly the diesels roar to life and the propellers began moving the boot quickly forward. I feel his wet hand slip from mine and look into his grief-stricken eyes as the distance between us grows.

"Herr Kaleun, stop! For God's sake, *stop!*" pleads Karl, but the boot is now sailing fast away from my crew.

I run along the deck as fast as I can, bound up the ladder and up on the bridge, I can hear my crew screaming wildly. I grab the engine controls and pull them down to all stop. There is no response! *U-49* is sailing fast, and the crew's cries are becoming faint.

A soft whisper comes from the speaking tube, "Herr Kaleun." I lean my ear closer to listen.

"Herr Kaleun, what have you done to your crew?" whispers a small voice.

I do not answer. I turn around, take two steps and drop down the conning tower ladder. My feet hit the wet control room floor with a slap as I simultaneously draw my Luger.

Standing where the Chief should be is young girl. Her deep blue eyes are friendly and she is smiling very sweetly. She is wearing her Sunday best, a white lace dress with a high collar and broach, white knee socks, polished black leather shoes. Her hair is long, light brown and gently curls around her shoulders. She is twirling a few strands of her hair around the small fingers of her left hand and has an impish grin on her delicate face.

She is giggling as she again asks, "Herr Kaleun, what have you done to your crew?"

I am furious and carefully aim the Luger. I can see the front sight blade right in the center of the rear groove. I can see the part in her hair, now lower and I see right between her blue eyes.

She begins to walk toward me, her heals clicking on the metal deck. She laughingly says, "Herr Kaleun, you could not kill an innocent little girl, could you? I did not abandon your crew. No, Herr Kaleun, it was you who deserted them."

I lower the gun and let it slip from my hand — it lands on the deck with a dangerous clank.

She is standing near me now and I can smell the scent of spring lilacs in her hair.

"Herr Kaleun, take my hand," she reaches her hand out to me.

I hear a voice very close to my ear. "Herr Kaleun, Herr Kaleun, wake up."

KTB Log May 21, 1940 00:02; Karl has awoken me with news from the KDB (German news service) It seems the Brits are starting an offensive in Norway, trying to capture the Narvik area. I told Karl to turn in for some rest and have Jurgen take his place. I am going topside to have a pipe of tobacco and to help with the twilight watch.

KTB Log May 21, 1940 00:37; WO sights two British planes, I order to dive the boot, ahead full and hard to port. I think to myself, why is the RAF flying this early?

00:40; SO reports depth charges in the water. We can hear a loud explosion on our port side, aft. I order hard to starboard. After three minutes we all stop and listen. I poked the observation scope up, but can't see any airplanes. We resurface a quarter of an hour later.

05:58; WO sights aircraft heading toward us fast. I clear the bridge and take a look over my shoulder aft, and I can see a plane setting up for an attack. I drop in quick and yell to the Chief, "Get us under and hard to port, ahead flank," as I spin the hatch wheel tight. I no sooner hit the lower deck that the SO reports depth charges in the water. We hear the percussion of two charges off to our starboard side. If these Brits aren't careful, they are liable to kill some innocent fish, poor little Flipper lives in a sea full of war, dying there under, under the sea....

KTB Log May 21, 1940 12:00; We are listening to the radio for news from the continent. Karl has tuned in the BBC and they report that on May 20, PM Churchill sent an urgent telegram to the U.S. President Roosevelt asking for Amerika to send destroyers.

BBC reports that the Royal Navy has released information regarding their operations near Norway (*HMS Royal Oak* and *HMS Furious* are operating there). BBC reports that in France, Guderian's panzers captured Amiens, then took Abbeville by nightfall of the 20th (from Abberville it is possible to see the ocean and to smell the fresh sea air).

BBC reports the French have sacked Gamlin and replaced him with General Weygand (he will not command well, he is out of touch with what is going on).

KTB log May 21, 1940 13:02; The bloody Brits have forced us to stay under the surface all day. We surface at 13:00, dive at 13:22, planes spotted, surface 14:05, dive from planes at 14:48. I have never seen such heavy RAF air coverage (BdU, I am not making up the part about the RAF. We can't break the surface without being driven back under by the planes. *Bloody hell* to the RAF, and where is the Luftwaffe to splash a few for us?) I will have Karl send in a plea to BdU to contact the fly boys to dog fight the Brits over the Channel, bloody RAF flying unmolested and in big numbers too.

KTB Log May 22, 1940; I awake early and we surface *U-49* 00:01 Current location, grid BF33, just off the Dutch coast. The night air is crisp, but the nights are too short this time of year and dawn will be too soon.

00:23; The RAF is up early, they must be going to the bakery for some scones, and we are forced to submerge. I don't want them to take our strawberry jam. We resurface and we taunt the British, the watch is using their best foul mouth language to vent some frustration.

KTB Log May 22, 1940 02:54; The Royal Air Force has come to say "Hello", but we chided them and submerged. I think they want our kippers! Where is the Luftwaffe? Must be sleeping in the barracks after drinking all night. 02:57 SO can hear two depth charges go off, sounded to me to be in front of us and to starboard. I lean back from my cubby and yell to the Chief, "Hard to port, ahead full." We don't hear any more DCs. We resume our patrol and return to our course.

03:10; I am manning the hydrophone and can hear a very faint sound of a screw at 117 degrees. I listen for a few moments and then we surface and began to try to get nearer for a better sound check. 03:24; Bees have buzzed us again, forcing us under. 03:29; The screw sound is now at 003 degrees and faint, but we are closing the gap. We resurface and, you guessed it, the RAF wants to play at 03:39, but I don't so it's dive time again.

03:52; SO reports warship at 012 closing fast, long range. That's not fair play on the part of the RAF, how dare they report our position, I think they want our glucose water.

KTB Log May 22, 1940 03:56; We rig for silent stalking, everyone grows quiet. I tell them to be ready for battle stations within a half-hour. The line at the head has formed. The torpedo crew has begun moving supplies forward (extra water, kommisbrot, a bucket to urinate in).

03:57; SO is waving his hand to me, he has a huge grin beaming from his face. I think the warship has tacked away from us.

KTB Log May 22, 1940 03:57; SO reports merchant moving slow, closing bearing 012.

04:00; SO reports merchant moving slow, closing bearing 066

04:00; SO reports merchant moving slow, closing bearing 072.

04:00; SO reports merchant moving slow, closing bearing 049

04:09; SO reports merchant moving slow, closing bearing 273

04:09; SO reports merchant moving slow, closing bearing 276

KTB Log May 22, 1940; HOOT! Man, we have come across a convoy, the feeling is exhilarating and invigorating. 04:17 we have plotted our course and we are making *U-49* battle ready. The crew is briskly and with great teamwork getting all stations ready. We will sail to our ambush point and wait for them to come to us.

SO reports merchants, bearing 266, 265, 269, 273, and three warships.

04:47; We are at our ambush point and the game is slowly making its way toward us.

05:10; We have moved approximately 200 meters forward, and we have marked the map with the course of the four columns of merchants. I can see a tribal DD leading the pack and the SO reports a warship aft of the pack, and one to each side of the convoy. I can see several small merchants, several C2 freighters, and I have identified two Type 2 tankers, one tanker is right in front of the other. These will be are primary targets.

KTB Log May 22, 1940 05:27; We sail 12 meters submerged, ahead 1 knot. Our target's course has been marked and we will be moving forward so we can fire from 800 meters. We are already in a position to have two rows pass us aft and two rows pass our bow. The DD out in front is moving slowly and sailing straight forward along its course.

05:37; We are setting the TDC, we will fire tubes one, two, and three at the first tanker. The eels are set impact, five meters, fast. We will fire at roughly 90 degrees, it looks as close to 90 as can be.

Tube four is set magnetic, 10 meters deep, fast. We will fire this at the second tanker.

Tube five is set impact, surface, fast. We will fire this eel at a secondary target (one of the small merchants to our stern).

KTB Log May 22, 1940 05:42; The tension has built over the past hour, the DD has slowly gone past and we are finalizing the TDC data. We have adjusted the position of *U-49*. I order the crew to open the tube doors. Our targets are very near and the leading ships have already begun to pass by. You can hear the hum of the engines through our hull and it adds a strange background noise to our hushed talking.

0546; I can see the tanker and we will be in a near perfect firing position.

05:51; The crosshairs have been positioned on her amidships just in front of the bridge area. The ship looks very large this close up and I feel we can't miss at this range. I see a boatswain coiling a line right above the point I am aiming at.

05:51; WE is given the order to fire tubes 1, 2, and 3. The compressed air sends our eels on their course. The path is straight and true. Twenty-seven seconds later I see the first one hit, quickly followed by number two. There is a huge splash of water and the tanker is rocked hard by the sudden impact. Number three hits and the ship is doomed. It explodes, spilling oil that becomes a burning sea of death. The crew that tries to get off are covered in the oil and drown or burn. I turn the attack scope to aft and begin to target a C2. WE reports at 05:51 enemy unit destroyed.

05:56; I give the command to open tube five, and we fire at 05:56. I have targeted the C2. I swing the scope back toward the second tanker. 05:57 WE reports an impact, and seconds later reports EUD, C2 sunk, grid BF33.

The second tanker is slowly tacking (all ships are now tacking). I have a good angle for the magnetic eel and have aimed for its bow in hopes that the eel swims right under its keel. 06:02 I give the order to open the tube door number 4 and moments later, the ship in position, I order *fire*. 06:03 torpedo impact. The ship is listing slightly to starboard, but she is keeping up steam. The crew is fighting the deck fire.

Our crew is reloading the torpedoes and we are sailing along with the convoy, the warships haven't closed in toward the pack so I feel we are safe enough. We hope to fire one more time at the burning tanker. The tanker is lagging behind and we are already out in front of it. At 06:28 we are ready to fire an impact, 5 meters, at 525 meters, 90 degrees more or less.

The number four-tube door is cranked open and I have the crosshairs aimed right in front of the bridge area. 06:28 I give the order to fire, seconds later there is an impact and we all breathe a sigh of relief. We never cheer on the boot. When we hit or sink a ship, it is just a relief to have done our job correctly and with deadly accuracy.

KTB Log May 22, 1940; WE reports EUD at 06:30, T2 tanker sunk. We rig for silent stalking and we disengage the convoy. I order ahead slow and go to 40 meters.

06:40; Onkel Willie is serving a late breakfast of cold kippers in soured cream with onions, and he is baking rhubarb pies for later.

KTB Log May 22, 1940 07:00; We rise to periscope depth and I see the convoy has sailed about 8500 meters from us, so we slowly surface and begin to follow.

07:15; I have Karl send a contact report on the convoy and a patrol report for *U-49*. We are 7000 meters back of the pack, Grid BF33.

Report: two tankers sunk, one cargo ship, total tonnage 28,190. Eight torpedoes left. Fuel 60 percent, soured cabbage is running low, please send more cabbage or 56 sexy nurses who are in heat … " Helloooooo Nurse".

07:26; BdU acknowledges our report.

08:29; The glorious men of our Luftwaffe have arrived! I can see a large formation of Ju88s level bombers. They are already setting up their attack run. Through my binoculars I can see them scoring hits. A ship is burning and they have scored two more hits. This is an awesome display of airpower (ps: this is not embellished). The raid was brief, maybe two minutes. But when I check the map I see they have sunk three ships! Gott bless the wonderful airmen!

(Personal notes) 12:07 May 22, 1940; This was a great hunt, the long days of searching were rewarded with a very satisfying harvest. We took out the vital tankers and sank a cargo ship and the Luftwaffe sank three more. We found a convoy and we engaged it and we escaped without any real danger from the warship escorts. They never came into the convoy after the first explosion.

13:16; WO reports aircraft. I am on the bridge and I order to clear the bridge. I take one last look and I see a few Brits flying towards us. I yell, "Ahead flank, 25 meters, hard to starboard."

13:30; SO reports depth charge in the water and we hear several loud explosions aft of our boot. In the process, we lose the convoy.

⊕⊕⊕⊕⊕⊕⊕⊕

Ahoy, this is a long update, but it was also the most fun I have had on any patrol. I feel that this is what the game does best, it gives the player a chance to "hunt" with whatever tactics the "skipper" wants to use. I like safe and slow, others might have used the deck gun (the sea was like glass). Plus I like to use the WE and

WO when it comes to helping set TDC input data, after all, I like to be a tactician not a mathematician. But I wish I'd kept the free camera enabled, I would like to see some of the action from something other than the periscope.

⊕⊕⊕⊕⊕⊕⊕⊕

KTB Log May 23, 1940 00:24; *U-49* is bound for the St. Mere-Eglise Naval Base in Grid BF35. Our goal is to be there by evening.

00:54; WO reports RAF heading our way. I am below deck and yell to them to get off the bridge and secure the hatch. We go to 15 meters and we don't hear any explosions. At 00:55 all stop and the SO is checking for screws. He nods that he can hear a faint merchant sound moving fast, moving away, bearing 149.

02:26; The sonar operator has informed me that a warship is closing slowly at 010 degrees. I order to surface and we will cruse at a very slow pace towards the ship.

04:48; WO reports ship at 007. We sail slowly closer and then we stop and wait to see what type of ship it is.

05:07; We have identified a small armed trawler. I order the bridge secured and then we head below deck and secure the hatches. I order Karl to send in a contact report. I tell the Chief to take us down to periscope, and hope the trawler hasn't seen us.

05:20; We saw the ship's flag as it passed, it's a French ship the *Bon Zou-Zou*, it looks like it was a private vessel at one time.

KTB Log May 23, 05:42 1940; We are following the trawler at about 3000 meters. We surface and we are sailing at a pace to keep our distance. The Luftwaffe is on the ball this morning and have sent two Ju88 level bombers. I am watching from the bridge and I can see the planes circle and align for their bombing run. They approach the trawler from his starboard side. They let loose the bombs, but unfortunately they miss.

But the gunners on the trawler do not miss! Oh, mein Gott, one of our gallant airmen is hit. I can see the AA has shot off his port wing. He is corkscrewing towards the sea. His crippled plane is leaving a smoky trail and I can see fire burning on his fuselage and what is left of his wing.

Those of us on the deck begin to scream for the pilot to bail out, other crewmen from below are screaming, too (we always keep the crew below deck fully informed of topside situations).

But he doesn't make it, the plane hits the water on a fairly flat trajectory, and it skips along the surface for a moment and then flips over violently, shedding the remaining wing, and it then slips out of my view under the water. He might have died before he hit the water. Too bad, we might have been able to get to him before the trawler did.

I was angry and wanted to open up with our deck gun to avenge the killing of our pilot, but I refrained from the temptation to get the French ship. I had Karl report to BdU about the crash and death of the pilot.

I need to break this sad mood on the boot, so later we will listen to the radio. Yesterday was too busy for us to stop and hear the news or even music. I think I will have Karl play some soft music.

KTB Log May 23, 1940 12:00; Karl has tuned in the BBC. The British Parliament has enacted a "War Powers" law giving the government control of all business and personal property. Petrol prices have been raised to one shilling and eleven-halfpenny a liter.

The news from France is great, the BBC reports that the 1st Panzer Divisions have arrived at Calais and (Guderian would be in Boulonge with the 2nd Panzer Division. There has been tremendous fighting at Samer and Desvers, but the French and British counter attacks aren't able to slow the advance. Guderian will be able to close the pocket with his fresh 10th Division; they are resting, refitting, and refueling).

(Personal notes) I feel nothing can stop Guderian's troops from killing or capturing the whole BEF. The BEF has run out of places to retreat to. Dunkerque it is the only port area left. Only that murderous leader of ours could foul this up. News from Belgium is that we have driven the Army to the Lys River. The Belgium army is putting up a tremendous fight.

KTB Log May 23, 1940 13:05; WO reports four Ju-88 level bombers heading NNW. We on deck scream our best wishes, and we salute the wonderful Luftwaffe.

14:36; WO reports planes, this time it's the RAF. We dive to 12 meters, and at 14:37 we hear two explosions aft, followed moments later by the percussion of two more bombs. We will resurface in a while.

KTB Log May 23, 1940 14:50; WO reports airplanes closing fast. I am on the bridge and can see a whole hive of bees coming for our honey. I yell to get the deck cleared and dive to 35 meters, hard to port, ahead flank.

SO reports DC in the water. It sounded very near our starboard side. The crew has become rather tense. I have never seen them so tense and that has me nervous, too. I wish they would stand at stations properly.

SO reports depth charges in the water, he repeats this six times and suddenly we hear explosions all around the boot. I order rudder amidships.

14:52; SO reports more DCs in the water and a second attack wave has begun. I have counted at least 16 explosions. Damn RAF, they sure had ammo to spare. It was rather a bit of a "popper", reminded me of popcorn hitting the side of a kettle.

The raid ended at 14:53.

KTB Log May 23, 1940 16:50; We have been dogged by the RAF a few more times but now the darkness shields us from the pesky bees. We are about 6 to 8 hours from St. Mere-Eglise Naval Base, approximately 125 km WSW. We are sailing ahead full, and might run at flank speed later. We will then patrol up the coast heading towards the Dover Strait.

21:16; We are 1 km from the base area, but we have no contacts.

KTB Log May 24, 1940 00:05; *U-49* is patrolling the French coast heading NNE. I will be going topside to help with the watch.

23:34; We have 15 to 25 km left to sail to be past the British patrol ships we encountered on our way into Der Kanaal. We marked our chart on the way in so I am confident we can sail safely past the warships.

(Personal notes) I am at my cubby having a late night cup of fish head chowder. I have just returned from the bridge and the rain is steady and twilight is drawing near. Heintz and Jurgin are sitting at the officers' mess table finishing a game of chess. Heintz is reigning chess champion on the boot. Most of the crew still prefers "Battling Ships." Backgammon is also popular, and some play cards when they are off-duty. The crew is on duty for four hours then they are off duty for eight hours. The crew has a lot of free time, some sleep much of the time, others like to read.

The funker, Karl, will be reporting for duty soon, and we will see what the news broadcasts have to say.

(Personal notes) May 25, 1940 00:10; We have the BBC news on and it seems Churchill was in Paris on the May 23, trying to hold together what remains of the French government. The tone of things makes me think the French will be governing in exile very soon. They also report the loss of several French ships in Der Kanaal on May 24, The *Jaguar*, *L'Adroit*, *Chacal* and the *Orage* all were DD type ships sunk. The Royal Navy has lost the destroyer *HMS Wessex*, the Luftwaffe repeatedly bombed and strafed it off Calais on the May 24.

Karl has received a very rare message from BdU from the OKW; it is titled Führer Directive #13.

Our next goal is to hold the Channel coast and to wipe out any troops trying to escape back to England. *U-49* is told to patrol NNE of Calais to Dunkerque. We are told to watch out because the Luftwaffe will be bombing and patrolling the Channel, and lower Nord Sea area. We are warned they will be shooting anything floating, and will likely attack us if they see us.

02:07; We had an all stop and the whole crew was assembled and told of our orders. There was some groaning. I don't like the idea of patrolling the coast. The water is too shallow and when the Luftwaffe starts dropping bombs, the bombs won't care if this hull was laid down in Germany or not. I told them we might be able to sink a troop transport ship, so the risk will be worth it.

04:21; Mars, why do you give Poseidon water? He does not want rain today. He wants sunshine and lollipops, rainbows and happiness. We want sunshine and torpedoes, targets, and sunken ships.

⊕⊕⊕⊕⊕⊕⊕⊕

Ahoy, a brief timeout from the log. Well, how are you? I sincerely hope you are feeling well, and your family is also in good health. Everything is shipshape here. I am waiting to see if the WorldMod has a large rescue fleet. Last night was the first night I saved and exited before lights out at 21:00. Taps was played and it was time for me to go to sleep. It was odd not hearing the ship sailing, or the diesels beating. I also didn't let the boot sail while I was at work.

⊕⊕⊕⊕⊕⊕⊕⊕

KTB Log May 25, 1940 06:00; Gottfried reported for morning sick call. He has a knuckle buster, he was turning a heavy bolt and the wrench slipped and he smashed his right hand. His hand and fingers have swollen and he has deep bruising. Willie has had a hard time washing Gottfried's hands. Gottfried is always working on the diesels and his hands are black with grime and oil. Willie is soaking Gottfried's hand in hot water and Epson salts. He will be relieved from duty, but he will still go to the engine compartment anyway. He is always working on his engines.

It is hard for me to believe he is the same officer I met a few weeks ago at the barracks. He was clean in his perfectly pressed uniform, clean-shaven face, the perfect Naval Officer. Now he looks more like a Viking, his hair is matted and his beard is tangled. In his apron he would be at home at a fiery forge. Hammering glowing steel into swords and battle-axes.

KTB Log May 25, 1940 12:34; *U-49* has had no contacts as of yet. The rain is steady and the fog is as thick as London pea soup. We have just finished listening to the BBC and we will tune in a German broadcast from France soon. The BBC reported that in Belgium the Hun has broken the front at Lys. The Germans are driving towards Ypres, the British plan to counter the move. But by the end of the day the British field leaders have done nothing. Belgium is quickly being crushed under our boot heel. The BBC has reported the total fall of Boulogne.

But in a brazen tone of contempt, Dave Wademon reported on a Luftwaffe raid: "The bomb was at Wickford, Essex at 01:55 hours, the dastardly attack by the Luftwaffe claimed the lives of two chickens when a hen house has struck."

KTB Log May 25, 1940 16:00; The rain is making our patrol safe from the RAF and Luftwaffe, but it's making the British safe, also. We have spent most of today using the batteries to patrol. It has been a slow day.

KTB Log May 26, 1940 03:35; Karl has awoken me and we are listening to a civilian radio frequency out of England. The British Admiralty has put forth a proclamation to all ship owners, to sail with great haste to Dunkerque France. (Operation Dynamo has begun!)

KTB Log May 26, 1940 04:22; All hands on duty are told of the situation. The ones resting will need the sleep so they will be told later.

U-49 is getting ready for a hunt. Battle stations are being readied and all loose items must be properly stowed by 12:00. (Heintz will take care of Command and Adolph will do the torpedo rooms). All garbage must be jettisoned by 12:00 (Willie is in charge of trim). All waffen in the conning tower need to be field striped, cleaned and oiled by 12:00 (August, waffen and ammo). All tools need to be made ready; all repair supplies must be made ready by 14:00 (Gottfried, engines, and Berthold, repair). All hatches must be inspected by 14:30 (Wolf, hatchways). All pumps must be inspected by 14:30 (Frantz, pumps and gauges).

KTB Log May 26, 1940 04:24; NA reports heavy rain and heavy fog, wind 8 mps. Someone is praying for rain and it is not me.

(Personal notes) Karl and I have been talking about the radio call that keeps repeating. They are clearly ordering all ships at sea and all ships at harbor to set sail immediately to Dunkerque. I estimate the fastest ships will be near very soon. There are many fast yachts that will already be sailing. Some from harbors in France as well as England.

14:56; WO reports sails and smoke approaching.

KTB Log May 26, 1940 17:00; We witnessed many ships and boats, but we were unable to engage the enemy because of heavy seas.

(Personal notes) May 26, 1940; *U-49* is approximately 35 km off the Calais, Dunkerque coast. The crew and I were witness to an amazing spectacle. When the WO called me topside, it was rainy but we could see several large sailboats. Their sails were trimmed and the wind was blowing from their aft, they were doing 15 knots. These first ships we saw were private yachts. Later we saw a flotilla of fishing craft, we saw tugs, skiffs, dories, ferries, large ships and small boats.

I watched through my binoculars one small boat very closely, named *Lone Tree*. It was an old coastal fishing boat, its white cabin and bridge sat upon a faded red hull. It was rigged with nets and gulls circled as if the two men on board were doing their day's fishing.

These men weren't fishing today. The old man at the bow held a red lantern to mark their way. The red light was like a glow of determination from the old man's soul as he lit their way through the gray fog. He looked like a typical seaman, big rain poncho, and large rain hat, thick mustache and beard. He was leaning forward as if to will the boat onward. His gaze was unflinching as he held out the lantern.

I would guess it was his son at the helm, a veteran of the War to End All Wars. The skipper was probably going to Dunkerque to save his own son. Two generations of Englanders sailing to save the next. They will have to face a firestorm of shells, bombs, and bullets, and I think they know that. These sailors, these fathers, brothers, onkles, and strangers, have all heeded the distress call. It doesn't matter the reason, if it's patriotism or to save a loved one or to save a stranger, they are all risking their lives to save others.

I will carry out my orders, but I have no zeal for this war today. I don't even know why I am fighting. This isn't restoring what was "stolen from us by blind hate". I will fight to stay alive and to keep my crew alive.

(Personal notes) May 26, 1940; Tonight we listened to the BBC. In a strong Welsh accent and in a very calm voice a man spoke:

"This is Alexander Webber at the dunes of Malo-les–Bains. I arrived here several hours ago. My journey here was fraught with danger every step of the way. I became separated from the CO and have hiked and at times had to run to stay away from the panzers and their infernal infantry. I traveled at times with a group of Canadians mixed with several French troops; the mix was always changing. Everyone offered help if you began to fall back, no matter what division or nation you where from. We all just wanted to make it to the coast, and to stay one step away from the Germans. Some of the soldiers who have been here for three days have informed me that there is no food, and very little water.

"The sergeants move along the troops offering encouragement. They have nothing else to offer. They have told me ships are coming to take us off the beach and dunes. I cannot imagine there are enough ships to get us all out of here before the Germans finish the attack.

"There is a feeling of dread. We all know the panzers are at the Aa canal; they must be waiting till dawn to advance and annihilate us.

"I can see one area of low water where the engineers have been taking automobiles and trucks and driving and dragging them to make platforms for soldiers to walk upon to reach the deeper water, where ships with a deep draft will have to stop.

"A very heavy fog has hindered the Luftwaffe, but we can be assured they will be attacking as we board and once we set sail from here. I have been living no longer by the day or week, no, I have been living one moment to the next. I have seen too many soldiers die these past few days and I just want to live one more minute. If tomorrow comes for us I will, God willing, be able to report. God speed, this is Alexander Webber."

(Personal notes) May 26, 1940 17:40; After evening mess I took my fork and broke off the middle tine to make it look like a small trident. I went back to my cubby and began to pray and to ask forgiveness from Mars. He tried everything in his power to keep me from this battle and I ignored his warnings. He wanted me to go to the sea near Iceland, but I disobeyed. He gave us two tankers and a cargo ship as trophies of our hunt. I should have been satisfied and returned to port. Now he will have to guide us as the RAF and the Luftwaffe will both be menacing us.

I will take the trident topside and say prayers to Poseidon and Neptune, and toss my humble tribute into the abyss.

KTB May 27, 1940 05:57; *U-49* is 1000 meters from the coast, heading NNE along the French coast. The rain has stopped and the sea is calm, but there is a heavy morning fog.

At 07:11 we plot a course SSW back along the coast. We sailed to approximately even with the Belgium-France border.

13:00; Karl has a French news broadcast on and they are reporting that hundreds of ships and boats are in the Dunkerque harbor. They report stated the Luftwaffe is bombing and strafing the harbor and beaches.

KTB Log May 28, 1940; A little after midnight we heard wonderful news that Belgium has surrendered. The Belgium Army could no longer sustain the losses. At midnight the truce is signed and the guns fall silent in another country.

KTB Log May 28,1940 02:34; *U-49* has sailed all the way to Ostend Belgium. We have plotted a return course back along the coast.

KTB Log May 28, 1940 04:11; Hoot, man! The rain has stopped and the fog has lifted. At least the WO and I can now see more than 10 meters in front of our boot.

I was on the bridge singing an old Amerikan song from the Great Depression, "Happy days are here again. Das skies are blue und clear again. So let us sing a song of cheer again, happy days are here again" the rest of the watch chimed in and in our best-broken English, we sang.

KTB Log May 28, 1940 04:31; WO yells, *two planes approaching fast, bearing 016!* I yell to clear the deck and to start getting us under. I am last off the bridge, but I can't see any planes, I just see many seagulls. We are 2,800 meters from shore in very shallow water and about 60 km from the Dunkerque area.

04:32; SO reports DCs in the water, we hear one explosion to our aft. I push the observation scope up just in time to catch a glimpse of a plane as he passes over us. To me, it looks like it is wearing a large cross, not a circle.

KTB Log May 28, 1940 04:35; Why did they not issue my boot a flag? I need to be able to identify *U-49* as German. If we had paint, I would paint a red field, a white circle, and a broken cross, on our deck. I feel the Luftwaffe is just as dangerous to our boot as the bloody RAF.

04:44; we resurface and continue down the coast.

05:30; WO reports aircraft approaching and we dive the boot. This time we could hear four explosions. We resurfaced at 05:40

KTB Log May 28, 1940 06:47; WO reports aircraft bearing 025 and 021. We dive and hear one explosion at 06:48. A second explosion at 06:50 startles me. Sounded close to our port side aft. Darn RAF made me spill my breakfast all over the table. I guess they are mad because we haven't been sharing our uber mess. We will resurface soon.

KTB Log May 28, 1940; RAF, let me eat in peace! 07:03 WO yelled, aircraft approaching fast, bearing 027, 034. We dove and this time almost as soon as we'd started our dive we could hear several explosions. I could not tell where they were but they sounded too close for comfort. We will wait a quarter of an hour before resurfacing.

KTB Log May 28, 1940; 10:02; Karl has received an update from OKW and it is more detail about Führer Directive #13 (leave it to BdU to forget to tell me.)

The limited use of weapons in the area around England and France for U-boots is canceled. Skl. orders the full use of weapons against darkened ships, tankers, and Greek ships is expanded, except a broad area south of the Irish Freestate. The Skl. tells Hitler that mistakes can happen and U.S., Irish, Italian, ships might be sunk. Skl. Chef says that neutrals and allies need to be informed about the enlargement of the war zone. Skl. Chef also said in the directive the joining of forces of Schiff 33 with a U-Boot.

(Personal notes) And on May 27 1940 enemy U-boots sighted in the Skudesnes-Fjord. The British H.M.U-boot the smelly old *Salmon* and the crapper *Snapper* are sailing around Norway.

The Irish Freestate. Thank Gott I don't have to fire upon any Irish! I sure respect the Irish. They love football just like we do. They like ale just like we do. They like dancing just like we do. And what sailor could resist a fine bonny lassie named Meagan O'Sofine, as she stands at the pier when you dock. Her curly red hair is a shimmering tassel of henna and sunshine. Her rosy cheeks and her pink lips that form into a warm smile.

KTB Log May 28,1940; 11:10 all is well. So far today our only threats have come from either the RAF or the Luftwaffe.

⊕⊕⊕⊕⊕⊕⊕⊕

Hail and victory! I am behind one day in game sailing. Yesterday when I came home from work, I mowed my lawn and that took two hours. Then I helped my brother with a small plumbing job and that also took a few hours. Then after dinner I started playing Viet Cong, and next thing I knew it was already past midnight.

This morning I woke up and resumed sailing.

⊕⊕⊕⊕⊕⊕⊕⊕

KTB Log May 28, 1940; At 12:13; WO reports land units spotted, bearing 010. Land units? I didn't know there were any. I train my binoculars on the coast. I see something I think is a panzer. But it is too big. I am able to identify concrete bunkers. They have a large cannon protruding from them; it could be an 88. I wonder if they are functioning. (These weren't here before, the Wehrmacht has great engineers and sappers!)

13:04; WO shouts, aircraft approaching fast! We are in very shallow water, so I have the WO ready the ack-ack. Arend yelled through the speaking tube to send Falke topside on the double. I climb down into the conning tower hatch and help Falke on with his vest and helmet.

Fritz Falke scurries up the ladder. Arend has prepared the AA and the first clip is being loaded just as Fritz and the planes arrive.

I was topside with my binoculars and had a good view of two RAF aircraft closing fast at about 144 degrees. I yell to the WO to open fire.

WO yelled to Fritz, "Fire at will!"

Fritz shouted back, "Which one is Will?"

I yelled, "Stop 'fudging' around and fire that ack-ack!"

The first plane dropped his bomb about 30 meters to our starboard side. It exploded, sending a fine mist of water our way.

(That is when out of game instinct I pressed the "thumb mouse button", I always map that for ducking. I saw the WO was ducked down, but I couldn't duck).

Fritz missed the first plane, but golly he hit the wing of the second. Smoke was pouring off the port wing. The RAF plane flew inland and crashed just beyond the beach. Our soldiers who were manning the bunkers ran to the site. I did not see a parachute so the pilot was probably killed in action. What a sight it made as it hit the ground, it looked as if a springtime flower of fire had suddenly sprouted.

It left a smoking crater on my mind, and I was blown away. The RAF came by and tried to bomb me for smilin' on a cloudy day.

That was *U-49*s first RAF kill, and Fritz has earned himself a special lunch that Onkel Willie is preparing, as well as a medal, if Gott willing, we return to port.

13:49; We identify five concrete bunkers. They are in Grid BF33, on the French coast, approximately 111 degrees NNE from Dover.

U-49 will sail into the channel, but we will stay away from the known patrol area of the British warships. Although I would like to lure a warship near the coastal gun battery to see if they open up on the warship.

KTB Log May 28, 1940; We will patrol the inner channel today and make our way back up the coast sometime late tonight.

14:07; Lunch mess is being served. Fritz is having his hunter feast. Willie baked a small loaf of fresh bread just for Fritz. The bread smells delicious. Willie also made him a sausage potpie, with cream and egg noodles. The crew and I have smoked trout and the usual vegetable and fruit fare.

(Personal notes) Willie knew how we would all want some fresh bread, but there isn't enough yeast for 50 loaves of bread. But he did bake us all each our own pretzel. We are going to have them as a special desert, with mustard and I will open a case of ale and the crew can each have half a pint of warm ale to go with the pretzels.

KTB Log May 28, 1940 14:40; We had hardly sailed more than 15 km into the Channel proper, when the SO reported warship closing fast at 334 degrees. I quickly ordered the boot turned around and ahead flank speed. Our plan is to get back to the area where the coastal bunkers are. We are only 20 to 30 minutes from the shelter of the guns.

14:44; *U-49* is about 600 meters from the shore and we are in the middle of the gun emplacements.

14:56; We see the unknown warship. I have Karl send the radio report, and he will send a second once we identify the ship.

15:11; We can see a DD, type V&W, we send in the report.

KTB Log May 28, 1940 15:19; The DD isn't cooperating, he isn't sailing close enough to us and the guns. I want to lure him near, he is 3,330 meters to our port side, not yet abeam with us. I order the NO to ping the bottom. He does it but the DD doesn't alter his course. I have the NO ping six times in a row, but the DD just keeps sailing by.

When he turns it is 90 degrees away from us, heading back towards Dover. Now I'm anxious to see the guns fire, so I order the deck gun manned. I man it myself and I fired a HE shell — BOOM – the cannon roared! I see what looks like a miss. I order to dive the boot fast! I am shaking with adrenaline as we dive under and brace ourselves for trouble.

But trouble never came, I pushed the observation scope up into the early evening air. DD was getting smaller. He did not even turn after I fired at him. I was really hoping the coastal guns would fire on the DD. The DD passed within 7000 meters of the big guns. The gun crew must have been having their evening mess, and had no time to help us.

21:30; The rest of the evening was uneventful for *U-49*. After we sailed past the coastal guns we continued into Grid AN87 and have begun to sail back ESE towards the Belgium-France border.

Karl has tuned in the BBC and we are all listening with great interest to the news reports. There is huge news out of England. The "Apostles of Appeasement" almost won, and England would have entered negotiations on surrender terms! But Churchill calls for a temporary adjournment and meets without the war council, and rallies the other Parliament members, that England and he plan to fight and not even think of surrendering. It works and when Parliament is reconvened on the evening of the 28th, Churchill has his way and there will be no surrender.

(Personal notes) I lay down for a bit of sleep, and I had a strange dream. I was walking down the road that leads out of Wilhelmshaven, there was a warm sea breeze, and it felt good to be back on shore. I walked along and up the bluff and instead of seeing the open country, I saw endless fields covered in red poppies, which swayed gently in the warm breeze.

I could see countless grave markers. I saw Gaelic crosses made of marble and large monuments of stone, Crucifixes, Stars of David, and Crescent Moons, I saw oriental prayer flags and incense burning as prayer offerings.

The air was filled with sandalwood and frankincense, but my heart felt the heavy weight of grief. I saw a procession of people. They were silent, and all I could hear was the tramping of their feet. I could see men in all types of uniforms, and all nationalities. I could see civilians too numerous to comprehend or count. There seemed to be people from all over the earth. There was no marshaling music, no cheering crowds, just this endless mass of humanity marching onwards.

KTB Log May 28, 1940 23:47; I awoke feeling very sleepy and am going topside for some fresh air.

KTB Log May 29, 11:49; *U-49* has had an uneventful morning. We tried to engage several planes, but we unfortunately missed them and they thankfully missed us. I was manning the ack-ack myself, all I did was waste ammo. I tried using proper deflection, but my fire control needs work. Also I couldn't see the planes that well, and I kept mixing up the seagulls and the planes on the horizon. A few planes dropped their payloads from high altitude. They seemed like RAF heading back to England that still had a few bombs to drop. The planes had too much altitude for me to see them.

15:33; Karl and I are eavesdropping on a nearby Luftwaffe Flugplatz (airfield). It seems that the Luftwaffe had a großer tag[122]. The RAF has had a bad day by the sounds of the chatter. Sixteen RAF lost in combat to our glorious Luftwaffe, and AA shot down another 8.

(Personal notes) I am disappointed in myself for not contributing a RAF kill today. I think I fired two hundred rounds and didn't so much as scratch the paint on the bombers.

KTB Log May 29, 16:00; Karl, our marvelous funker, has tuned in the German news being broadcast out of France.

(Personal notes) I wish I knew who the newsreader is, she sounds very pretty. She has a wonderfully powerful and domineering voice, I guess by her accent she is from Berlin. She has that metropolitan way of pronouncing her words, she emphasizes her speech by stretching the words as she speaks. When she talks it isn't in a staccato fashion ... "und das Luftwaffe"... it is more sexy and slow ... "unnd dasss Luffft-wafffe"....

She reported that the Wehrmacht has taken Lille, (I am not sure where they took her, it might have been to dinner or maybe to a show) ... oh, she means Lille, France; and we have also captured two more cities in Belgium, Ypres and Ostend.

And the best news was for last. Our fellow sailors have been very busy today. *U-62,* while patrolling in Der Kanaal and approximately 6 km from Neiuport, torpedoed the *HMS Grafton* DD. It didn't sink but was scuttled later in the day by the *HMS Ivanhoe,* after the troops had been evacuated. The Luftwaffe near the beaches of Dunkerque sank the *HMS Grenade.* The Luftwaffe also sank the *HMS Waverley* and approximately 400 of 600 survived that attack.

The V&W DD *HMS Wakeful* was sunk with over 600 soldiers below decks; only one survived the attack by our Schnellboot *S30.*

Tragedy strikes the British again when the *HMS Lydd* attacks the trawler *HMS Comfort* loaded to standing room only. The *Comfort* had earlier suffered some flooding and was attempting to return to Dover. *HMS Lydd* sees the ship and thinks it is an S-boot! The *Lydd* opens fire with its Lewis guns and its four-inch guns and is scoring hits.

---

[122] Big day

The *Comfort* goes dead in the water and the *HMS Lydd* then rams the *Comfort*, splitting the ship in two! The few who survived the ramming attempt to swim to the *HMS Lydd*, but the soldiers and crew kill the survivors with small arms and the AA from the *Lydd*.

(Personal notes) It seems as if *U-49* is destined to be left out of the fray. I was sure we would see action today (I guess WorldMod doesn't have action here at Dunkerque, May 29, 1940 was the peak day.).

KTB Log May 29, 1940 20:00; All is well aboard *U-49*, all hands are in good health and their morale is still holding up. A few were hoping for action today, and like myself are somewhat disappointed at our lack of being able to engage any troop transports. We patrolled very diligently; there just were no ships to be preyed upon. Our fuel load is just slightly above the 50% mark, 8 torpedoes left.

KTB Log May 30, 1940 00:07; It is a new day, we shall see what it brings, we are just south of Calais and will patrol back up the coast (I have lost count on how many trips back and forth along the coast we have done).

00:10; I will be going topside for a pipe of tobacco. Jurgen is on watch and I will have him regale me with his tales of Paris. He is the only one on das boot who has been there. He has some rather saucy stories of Französische Frauen und Nächte verbracht mit zwei Damen für den preis von einem!

02:36; I make the recommendation that if *U-49* is ever stationed or weighs anchor at a French port that Jurgen and I be allowed to travel to Paris. He has informed me that in his past travels to Paris he was able to make the acquaintance of several French citizens. It is my belief that these women and dancehall girls hold many secrets that need to be investigated.

06:00; Morning sick call had seaman Erwin Hansen reporting for the treatment of several cuts and scrapes he received while he was aft removing seaweed that had collected around the rudder and its housing. We were at an all stop and under the cover of our coastal guns. He and another seaman had to enter the water and cut away the seaweed. He had finished his work and was attempting to board the aft deck, he slipped and he slid down the side of the boot, the barnacles cut his hands and chest badly. He must have tried to grip the hull as he slid down. He was quickly pulled back on board. Onkel Willie is now making a homemade salve to help heal the cuts.

KTB Log May 30, 1940; BdU has sent an urgent update to all U-boots. Recon has reported the *HMS Ark Royal* has left Grennock at 05:45, being escorted by the DDs *HMS Acheron, HMS Ardent, HMS Acasta*. The task force is believed to be heading toward Norway.

KTB Log May 30, 1940 14:34; The only sound contacts are the normal British patrols in the Dover Straits. We have spent the past 8 hours running 12 meters deep, ahead two knots.

BdU also sends another radio message that in Holland the Southwest Ostende harbor and facilities are operational and secured, although the lock at Zeebrügge has been rendered useless by the cowardly saboteurs. (I wonder if Holland has locks on any dykes? I wouldn't mind handcuffing a few myself.)

19:00; We had no RAF bees today. I was unable to practice my aiming.

KTB Log May 31, 1940 01:00 I am going up on the bridge to hear of Jurgen's further adventures in Paris. The Watch and myself became distracted by Jurgen's tales of Paris. WO reports aircraft close range, bearing 279. Bloody hell! Why are they up flying so early? Bloody hell! WO gets his AA gunner on deck very quickly, and just as he begins to engage the planes the first bloody RAF'er, drops his bomb and it seems to detonate right on the surface. Very near our boot!

I yell in my best pirate voice, "Shiver me timbers! Avast ye maties! Man yer bat'el stay-shuns." All hands are scrambling below deck. The boredom of the past few days fades into a distant memory.

KTB Log May 31, 1940 01:40 CE reports Flak gun damaged.

01:40; CE reports Watchtower damaged.

01:40; CE reports Main pump damaged.

01:40; CE reports forward deck damaged.

CE is ordered to get the repair crew moving, on the double.

01:41; CE reports Main pump intact.

01:42; CE reports repairs on Flak gun complete.

01:43; CE reports unable to bang the dent out of the watchtower, he considers the damage minor.

01:43; CE is unable to repair the forward deck and considers the damage minor.

01:45; CE reports the hull is still 100%.

02:00; I am fortunate the gunner wasn't injured or killed in action.

02:27; WO reports plane, bearing 179. The gunner fires two clips but is unable to score a hit. The RAF bomber misses by a large margin.

(Personal notes) Bloody RAF was trying to stop Jurgen from exposing die Magie der Nachtmädchen!

KTM Log May 31, 1940 03:10; Tragedy has struck *U-49*. I was up on the bridge helping with the watch, when at 02:48 our highly decorated officer, Arend Akermann dropped his binoculars, looked over at me, his eyes rolled back in his head and collapsed. He hit the deck hard. We rushed him below, but there was nothing Willie could do.

Arend was dead; he must have died from a "bug" bite or a massive heart attack. I know for a fact he wasn't wounded during the earlier RAF air raid.

He was the only officer I have had on board since the first patrol. He was my main watch officer, the repair officer, and my torpedo officer. He will be hard to replace.

KTM Log May 31, 1940 04:02; I had Karl radio BdU for instructions.

04:20; We were told to rendezvous with *Schiffe 31*, it is heading toward our present location, and is acting as a temporary morgue.

(Personal notes) I am glad I don't have to have a burial at sea. Being this close to shore, his body would probably get washed up. The loss that will be felt by his family will be tremendous. Perhaps they will find some solace if they can bury his body at their family cemetery plot.

KTM Log May 31, 1940 04:42; I will assemble all his personal items (medals, all his gear,) and send them with his body back to Germany.

(Personal notes) I have gone through his personal possessions, and it was rather depressing. He must have loved his family dearly. He had written letters to his wife, his son and also his three daughters. Perhaps they will be able to "hear" his voice in the words he wrote them. He also had brought on board a small collection of photographs. There were a few studio pictures, the whole family standing so proudly in their finest clothes. His face seemed so different than the one he always showed us. I never saw him with such a glow of peace and tranquility, as he has in these photographs.

The other pictures must have been taken by Arend and his wife while on holiday with his family. They were at the seashore. Pictures of the whole family playing on the beach and picnicking on the bluff. He looked so at ease and enjoying himself immensely in the company of his family.

I have known Arend little more a year and he never talked of his family. I wonder if it was just too painful for him to speak about them. I will have to write a letter to his wife and send it along with his corpse.

✠✠✠✠✠✠✠✠

Last night around 22:30 I did a save and continue. I did the save while on the bridge with Arend and the regular watch topside. I have done this dozens of times. After clicking "OK" to overwrite the past save, I was again, like always, back on the bridge, only this time Arend was there one moment and suddenly he was gone.

I though he had become a wizard from Morrowind and had cast an *egress spell*.

I clicked on the crew management screen and sure enough, Arend was still on the bridge just all red and red means dead.[123] I know for a fact he didn't get hurt during the air raid, so it must have been a bug or glitch, or he died of "natural causes". Either way he is gone, and ain't nothin' gonna bring him back.

✠✠✠✠✠✠✠✠

KTB Log May 31, 1940 08:16; I have tied an identification tag with name and serial number to Arend's foot and I wrote in ink an identification tag that will be secured to the outside of the shroud. A boatswain has used his best knots to secure a white sheet around Arend's corpse as a shroud.

10:04; We have arrived at the rendezvous point and are waiting for *Schiffe 31*. CE reports repair completed to the watchtower and has repaired the forward deck.

10:54 *Schiffe 31* has arrived and preparations are being made to send the corpse, some posts from the crew, my letter to Arend's wife, a sealed packet containing his medals and wedding ring, gold watch and bob, photographs, to the ship.

11:07 *Schiffe 31* has sent a launch to retrieve Arend's body and his possessions. The launch has brought us a fresh supply of flour, sugar, three cases of eggs, canned milk, fresh carrots, beets, radishes, Swiss chard, five cases of lemons, five cases of oranges, cigarettes, and pipe tobacco.

---

[123] In Silent Hunter III, the crew management screen displays small images of the crew and where they are stationed on the sub. When a crewmember is killed, his image turns read.

17:03; *U-49* has spent the past few hours using our batteries for propulsion. We sailed as far as the Netherlands-Belgium border and we have plotted a return course, our plan is to exit Der Kanaal and patrol Grid BF35.

KTM Log May 31, 18:00; The BBC reports that yesterday 53,000 more soldiers made it off the Dunkerque beaches and harbor. Today 68,000 are evacuated back to Dover.

(Personal notes) The German leadership has undone the greatest military maneuver conducted in the 20th century. The main architect of operation "Cut of the Sickle" is Lt. General Erich von Manstein. His overwhelming victory is left unfinished when the BEF is allowed to escape. The thrust through Holland, Belgium, Luxembourg and France started on May 10, 1940 and by now it should be over with a complete victory for Germany over the BEF and the 1st French Army.

Had the BEF been annihilated, Churchill very well would have lost the support of the Labor party and the Apostles of Appeasement would have entered into negotiations with Germany. Germany could then secure the eastern front and Russia could be left alone. Amerika would stay in the Far East and we could have peace here on the Continent.

(Personal notes) May 31, 1940 21:00; *U-49* has become enveloped in an air of silence and retrospective. Each person has kept to himself today. After we transferred Arend, we ran slow 3 knots, and made our way SSW back towards Der Kanal. I will have to resume patrolling when we exit the western side of Der Kanal. There will be no time for moral contemplation. I will have to see that the crew makes it through this calamity.

KTM Log June 1, 1940 00:07; We are sailing on the surface and I must fill in as WO and be ready to have Fritz man the AA if the pesky RAF are doing their early morning patrols.

01:16; I have the CE dive the boot to 12 meters and ahead 4 knots.

05:09; We have been sailing while submerged and will resurface to replenish our air supply and to recharge the cells a while.

(Personal notes) June 1, 1940 06:00; I reported to sick call, I wanted to have a talk with Willie. I feel I am responsible in some way for Arend's death. I know in combat people die, and we are in combat even when it feels like we are in a lull. So I shouldn't be shocked, but I am. I was sure Arend would be around on all of *U-49*'s patrols. He was a tireless officer and he could fill so many roles on the boot.

Onkel Willie was too blunt. He looked me eye to eye and said, "Captain

Wrratt, you *are* responsible for Arend's death. You are responsible for our lives and therefore if we die, it is your responsibility. You are the Skipper and with that comes all the weight of accountability.

"Maybe Gott called Arend home to spare him from this horror we have become part of, maybe Gott will call all of us home, too."

He really wasn't very helpful, I feel worse now than I did before.

He was right; I have to be responsible. Maybe I worked Arend to an early grave. With all he did on our boot, I can't even begin to imagine how vital he was to his wife and family.

KTB Log June 1, 1940 11:32; I am on watch and although the sky is clear, it has begun to rain. (This is the first weather change I have witnessed while on the bridge). Thunder and lighting have begun to form and the clouds are thickening quickly. The wind is 7 mps blowing at our backs. We are heading SSW past Dunkerque.

12:15; The rain is heavy, fog heavy, wind 7 mps, although the sea isn't yet choppy.

12:17; We dive to 30 meters and will cruise slowly for the rest of today.

(Personal notes) That way I won't have to do watch duty in the rain.

13:00; The boot is reestablishing its routines.

KTB Log June 1, 1940 14:58; SO reports sound contacts of the warships that are routinely patrolling the Dover Strait. Onkel Willie is serving tea! He has made several pots of tea and says all hands must have some Irish tea. He has baked several loaves of Irish Soda Bread, and we are enjoying the fresh bread immensely, it tastes fantastic, and smells marvelous, too!

(Personal notes) Bless his soul, Willie has hidden away two records of Irish music. He is holding an Irish wake for our dear, departed friend Arend! Music, laughter, Irish whisky, Irish tea, the good fellowship of comrades. Everyone has wonderful stories to tell of Arend. I never heard any of the tales from the torpedo room, or of his work doing maintenance on the boot. All hands swear they will pick up the slack now that Arend is gone. And I will do my best to try to keep our boot out of harm's way. Willie has righted the mood on our boot. He has gone above and beyond the call once again.

19:00; We are surfaced, sailing 4 knots (I am trying desperately trying to make the most of each liter of diesel fuel. The CE hasn't yet reported 50 percent fuel remaining, although we are very close.)

KTB Log June 1, 1940 20:00; The reports out of Dunkerque are hard to comprehend, there must be 600 ships sunk or burning (I wish I could see it). The Luftwaffe yesterday and today has made hundreds of sorties. Many from makeshift airfields near the front. They are pounding the thousands of boats and ships near the Dunkerque harbor and beaches all the way to LePanne Belgium. (I wish I could see that, too!).

KTB Log June 2, 1940 02:37; All is well, the rain is still heavy. *U-49* is in Grid BF33 heading slowly SSW. Today's rain has given us a respite from the RAF bees and we are sailing slowly through the fog and rain. The plan for today is to cruise at 12 meters ahead slow and conserve the diesel fuel.

KTB Log June 2, 1940 11:36 We surface for some fresh air. The rain is still falling at a steady pace and there still is a heavy fog.

KTB Log June 2, 1940 16:17; The rain is steady and all is well. *U-49* is sailing SSW in Grid BF 33.

BdU sent a U-Bootslage for June 2, 1940:

U-37 Kplt. Oehm has reported to be sailing back to base. Out of aals,[124] with 39,368 tons sunk. (Personal notes) Great job! Outstanding work by *U-37*!

U-46 is sailing near center Nord Sea.

U-101 has reported heavy British patrols at east end of das channel, surface sailing must be avoided. BdU orders available U-boots und S-boots to attack all schiffes in the channel.

U-29 is sailing north of Ireland.

U-58 advised of enemy U-boot search.

U- 43, U-56 patrolling in the Atlantic.

U-13 has reported to have torpedoed the trawler *Blackburn Rovers*. We are informed that das Kaiser-Wilhelm-Kanal und das Flensburger Außenförde are operational once again after they were swept of mines.

20:03; We surface and we are moving the fore external torpedo. It will be dangerous in the rain, but the sea isn't too choppy.

20:5l; The aal is in its holder and the crew is taking a short rest before moving the aft external aal.

21:20; The crew is in the process of winching the aal below deck.

---

[124] Aal: eels, German nickname for torpedoes.

21:55; All our torpedoes are now available for battle. Six steam forward, and two electric aft.

(Personal notes) Heintz has started a chess tournament, but you must make your moves within 30 seconds. It is fast paced and only the quickest players who have set moves are doing well. I was dropped after my first match against seaman Gorge Rausch. I forgot to castle the rook and king before starting my fianchetto. Gorge asked me what kind of opening I was using and I told him it was the "Glockenspiel fudge up, perfected in 1734 by a Viennese barmaid named Frienredi." I lost in 14 moves. I think I will stick with "Battling Ships."

KTB Log June 2, 1940 23:37; *U-49* is sailing on the surface and we will be diving soon to use the batteries for the next few hours.

(Personal notes) I dreamt I was hiking along an alpine chaparral. The small pine trees were low and twisted, and the path was rocky. The air was warm and dry, and although I couldn't see the sun I could feel its warmth.

I had walked along for some time and had come upon a fine vista that overlooked a splendid river that flowed lazily along. I was looking as a small canoe came drifting downstream. I began watching its slow course as it bobbed and rocked along. I couldn't see anyone piloting the canoe. Yet it seemed to be sailing straight and keeping to the center of the river.

It passed right in front of me and I could see *U-49* painted in green on the canoe's bow. I was suddenly sitting in the canoe. I thought to myself, if this is my boat, and I'm not in control of where it sails, then who is? I was pondering this and wondering what fates control the boat's destiny? Was it my skippering? Pure luck? Some greater force?

KTB Log June 3, 1940 04:26; Jorge the SO woke me. He has a warship at 111 degrees, moving fast and closing.

04:27; I order rig for silent running and we stay our course. I return to sleep with orders for Jorge to wake me if the warship doesn't tack away.

05:32; BdU sent an urgent update to all schiffs, to be aware that passenger ships are leaving from the Continent and Ireland flying the Amerikana flag, and we must not fire at any Amerikana ships by an order from Hitler himself.

06:00; Two of last night's watch have reported for sick call for some of Willie's "shoo flu brew", of hot lemon juice, whisky and bitters.

07:08; Last night's chess tournament came down to a match between the reigning champion, Heintz Geisler, who plays a traditional European style, and a new crewmember (this is his first patrol), Oskar Albrecht.

Oskar made it through the rounds easily; he plays a fast and fluid offense. His final with Heintz was dramatic as Oskar reacted swiftly to Heintz's moves. Oskar countered Heintz with unorthodox moves that soon had Heintz frustrated. Heintz forgot to stick to his style of play and 11 moves later, Oskar was crowned new chess champion. He not only won bragging rights, Heintz has to treat Oskar to a night at Fru Adrienne's Haus der Geschlechtbezirk-mädchen when we return to Wilhelmshaven.

KTB Log June 3, 1940 11:21; I feel like *U-49* is caught in a freak rainstorm. The weather is still bleak. We have had less then 24 hours of clear weather since Arend died. The only positive outcome of the rain is that we are sailing mostly submerged. We are saving diesel fuel (we are still a few liters above 50% remaining).

11:22; We have encountered no merchant ships while cruising SSW through Grid BF33. We are now in grid BF 31, cruising ahead one-third. We will surface soon to recharge the cells.

11:49; *U-49* is surface sailing in heavy rain, fog and winds 8 mps, shifting in direction.

15:26; We surface and the rain has abated. The sky is still very gray and the fog is still thick. I will stand watch myself.

18:15; All is quiet, the sea is very choppy, but no waves break over the bridge. We were all tethered to the rails as a precaution.

(Personal notes) While on watch, Jurgen was regaling us with more of his adventures while on holiday in France. He was talking about a Madame who lives in Rennes. LaMorn is her name. Her beauty and grace is well known. She is a famous model and dancer. She poses for many artists, she makes a most attractive and desirable study. Her dark red hair contrasts wonderfully against her fair skin and her years of dancing has made her body curvaceous and her muscles well-toned. She is now the proprietor of the finest Maison du Plaisir in France. Her filles et dames are from all over the world. You can have anything from a dark-skinned lovely from Afrika or a delicate beauty from the Orient, or a platinum blonde from Hollywood. Jurgen makes it sound like taking over the French will be worth it if we get to visit some of the exotic women of Madame LaMorn.

KTB Log June 3, 1940 18:23; CE reports cells recharged.

18:23; Gottfried would like to check the port side diesel with the assistance from the CE, during our next dive. I tell him we will dive soon and he should advise me of any needs that arise during his inspection.

19:27; We dive to 30 meters and will cruise ahead one-third.

(Personal notes) After mess I will go see and what Gottfried is doing in the engine room. I will bring him some coffee and his dinner.

KTB Log June 3, 1940 21:12; Gottfried is finished and reported that the diesels are in excellent condition. He has returned to the electric motor room and we are cruising ahead standard.

22:04 All is calm, most of the crew has turned in early.

KTB Log June 4, 1940 00:52; We are sailing on the surface ahead one-third. The sky is gray with clouds and thunder can be heard off in the distance. The fog is still fairly thick and I will have to stand watch while August Meyer has some much-needed rest.

KTB Log June 4, 1940 01:08; CE reports 50 percent diesel fuel remaining. The CE Heintz has assured me that we have used the first half of our fuel load very efficiently. He said Gottfried has been getting each kilometer's worth from each liter of diesel.

KTB Log June 4, 1940 03:00; August reports for duty, and I am relieved of watch for a while. I am soaked and want to change my trousers and sweater.

04:00; Karl has been listening to some British ships; it sounds as if the Dunkerque evacuation is complete. The last ship to leave is the DD *HMS Shikari*; it left at 03:30. Approximately 340,000 troops have been rescued; the rescue fleet was comprised of over 800 civilian craft and 222 Royal Navy ships/boats. The RAF lost 106 aircraft and our Luftwaffe lost 150 airplanes.

(Personal notes) Germany is going to rue the day we allowed the Brits to escape from France. No matter how many land and air battles we win, none will ever relieve the burden of this debacle.

07:00; NA reports depth under keel is 55 meters. *U-49* is in Grid BF 26, approximately 25 km NNE from the channel island Guernsey. The sky is still thick with gray clouds. The watch was tough this morning; the boot was pitching and rolling in the waves. We had a good dunking from time to time, but fortunately we didn't have any waves break over the bridge.

09:52 Karl has tuned in the German news and they have announced that the Wehrmacht's 4th Army has entered Dunkerque at 09:30, to no resistance and has captured the rear guard (about 40,000) and 800 civilians. Von Kluge rode into the town center and Dunkerque has fallen.

(Personal notes) No one cheered and no one laughed and smiled as we heard the news. From young seaman to Onkel Willie, we all know the Englanders resolve will be renewed with the "miracle of Dunkerque" having saved the backbone of the British forces to fight another day.

10:17; *U-49* will continue on our course and I will hit my bunk. This sad news makes me tired.

11:40; The SO has awakened me to inform me about a warship contact, moving fast, moving away, bearing 271. I go topside, all is calm and the watch and crew is rotated. I return to my bunk and go back to sleep.

Log June 4, 1940 16:07, I awake from a good few hours sleep, and make my way to the head, then to the galley for some juice, and go topside to see what there is to see.

Log June 4, 1940 18:00; *U-49* is heading SSW and we are approximately 95 degrees and 70 km from the island Guemsey, Grid BF25, depth under keel, 90 meters. I like deep water.

(Personal notes) Otto was expressing his concerns for his older brother, Herman. It seems Herman is attached to the 6th Army and has been seeing a great deal of combat, since May 10,. Otto told me his brother drives an armored panzerwagen SdKfz251. It's fast and usually travels in groups of two or three, along with several BMW motorcycles, and perhaps some light tanks for cover.

I wonder what it would be like to be part of an advance scout team, the eyes and ears for our division. We get our orders and we head out, the roads we travel are mostly lined with the shattered remnants of stores and homes. Most have had the roofs burned off or caved in; the windows are shattered from the Luftwaffe and its earlier air raids. They look ghostly, as some still have gray smoke drifting up from their skeletal remains.

We have to keep a careful watch for any snipers or traps. The retreating Englanders and French forces have removed all the signposts. It doesn't hinder us at all; we have very gute maps. The road is passable, there is some rubble scattered about, a burned tram, a twisted frame of a bicycle, a few carcasses of dead dogs and horses, a few dead people, the air is heavy with smoke and the smell of death. Our road to victory is paved with gory.

We don't see many civilians and the ones we do see, once they see us, they run and hide. We don't engage the civilians. We sometimes have targets of opportunity that we are allowed to act upon. Such as what happened four days ago. Our group was scouting the roads leading towards Dunkerque when we came upon a British blocking force, they fired at us and we withdrew behind a ridge. Gus "Grim" Deitch, is our sniper, and he and I stealthily make our way to the top of the ridge, sticking a few pieces of tall grass into the webbing on our helmets, as we advanced. Upon reaching the top of the ridge, we could see their captain, prancing along the line as if he was in some sort of fantastic battle. He waved his arms and pointed his swagger stick at an imaginary Hun, yelling for his troops to fire. The troopers were half-heartedly responding to his commands.

Gus and I lay flat on the ground and crawled to get behind a few small shrubs, and a fallen old tree. I acted as spotter, we were in a fine position for the hunt I looked the line over and checked for any overwatch to its rear. I scanned for the captain and could see he was standing still. The captain had paused to survey his situation. He never looked our way. I gave Gus the all clear to fire.

Gus had been tracking the captain, his crosshair was right on the side of the captain's head. "Grim" gently eased back the trigger — Crack!

I was watching with my binoculars as the bullet struck the captain's head. There was sort of a blur of blood and brains. He stood for a few seconds, or so it seemed, and then he dropped right where he stood.

Bloody Brits cheered! Gus and I laughed as we made our way back to our panzerwagen! I guess he wasn't very popular.

Log June 4, 1940 18:20; All is calm, but the weather remains unchanged so the boot is pitching and rolling. We will dive to 40 meters and cruise on the cells for a while at ahead standard.

KTB Log June 5, 1940 03:02; CE Heintz reports that the battery is down to 50%, that was a fair cruise on the batteries. We surface and begin to recharge the cells. Weather unchanged, wind 15 mps, rain none, cloud cover heavy, fog medium, donner and blitzen all around the boot, but off in the distance. I will stand watch myself for a while. Today has been uneventful, and the monotony of the weather is more fitting for winter than late spring.

(Personal notes) I think I will have Karl play some music softly, perhaps some Paul Robeson, perhaps the Joe Hill song. In a deep bass voice, Paul's voice fills our boot and refills my spirit. I dreamed I saw Joe Hill last night as alive as you or me. I love this song! I want the crew to learn it well enough to sing.

Then we will listen to the *Peat-Bog Soldiers* a.k.a. the German song *Moorsoldaten* by Paul Robeson.

KTB Log June 5, 1940 07:00; We have been monitoring the enemy's radio waves, the Englanders sound like they are preparing for us to invade. The government has taken control of everything, all business and employment, prices, all aspects of property ownership. Some of the civilians are making arrangements to have their children sent to the safety of Canada and Amerika. Churchill made a big speech last night, he said that the Brits will fight us in the air, on the sea, and even at the landing beaches, that the Englanders don't plan on surrendering.

KTB Log June 5, 1940 SO reports at 10:01 warship, moving fast, moving away, bearing 304. We plot an eastern tack and put the warship to our aft, and the crew farts in its general direction.

KTB Log June 5, 1940 12:27 SO reports merchant closing slowly at 010, long range. We all stop and listen to what angle the merchant is sailing in relation to our bow.

12:32; SO and the NA and myself have determined she is sailing SSE to our location. I order battle stations; all hands have been ready for the past hour. We surface and plot our estimated interception course. Seventeen kilometers separate us. I go up on the bridge and take Jurgen with me and I have him bring his harmonica. He will play a few tunes as we sail to our first hunt in several days. My spirits are very gute and I am looking forward to a hunt. The lighting and donner are an excellent backdrop for the carnage to come. Jurgin is playing some old time sea tunes *(Blo' das man Down, Spanish Ladies, Away Rio)*.

(Personal notes) Next time at the barracks I will have to see if any recruits can play the fife, banjo or a concertina, that way we could have some live music and tunes while sailing. I think it would be good for morale. And I know Jurgen would like to have some accompaniment while playing.

I feel like a New England whaler, the music has made me calm; I have never had my nerves so settled as they are now. The whole watch is in unison with those below deck. This early stage of the hunt is usually the start of the tension. But Jurgen's tunes have us relaxed and going about our duties with clockwork precision and steadfast determination.

KTB Log June 5, 1940 13:38; *U-49* dives to 12 meters for a sound check. The ship is still on a course to cross our bow. We surface and resume our positioning for an ambush.

13:46; We are less then 1000 meters from her course. She is still down range, closing slowly. I am attempting to ID the ship. She is really pitching in the high seas.

13:48; We have a C-2 cargo of unknown origin (we must not fire if the Amerikana flag is sighted).

13:49; The WO IDs the C-2 and gives me its estimated bearing and speed. We dive to 12 meters, ahead slow. I push up the attack scope and begin to plot the final approach. Plan is to fire at her 180, 525 meters out, two aals, impact, 5 meters deep, fast.

13:53; TDC is set and I check her approach, she is rolling and waves are obscuring my scope. All stop and open tubes one and two. The ship is large and I can clearly see the British flag acting as a death beacon for the ship.

KTB Log June 5, 1940 13:56; Our game is at the appointed hour, and I order tube one fired, tube two fired! I have targeted amidships, just in front of the bridge, range 928 meters.

I can see the ship very gute, and at 13:56 WE reports impact, the ship's bell has tolled. She split apart with the first aal, she must have been transporting ammunition. The explosion was huge, I saw cargo go sailing into the sea. The second aal struck the stern section as it went bow up. Seconds later all remnants of the ship were gone.

13:56; NA reports enemy unit destroyed. The forward torpedo room is reloading tubes one and two. Great work by the crew with no officer to oversee them (Arend would have done that). We surface and have plotted a patrol course due west, depth under keel is 90 meters, ahead slow. We send our patrol report; Karl is encoding it so the Brits can't figure out our location in case they intercept the radio transmission. *U-49*, ship sunk, C2 cargo, 6445 tons, Grid BF24, 6 aals remain, four stern, two aft. Fuel 45%.

KTB Log June 5, 1940 14:09; The tubes are ready for our next hunt, and the crew is told to get some mess. Willie has made potato soup and sausage, with strawberry jam and biscuits for dessert.

14:32; The crew is rotated and then everyone will have had mess. The morale is very gute, and we are enjoying some good fellowship, and tales of our hunt are being told. I always get a kick out of how the torpedo crew "sees" the battle. They say that they know as soon as the aals swim if it will be a hit or a miss. They told me our boot tells them. The 'woosh' of a hit will be strong and forceful, but the times we miss the 'woosh' is weaker and less confident. (It makes me want to do a compression test. I wonder if the gauge is off or if there is an air leak in the compression lines. I will have the CE double-check all lines, fittings, valves).

KTB Log June 5, 1940 15:56; SO reports warship closing fast, bearing 229. We dive to 12 meters and the SO tracks the warship.

16:08; He is traveling WSW, nearly parallel to our current course. *U-49* adjusts our heading NNW away from the unknown warship. Ahead one-third, and we resume patrolling.

18:39; We dive to 25 meters, and I order the CE to all stop. I order the SO normal sweep, and I turn in for a few hours sleep.

KTB Log June 6, 1940 01:49; I awaken and order the boot to be surfaced, ahead slow. We will turn and head due south. *U-49* has reach the farthest west we can sail and still have 25% fuel remaining for the return sail to Wilhelmshaven.

KTB Log June 6, 1940 02:27; SO reports warship, moving fast, moving away, bearing 029, long range. We will maintain our present heading in Grid BF24, ahead slow. We are actively patrolling (diving every 90 minutes for a sound check).

06:02; *U-49* is 12 meters deep; I am manning the hydrophone while Otto goes to the head. I hear a sound at our 180, we are at an all stop. I order ahead full and we turn the bow to the sound.

KTB Log June 6, 1940 06:03; All stop, I can hear something that could indicate a convoy or military task force, (although none have been reported in our area). The sounds are widespread, and very indistinct, very faint.

We surface ahead full and we will see if *U-49* can close the gap.

KTB Log June 6, 1940 06:24; SO reports warship moving fast, moving away, bearing 349. We dive to 12 meters so I can have a listen to see if this is what I heard at 06:02.

06:27; I can hear the warship, but I also hear indistinct sounds at 349 through 011. *U-49* surfaces and ahead flank for 30 minutes and we will dive again for a sound check.

KTB Log June 6, 1940 06:56; Jackpot! SO reports merchants bearing 349, 346, 348, 349, 345, 349, 349, 345, 349, 349 … everywhere.

06:59; Surface, ahead flank, we plot a course to sail in a straight line toward the convoy. Then see what the warship coverage is like, mark the targets on our chart, and plot their probable course. Then flank them and it will be time for a hunt. This is our second found convoy.

KTB Log June 6, 1940 06:55; We are making 16-17 knots, we should catch up easily.

07:03; SO reports merchant, bearing 345, moving away. We alter our course.

KTB Log June 6, 1940 07:08; the convoy is heading WSW and we easily track parallel to its course. We see smoke from several ships 3.8 km out. We will easily overtake the ships and we will set up for an ambush. I see what looks like tankers in the second row. We can see a line of three ships; then a second, and the SO can hear a warship at the front and one off to the far side. Could there be only two escorts?

07:23; The fog and the heavy sea have made the ships hard to see, we have passed the pack and they are fading fast.

08:18; We are in our ambush point, we can see a DD tribal about 625 meters to starboard, 110.

08:38; The DD has passed. We have targeted the second tanker and will fire all four of our stern aals. The TDC data has been processed by watching the leading ship in the convoy. Distance 475, speed 4 knots, angle 180 to the target's course.

08:50; Tubes one through four are opened.

08:52; I give the order to fire all four tubes.

08:54; WE reports three impacts, one miss.

08:58; NA reports enemy unit sunk, tanker T2, 10,867 tons.

KTB Log June 6, 1940 09:13; We report the convoy sighting, heading 241, 4 knots, 12, merchants, 2 warships. We send our status report: *U-49* Grid BF24. Two aal left, aft. Five ships sunk, three tankers, two cargo, total tonnage, 45,502. We will cruise submerged for a while. Our hunt was fine this morning. The crew has done Arend's memory proud.

KTB Log June 6, 1940; *U-49* is plotting a return course; we are approximately 1,700 km from Wilhelmshaven. It is getting too hot already not to be wearing our summer uniforms. I give the order for all maats to wear the white shirts, with the blue collar with 3 stripes, scarves optional, hats and cap optional, unless on watch. All seamen will wear the Arbeitbluse with Kleiner Dienstanzug, hat and caps are optional. But all on watch must have a cap or hat on. Officers and maats, jackets are optional, but Officer of the Watch will wear his Jacke. Exceptions are crew in engine room and torpedo room, repair and maintenance crew, they shall continue as before.

13:00; I will hold inspection. The crew is told to be ready by 13:15.

13:15; I step into the command room and the first maat shouts, "Attention!" The crew responds quickly.

(Personal notes) I walk in with my full dress uniform; I am even wearing my shoulder straps and epaulettes. All eyes aren't looking front, and I order, "All eyes front," and I yell to Karl to play *Die Fahne Hoch*.

I walk as sternly as I can, my shoe heels clicking on the metal deck. I closely inspect the crew. The crew is unsure of why the sudden Nazi zeal on the boot. I am trying to look angry and stern. I give them my best Hitler imitation, "Mein comrades, *mein dog has no nose!*"

Jurgen shouted back, "How does he smell?"

"Awful!" I yell back.

Things are back to normal and I shout to Karl to play some Maurice Chevalier songs.

14:30; The crew looks in excellent fighting trim. The crew's hair has grown back and they look sharp in the clean uniforms. We no longer look like a ship of convicts. We look like a crew on a mission. It could be that we sank our fifth ship of this patrol yesterday, or it could be the clean clothes. But something has made the crew walk a little brisker, have a sharper glint in their eyes, a renewed sense of purpose and determination.

General shape of the quarters is gute. Later today the repair and maintenance crew will inspect all electric components, (wiring, fuses, fuse panels, switches, generators, pumps, cells), and check all torch cells.

16:00; All hands have done an outstanding day's work. The hunt only claimed one ship, but it was a clean kill. I will turn in for some needed sleep.

KTB Log June 7, 1940 02:17; BdU sent out an update:

All ships and boots are warned to stay away from western entrance to Dover; it is now heavily laden with mines. All ships and boots are ordered to increase our surveillance of all shipping traffic as we exit the northern side of Der Kanal. Record and report all ships spotted

KTB Log June 7, 1940 04:07; *U-49* is making its way ENE through the Channel; we are in Grid BF 25.

KTB Log June 7, 1940 06:00; August "Gus" Meyer has reported to sick call, he was sharpening a fishhook and he has pierced his finger and the hook is stuck. Luckily it passed through and Willie will cut the barb off and pull the fishhook out.

(Personal notes) August said he was getting ready in case the weather clears enough for us to get in some fishing. He said that we are in gute fishing grounds, that the steep drop off and shelves are bound to have game. I sure would like some fresh fish for dinner, I wonder if we could catch a few sea bass or some blue fish? I tell Gus to go get his fishing pole and tackle; we will see what the weather is like topside.

(Personal notes) Fishing fever has gripped the boot. Gus has caught a 27-pound sea bass, on a hook and some bacon rind. Now, most of the hands are rigging a pole that we will attach to the aft rails. The pole will jut out and the crew will tie several lines (no hooks) and white strips of cloth that will float and create "action" on the surface about 300 meters behind as we troll at about 2 knots.

August will then troll below the "action" with his line and hook. Two other crewmen are fastening lines and hooks onto makeshift fishing poles. They will have to reel in by hand and glove, Gus can reel in using his tackle and reel.

KTB Log June 7, 1940 07:01; We are sailing slowly to conserve fuel. Several of the crew are topside for fresh air. There has been an emergency; our food supply has to be supplemented immediately. We are going to have to catch some fish.

(Personal notes) I want to get back up on the deck. We have caught seven more fish. Gus has been sharing his fishing pole and all hands want to have a go at trying their luck. Onkel Willie has made a gaff out of a prong that he uses for cooking. He looks like the old salt of the sea, as he swings his gaff and hauls up a wriggling fish. Everyone cheers with delight as each fish is caught and they yell with anguish when a fish slips off the deck and back into the sea. The sky is still very gray and that should keep the bees away. We will troll for an hour and I will go back topside to help with the watch. With 12-15 crewmembers on deck, it would be a vexing situation if we should have to dive quickly.

KTB Log June 7, 1940 08:11; We have resumed sailing at ahead standard. *U-49*'s crew have taken great risk to procure a fresh food supply. We caught 15 sea bass, 8 blue fish, 1 mackerel. But we have ruined our fresh uniforms, so we are back to wearing whatever the crew has handy.

12:25; SO reports warship, closing fast 075, long range. We alter *U-49*'s course 10 degrees and continue sailing ahead full.

13:47; I order the NA the resume our plotted course.

14:25; SO reports warship, moving fast, closing, 128 degrees, long range. We continue on our course and I will go topside to do watch duty.

KTB Log June 7, 1940 15:45; Sailing ENE. *U-49* is in the western most edge of Grid BF 35. WO reports ship spotted, medium range, 228 degrees. An Amerikana passenger liner has been sighted and we send in a report of its heading and time we sighted it

(Personal notes) I didn't even notice it, it was close about 3,500 meters, and it was big! A passenger liner was leisurely sailing by as if on holiday. I ordered all stop and we watched an Amerikana ship pass by. Its closest point was 823 meters. This was our first sighting of a Stars and Bars. I imagine it was a boat of evacuees from Europe heading to the safety of the States. Hitler told us no shooting of any Amerikana ships would be tolerated, and I don't want any Gestapo paying me a visit when we dock, so we will just observe the ship.

Some of the crew requested to come topside to see the ship, permission was granted and we rotated the crew and had a few extra hands on the bridge. The crew was anxious to see their first Amerikana ship. I was hoping it would be the first and the last. I didn't like how slow the captain was sailing by, he should have had his boilers raging and doing 30 knots, and instead he was sailing as if he was trolling for fish.

16:00; *U-49* resumes sailing ahead full on our plotted course.

17:00; We dive to 20 meters, depth under keel 35 meters, Grid BF35.

17:03; Onkel Willie has prepared our emergency rations.

(Personal notes) *U-49* is having a modest fish fry. Willie and a few seamen have been cleaning and preparing our fantastic catch. Willie has fried the fish rolled in corn meal and spices, and the boot is filled with the wonderful smell of frying fish. We are dining in relative luxury, Willie turns out meals that not only please the taste buds and fill our bellies, they also give us a sense home and well-being. As we dine, Karl has tuned the wireless and the boot is reverberating with jazzy music from France. The crew is enjoying their fish; it must taste better because we caught it ourselves. The crew's morale is high, and there is a strong feeling of good fellowship on the boot.

KTB Log June 7, 1940 23:33; *U-49* is making gute progress as we make our way out of the Channel. We are approximately 460 km from exiting Der Kanal, and we have sailed approximately 530 km since sinking the C2 in Grid BF24. All is well and we will keep on keeping on. If Sally sells seashells by the seashore, and seduces seaman by the seashore, would she be known as Sally the seashell sellin' seashore whore?

KTB Log June 8, 1940 03:07; BdU daily update for today: Radio intercepts indicate that the *HMS Repulse* and the *HMS Sussex* and four DD and several auxiliary cruisers are in the Island area. The British still show no signs of knowing about Admiral Marschall's task force.

Schiffe 16 reports six British ships missing off the coast of West Africa during mining operations.

BdU reports convoy sited 7th of June, 23:24, near Foreor, Norway heading 240.

BdU repositions most U-boots at sea:

U-29 will patrol Grid BF75 and BF76.

U-32 and *U-47* told to sail to the northern end of Der Kanal.

U-38 sailing towards the area north of the Shetlands.

U-25, *U-30*, and *U-65* all patrolling southern area of the Nord sea.

U-60 patrolling Nord sea.

Oberkommando der Wehrmacht stresses no U-boots or Schiffes are to fire upon any neutral ships, especially Amerikana ships. O.K.W informs us that Italy will be joining the war effort very soon.

KTB Log June 8, 1940 13:53; Grid BF 33, we have spotted another Amerikana liner heading WSW, sailing slowly. I had Karl send in the contact report. We dove to 12 meters, ahead flank. We have pulled abeam with the liner and we will surface and give the tourists a bit of a show. I fancy that they have never seen a U-boot surface before.

KTB Log 8th of June 1940 14: 32; We are 325 meters off her port side. The WO is in uniform and watch crew is in clean uniforms and I have put on my clean Jacket. We want to look like sailors and not a lot of scurvy sea dogs. I will guess this will be the talk of the voyage for the Amerikans and the Europeans who are sailing to Nord Amerika.

14:37; The crew is given orders to briskly get to stations on the bridge when we surface. They are to stand watch as professionals, no taunting or yelling toward the ship will be tolerated.

14:38; The WO will not hail the ship to stop, we will sail alongside it, then we will briskly clear the deck, and we shall disappear like an apparition back beneath the sanctuary of the waves. I am sure the captain will radio our position and heading, so we won't stay surfaced for long. Two crewmen are told to stand by in the conning tower in combat gear, weapons out of the racks and ammo ready to be loaded if the need arises.

KTB Log June 8, 1940 14:40; CE Heintz, is told to surface *U-49*. The watch and myself took our positions on the bridge, the port rail of the passenger liner was already being lined with people curious to see our excellent craft. The crew of the liner was very excited; they were running about on deck like a Chinese fire drill. Some of the passengers also ran, but it was to the rails to have a look. We sailed next to the liner named *Maryland's Pride*. After two minutes sailing, I ordered the bridge cleared and we secured our hatch. CE took us to 12 meters and we turned 180 degrees and began to separate from the liner. I kept the scope down; I didn't want the ship to know where we went once we submerged.

(Personal notes) I saw some of the passengers quite clearly; I could see that there were many women and children on board. This must be another evacuee ship. They looked like my Onkel in New York. He said Amerikans look just like us Germans except, he said, there are whole thriving neighborhoods of every sort of ethnic group you can think of. Not always getting along so well, there is a great deal of racism amongst some of the Amerikanas. He said some places dark-skinned people aren't allowed the same rights as white-skinned people. That they hang blacks in what they call lynching style. It is basically the same hate philosophy as the Reich preaches, one race superior and another inferior. I guess there is a great deal of hypocrisy in what a government says it stands for and what their real aims are.

14:07; *U-49* is approximately 115 km from exiting Der Kanal, The sky is still gray, winds 15 mps, depth under keel 30 meters.

KTB Log June 8, 1940 19:05; We arrive at our gun battery at Calais. The weather has finally changed; the wind is down to 6 mps, although the sky is still gray, with intermittent lighting. Perhaps we can find a schiffe to use the deck gun on. Plus we still have the aft aals. *U-49* is patrolling due north, we have reduced our speed to ahead one-third for the rest of today, and we dive to 12 meters.

KTB Log June 8, 1940 22:00; Karl has reported for duty. 22:09; BdU sends a second update, this one is double encrypted. The news is brief and states that Admiral Marschall has engaged several RN ships near Norway. His task force has sunk the fleet carrier *HMS Glorious*, DD *HMS Ardent*, DD *HMS Acasta*.

(Personal notes) This must have been a tremendous battle; I will have to wait till we return to port to learn more of the details. It is gute that the RN never got wind of our task force, the *Scharnhorst* and the *Gneisenau*, must have hit them hard and fast.

KTB Log June 9, 1940 00:04; *U-49* is patrolling north in grid AN79, all is well on the boot, and soon I will turn in for the night.

00:31; Sonar Operator reports sound contact, warship, moving fast, closing, bearing 348, long range. I order the SO to follow the nearest contact.

00:55; The warship isn't on a menacing course, so we will soon surface.

01:39; *U-49* breaks to surface and the day's first watch is assembled. I tell them to be extra vigilant of bees. I will watch for ships, but they must keep a sharp eye out for the pesky RAF.

KTB Log June 9, 1940 01:40; NA wants to gives me his weather update, I told him I don't need a weatherman to know which way the wind blows. He told me anyway; rain none, fog light, wind 6 mps, direction 181.

02:14; We are at 12 meters, and I am manning the hydrophone while Otto has breakfast. I hear a very fast screw at 300 degrees. I don't signal the crew yet, no need to get them nervous.

02:20; The sound is growing louder and is steady at 300, so I have the NA alter our course 28 degrees NNE. I tell the crew a warship is about 10 km west of us and we will attempt to put some distance between us. We surface ahead full.

02:27; I am on watch and we can't see any smoke from the warship.

02:50; We dive to 12 meters for a sound check. SO reports warship, moving fast, moving away, 099. Gute, that must have been the warship that was closing.

KTB Log June 9, 1940 02:56; There are no other sound reports so we surface *U-49* and alter our course to sail towards the NNW corner of Grid AN87. Estimated time of arrival is one hour and twenty minutes.

03:11; Karl has received the BdU update: Kpt. Lt. Oehm has brought *U-37* back to base; they sank 10 steamers, approximately 43,000 tons. (*U-49* will have to get one more ship to be on par with the others who are out hunting.)

U-28 has put out to sea from Trondheim, Norway, it will attempt to attack troop ships leaving Narvik.

Ua, *U-25*, *U-51*, *U-52*, and *U-65* are all told to sail to the Shetland Islands area to attack troop convoys heading from Narvik.

KTB Log June 9, 1940 05:08; WO spots an RAF plane closing at 249. I order the bridge cleared and I yell to Heintz through the speaking tube to get us under. I drop down the ladder into the conning tower and we secure the hatch. I order ahead one-third and we change course to due east.

05:09; SO reports warship closing fast, bearing 036.

05:10; I hear an explosion to our aft, probably the RAF dropping a depth charge. Dirty RAF'er, he must have radioed our position to the warship. The crew is looking tense.

05:19; We rig for silent running and cruise at ahead slow.

KTB Log June 9, 1940 05:24; The warship is about 8,200 meters 254 degrees, moving slowly, they are searching, but in the wrong spot.

KTB Log June 9, 1940 05:56; The warship is moving slowly NNE, the distance has grown to 12,600 meters. I feel we eluded the warship. I wonder if it was coincidence that the ship came near, or did the RAF actually radio our position?

KTB Log June 9, 1940 06:00; Sick call had Hilmar Chausz reporting for a very painful earache. Willie has made an ear candle and says he will use it on Hilmar, and he should be relieved of his earache soon. He is fit for duty.

06:36; We secure from silent running, and surface. The warship contact has faded and we resume patrolling. *U-49* is heading due east in Grid AN84

(Personal notes) June 9, 1940 07:11; I went down to the deck casing for a morning pipe of tobacco. It was exhilarating as the sea spray off the bow blew back towards me into a fine, salty mist. I kept think about the Wehrmacht and I was wondering how the drive to Paris was going. The reports from the past two days have some of the crew with brothers fighting at the front worried. The fighting has been fierce as the French and remaining Englanders stubbornly attempt to hold their ground.

Our soldiers should have crossed the Aisne River by now, and be engaging the enemy. The next big river is the Seine. Will they cross in Ande, Courcelles, Portmont, where will the weak link be? We can expect all the bridges to be blown, but what if they miss one? Perhaps that will ease the burden on the sappers. Of course, our engineering corps is second to none. They will build bridges for our panzers and Wehrmacht, and the battle will grow ever closer to the capital.

KTB Log June 9, 1940 07:47; *U-49* dives to 12 meters for a sound check. All is still, so we alter our course NNE.

KTB Log June 9, 1940 15:46; *U-49* surfaces. Oh, fiddlesticks! The wind has picked up to 7 mps. The sky is still gray; we must have slipped into the largest low pressure system in the past millennium. How could the sun not be able to break through the clouds? I guess everyone complains about the weather, but they never do anything about it.

16:00; All has been calm today. Fuel load is still slightly above 25%. *U-49* has plotted a course NNE through Grid AN84.

KTB Log June 9, 1940 19:40; The news from France is wonderful, the Wehrmacht is on the march! We have begun to cross the Seine River; this will drive the French and the British toward the remaining harbors.

KTB Log June 9, 1940 20:14; BdU sends an encrypted radio message. Italy will declare war on France and Britain, effective at midnight.

(Personal note) With a friend like Mussolini, who needs enemies? Mussolini and the Italians are committing genocide against the Ethiopians. Hallie Salise has been disposed, and the Italians are wreaking havoc in Africa.

23:09; *U-49* has just entered Grid AN81 heading north, ahead one-third. Karl has been picking up a great deal of Kriegsmarine chatter.

KTB Log June 10, 1940 00:11; WO called down to wake me. They have seen a red star shell about 7,000 meters to our starboard side. I am dressing and will rush topside to investigate the possible distress signal. We will alter our course toward the possible distress signal.

01:14; According to the WO the red distress signal, I just witnessed was one hour after the first sighting. Nearly straight in front of us at approximately 007 degrees, approximately 4-5000m. A red star shell was fired from a flare pistol. We have readied the arms from the conning tower, I have loaded my Luger, and we are preparing to investigate the sighting.

01:19; WO has called down, he sees three German life rafts, they seem tethered together, at about 1,600 meters. I have Heintz turn toward the boats at standard speed. I order the armed marines on deck and I get my Aldis lamp. The WO has shown me where to look and I can see the lifeboats. They are German and we hail them. Their CO responded that they are from *T-21*, sunk two days ago.

01:25; I have Karl radio BdU of our contact with the crew of *T-21*. I will go back topside and help with the maneuvering to the boats.

KTB Log June 10, 1940 01:32; We have the three lifeboats secured and we are giving water to their crew. Their CO is here with me and we are discussing our options. My only concern is the discipline of his men, not one of them has ever been in a U-boot, and we can't stay on the surface the whole trip back to Wilhelmshaven. The CO says he has 22 crew, two officers, and himself. Twenty-five more people on our boot.

I order five of my crew to begin clearing out stores in the forward and aft torpedo rooms. We will have to jettison some crates. Willie and Heintz will have to figure out how to trim the boot. I will have to rely on their expertise in trim and ballast with so many people on board. The CO is explaining the situation to his crew, and I will do likewise.

02:11; Willie has moved his stores around and wants me to put the three officers on the deck near the radio room area. Fourteen men in the forward torpedo room and eight in the aft. I tell Willie I want the officers and CO in with his men, we will pick three of his crew that look like the fittest sailors, and they can work with my crew. We can't afford any trouble from his crew when we dive, that is why his crew must remain in contact with their officers. I will intermingle my crew to help ease the tension of the first dive. Jurgen will be with the crew forward. He can distract them with his lurid tales from France. Onkel Willie will have to work very hard feeding us all.

KTB Log June 10, 1940 02:14; The CO has transferred his war log and other ship's papers to my boot. The parade has begun. I was standing on the bridge as the officers came on board. They looked tired but very glad to be on board. Their CO was below decks getting them to their proper places. Next came the seamen, a gute, fit group considering the conditions. I gave each of them reassurance and a very hearty greeting.

02:29; All are on board and the CO has returned to the bridge, we discuss what to do with his boats. I can't fire my deck gun in 7 mps wind, and we have no explosives on board, so we can't scuttle them. We cast them adrift, and we go below to survey the situation.

02:45; Karl has informed me BdU wants *U-49* to return with the crew from *T-21*. We are now plotting a return course to Wilhelmshaven. NA will give me his calculations soon. I have explained to the CO and his officers that we must do a planned dive soon, instead of an RAF or warship-forced dive. They will get their men ready and will report back to me. I have to check with Willie about the trim of the boot. I will have to check with Heintz about CO2 and how long we can safely stay submerged.

KTB Log June 10, 1940 02:48; NA reports estimated time to Wilhelmshaven is 26-30 hours at ahead flank. We can't run the whole 680 to 700 km at ahead flank, we can't afford to have an engine failure now. Row, row, row your boat, gently down the stream.

04:47; We won't be diving with so many sailors on board. Karl has radioed BdU, and we will be meeting an escort at 12:00, in Grid AN83.

05:48; *U-49* has exited grid AN82.

06:00; Willie has been doing sick call with the crew from *T-21*, most have been treated for dehydration and they have been given extra servings of food. There were two wounded men from *T-21* who had temporarily lost their hearing.

KTB Log June 10, 1940 09:58; CE reports fuel below 25%.

KTB Log June 10, 1940 12:00; Noontime mess is a scene; Willie has somehow made a nice feast for our comrades from *T-21*. Everyone is enjoying the gute fellowship, and gute food.

KTB Log June 10, 1940; 12:40 we have rendezvoused with an armed trawler out of Holland. They have signaled that there is light-to-no enemy traffic from here to base. Their orders are to stay with us until grid AN83.

14:07; The captain of *T-21* was standing watch with me, and was amazed at how well our ship cuts through the sea. He said it was a fine craft. *U-49* is making gute headway toward home. The usual end of patrol routine has to be held off until we offload *T-21's* crew.

KTB Log June 10, 1940 16:07; The rain is now heavy and the boot is rolling too much with the additional crew on board. Heintz and Willie are readjusting the trim. We have entered Grid AN83, NA reports depth under keel is a mere 12 meters.

16:16; We have been released by the escort, I saluted them from the bridge and soon the fog had obscured them from view. This heavy fog should keep us safe until we reach our next escort in Grid AN96. We slow to ahead full. We are averaging 14 knots in the heavy seas.

KTB Log June 10, 1940 18:02 I order ahead flank, we have sailed into Grid AN95, depth under keel is 14 meters.

18:22; The familiar sound of seagulls can be heard off in the distance. Rain and fog is still heavy, wind 15 mps. It seems Arend's dying cast a gray sky over *U-49*. Since his untimely death, we haven't seen the sun.

KTB Log June 10, 1940 18:49; We have made visual and Aldis lamp contact with a small German fishing craft, the *Sea Shtick*. I asked the skipper what he was doing. He replied that he was on a three-hour tour with Killagain, the Skipper, too, a wealthy industrialist and his mistress, a striptease artist, the professor and Helga Ann. I wished them gute sailing, and not to become washed up on one of the uncharted deserted isles.

22:55; SO reports warship moving away, bearing 029, long range.

22:59; *U-49* turns SSE grid AN95, ahead full. I don't want to miss this escort in the heavy fog and rain.

23:20; We have lost sound contact with the warship.

23:20; I order depth 10 meters and all stop. The SO reports the sounds of a sinking ship at 330. I listen and I can hear the unmistakable groaning of a sinking ship. Ahead standard and we surface the boot. We alter our course toward the wreck.

KTB Log June 10, 1940 23:44; *U-49* slows to begin searching.

Approximately 6 km from where we first heard the sounds, we have spotted a lifeboat, with 8 German sailors on board.

23:53; We have contacted the CO of the sunken armed trawler. I have explained that our boot is full to standing room only already. The plan is to tether a long line to his lifeboat and tow them to Wilhelmshaven and then transfer them to the first escort we see.

KTB Log June 11, 1940 00:12; SO report warship, moving fast, closing, bearing 339. Karl has radioed to the nearby warship; we will meet them and transfer the crew in the lifeboats.

00:39 *U-49* has rendezvoused with *S-boot 12*. We are at an all stop and the trawler crew is being untied, and they will row over to the S-boot.

00:46; We have bid farewell to the S-boot, they will take the crew from the lifeboat to their base, and we have resumed our trek to port. SO reports warship moving slow and closing, perhaps this is our escort.

00:58; Wind is down to 6 mps, heavy rain, heavy fog. I hope I can spot the escort, and the lighthouses at our base.

01:27; We are following an armed trawler towards base. We have to stay about at 350 meters to remain in visual contact.

KTB Log June 11, 1940 02:14 I hear a dreadful sound.

02:14; CE reports bow torpedo damaged.

02:14; CE reports we have flooding

02:14; CE reports hull damaged.

02:14; CE reports we are taking damage.

02:15; I order the repair team assembled, but they are hampered by the overcrowded fore torpedo room. I yell to Heintz, *Emergency back!* There is one last grinding sound as we free *U-49* from the nets.

KTB Log June 11, 1940 02:17; *T-21's* crew has had a bit of a scare, the sounds and the flooding made them think the boot was sinking. Order has been restored and the repairs are complete. Hull integrity is now 89%. Damn, the harbors are tough. Dumbass escort.

02 23; I couldn't have missed the opening by much, so we back along parallel with the far away harbor lights. I measure on the navigation map; we are about 4.8 km from base. We turn 35 degrees starboard and resume sailing ahead slow toward the lighthouse.

KTB Log June 11, 1940 02:48; We have sailed past the lighthouse and ahead 3 knots as we make our way to an open pen.

03:11; *U-49* has been secured to the pier and the gangplank will be up soon. *T-21's* crew will disembark first. The CO has thanked me for the ride and his crew was equally grateful. My crew won't disembark for several hours, they must secure their personal possessions and secure their stations.

KTB Log June 11, 1940 03:38; *U-49* sank five merchants, 45,502 tons and downed one RAF plane.

KTB Log June 11, 1940 04:09; A runner from command has arrived and brought me gute news. A "Review Board" has found no negligence on my part for Arend's untimely death.

I was right, it was a "bug" that got him.

KTB Log June 11, 1940 04:10; The crew of *U-49* has won medals: one Iron Cross, First Class, two Iron Crosses, Second Class, one U-boot war badge, and nine U-boot front clasps. I will walk over to the officer's club for a shower and be ready for the base commander when he reports to his office at 06:00.

⊕⊕⊕⊕⊕⊕⊕⊕

KTB Log June 29, 1940; *U-49* is still in drydock from the last patrol. Hitting the net or wall at the end of the last patrol damaged the dive planes as well as the hull. *U-49* is also undergoing a motor overhaul, and the engineers are busy replacing the cells. Wiring must be overhauled; all the welds must be inspected. All the bearings in the drivetrain must be replaced. All bunk boards must be replaced. We need a new stove and all the old storage bins need to be scrubbed and repainted. I need the engineers to put a better lamp at the cubby; we need a few electric fans.

They issued us new uniforms that are British Khaki. Seems the Brits left them at Dunkerque by the ton. The High Command granted me and the crew extended shore leave, three weeks for doing such an outstanding job while we were in der Kanal. Two weeks have passed and we are still on leave. I don't have to report until the first Friday in July.

The first week in port I wasn't yet on leave, I had the standard after action reports, debriefings, and meetings about the latest developments in the war to occupy my time.

The second week I went to visit Arend's wife and family, and to pay my respects at his grave. It was in a beautiful graveyard, very peaceful. His family and I went together and had a picnic lunch there. It was very sad and at the same time I felt his presence was there, too, so I was somewhat comforted.

His widow said she and the family would be all right. She comes from a wealthy family and they have given her all the financial aid she needs. She said that she had bad premonitions during the week he died. On the day he died, she said that a small black bird flew into the house and landed on his chair at the dinner table. It sang a few notes and flew back out the open window. She told me she knew then that Arend was dead.

⊕⊕⊕⊕⊕⊕⊕⊕

Well, that was a sad week so Jurgen, a few of the mates and myself are heading to France! We kicked the French armies' arse and the Brits have been pulled back to England. The western continent is Germany's! A few days after reaching port, Paris fell. So we are off to do some R&R with those sexy French women!

You know you've been with Subsim a long time when...

...You've become obsessed with the idea of tracking each rodent in the attic by their unique sound signature.

SUBSIM ROLL CALL

I was looking for anything related to modern submarine warfare, and specifically Jane's 688i H/K, when a search on AltaVista brought me to Subsim; specifically, the "Fix My 688I" Campaign. I signed the petition: I was blown away that someone loved the game as much as I did, but was doing something about it to make it better. Subsim.com became THE place for me to go to learn about modern and historical naval warfare, new subsim releases, and read the cool articles. Thanks for many years of service and enjoyment, Subsim. And here's to many more to come!

James "Skweetis" McQueen

My story is quite similar to many others whose path to Subsim seems to be thanks to the nuke subsim Sub Command. I joined in around September 2002. It was probably as a result of a link to a review. I have been a member ever since and it is the only real forum that I visit regularly.

Benjamin "Konovalov" Sadler

I first visited Subsim shortly after I got Jane's 688(I) Hunter-Killer in the bargain bin at Wal-Mart. When I would search for a patch for Command: Aces of the Deep, the search always leads me to Subsim. Through the years I was able to remember the name of this site rather easily. I would check in now and then to see if there was any news. When Silent Hunter III was announced I made my way back to the forum. I remember when the big news broke that the campaign would be changed to a dynamic type. That was a happy day! One of the most fun times I can recall here at Subsim was the Recon-Test. I would try and check in everyday to see what the intel photo was going to be. They were all pretty challenging and that is why it was so enjoyable.

Jeremy "Syxx Killer" Craven

I came upon Subsim while searching for game tips. Soon after, I bought SHIII and have played nothing but ever since.

Jason "Camaero" Packard

I had picked up SHII and DC and was looking for some info as I was already a flightsimmer and figured there must be a place for subs and naval stuff. SHII had Subsim listed in the credits.

Joe "joea" Apostolidis

I've been visiting Subsim regularly since I got on the Internet, which would be around 1999. First time I got here was when searching for the expansion disk for AoD. I found 1.31 patch for SH1 (and how to make it work with the German version) at Subsim.

Martin "Viper the Sniper" Bisanz

I first visited Subsim ages ago, but only to read this or that review. Then I came to see the SHIII review. I've come to see Subsim as an excellent community. For me, of course, it's still mainly a sort of centre-of-gravity for the SHIII community, but I've come to realize that it's much more than that. Nostalgic times? It was when the "SH3 Mod Team" forum was active and the Operations/Real U-boat team was together.

George "CCIP" Ross

Subsim is a great place to come to chat about games and world events, a great community, latest info on what games are coming out, gaming tips/help from other subsimmers and to catch up on the odd bit of navy news happening from time to time. I think what really made the wow factor for me is when I realize some posters are actually sailors from real subs and ships and here I am in the same forum with them! Coming from New Zealand, that's a first for me. In 2000 I actually registered with Subsim, yet I wasn't in there often -- was hard at work at the time so only popped in very rarely. Now I'm addicted to the forum.

Frank "kiwi 2005" Slavich

I started to visit Subsim way back in 1999. I was one of the "Fix my 688" Campaign petitioners.

Willem "Fish" Peschier

I found Subsim whilst looking through a Google search. I had been playing Sub Command for a long time and was looking if there were any newer games coming out, and low and behold, I stumbled upon the gold mine of the Subsim world. I personally believe that Subsim is the hub of all naval games and I hope it remains so for a long time to come. There are a lot of excellent players, guys and girls here we all can learn from, and the best thing of all is there is not a lot of bitching, fighting, scrapping, or what have you (okay, okay, in fairness, I have started a few). I am proud to be a Subsim member. It's one of the very few decent forums on the Net, one where we can all come together and share our views opinions.

Blair "Kapitan" Shaw

I first came to Subsim, in January of 2002, after joining the WPL the previous month. The thing I recall most from that first visit was all the naval-related news stories on the front page. What I remember the most is the people that I've talked to over the years through the Subsim community.

Jim "Malefactor" Brooks

Blind Man's Bluff led me to Jane's 688(I). A search for more info about 688 led me to Subsim. I lurked for a long time as a stowaway. I found it to be a nice place with a nice and friendly bunch of guys. I especially fell in love with the Indy forum that gave me so much inspiration and encouragement. Now I can't do much more than work on my indy subsim.

Heinrich 'Deamon' Lang

I would like to say that Subsim has been for all these years one of my favorite homes that I always come back to. Mainly I like the environment here. We don't tolerate flamers, trolls and other people who make most other online communities a living hell (look at any MMO company forum for an example of forums at their worst). Partly I think it's this way as we tend to attract older gamers, but mostly I think its because the Subsim staff and moderators (myself included) try hard to keep this place clean, helpful, and friendly. As long as this place exists I will stay.

Bronwyn "NeonSamurai" Despres

I visited Subsim the first time roughly half a year before Sub Command came out. I was busy with Jane's 688i, had questions, and stumbled over the place. From there, I received some well-meaning e-mails by someone, which did not help at all, unfortunately. He finally said (wanting to send me back to the forum where we had met): "Ask Neal about it." And I replied something like, "Who is this Neal, and where can I find him?"

Skybird

I've been with Subsim since sometime in '97. At the time I had dusted off a copy of the old DOS game Seawolf. I had never beaten the last mission, and while looking for hints, came across Subsim on the Internet. Wild Bill told me to stop using my torps in high-speed mode, and that was what I needed. Jane's 688(I) was what most of the people were playing at the time. That era was my Subsim heyday, sparring with Kraut/Akula971 and Sixpack.

Brian "Terrax" Logan

I learned about Subsim after playing Silent Service II and Silent Hunter. I was very interested in Gato subs and wanted to check RC models. I found Subsim.com in 2000 and used it as a reference. After a while I started to check the news section, since 2002 on an almost daily basis. I was impressed by Drebbel's Silent Hunter III preview of in the fall of 2004 and decided to send that fellow Dutchy a compliment by e-mail. He responded graciously and asked me if I ever posted on the forum. "Nah, nothing for me," I replied. But Drebbel didn't take 'Nah' for an answer and convinced me that the Subsim forums were real fun.

Abraham Zeegers

I found my way to Subsim some time during the previous millennium. The first great memory I have from Subsim was when I corresponded with a user that had signed up just 4 months before me; his name was Abraham. We talked about Subsim, SH2, SH3 and about Drebbel's Dutch Submarines webpage. Back then both of us were two unknown members of Subsim, but in the end, both "bilge rats" ended up as moderators, at the same time. Subsim truly works in mysterious ways.

Sebastian "Polak" Kowalski

I've always been an avid military historian with a wide variety of interests. I've been playing war games and flightsims, since the days of PT 109 back on the old Tandy 386SXs. I discovered Subsim with SH II.

Jeremy Scott

I first got into playing simulations back in 1993. I went through SeaWolf and 688(I). When Aces of the Deep came out I got deeply involved with subsims. I started looking for information on the Internet and discovered Subsim. I had the SubGuy website up in 1995. Jim Wolford (Australia), who was also playing Silent Hunter, assisted me in putting up that website.

Robert "SubGuy" Johnson

I found Subsim in 1998 after I picked up a copy of Silent Hunter: Commander's Edition. I'd always been interested in sub games, having started back in 1985 with Silent Service on the C64. But it wasn't until Silent Hunter III was released that I joined the Subsim forums; quite a few talented modders had assembled there and I didn't want to miss a thing!

Jaesen "JScones" Jones

I found out about Subsim from the SH2 manual. I then downloaded the mod that changed SH2 to US subs in the Pacific. I would like to mention the patience shown to newcomers who ask the same questions over and over again. I used to be one and I now try to show the same helpfulness and respect that I received.

A. Khayman

Got back into Sub Command after a long hiatus and was searching the Internet for mods, downloads, and patches. I saw an article that said Sonalysts was making a new game. I kept searching for stuff and found SCX and Subsim. After being impressed with SCX and the highly anticipated DW I registered as a forum member. That was a little over two years ago. I'm a member of two other forums. One related to jobs in the aerospace industry, and another to dirt biking. Subsim is by far the best one and often times, the most interesting one.

John "Sea Demon" Churan

It's a snowy Saturday night in 1960 in Winnipeg, and a very young John Channing is curled up on his father's lap watching Silent Service *on an RCA Victor black and white television as the fireplace crackles softly in the background.* Skip ahead to 1995 and I had just bought my first computer, a Compaq 486 sx25 with 4 Megs of RAM. I had just come across EA's SSN 21 Seawolf. This was quickly followed by Wolfpack, Aces of the Deep, Silent Hunter, and Fast Attack. It was the Golden Age of Subsims. With the help of a 14.4 kps modem, I typed "Submarines" in webcrawler.com and came across "Neal Stevens' Deep Domain." Off I went…and I never left.

John Channing

I was looking around Google for sub games and ta-da! I have found the subsim Mecca of the world here. I got here in time to witness the record for most people online in the forums.

Minan Stewker

I first heard about Subsim.com way back in 1998, while looking for tips for Silent Hunter. When I retired, I bought my first computer in 2001.

John "donut" Glaze

I am a real "Newbie" to Subsim. I am 71-yrs old and was a submariner for twenty years (1954-1974) in the Royal Navy.

Rev. Anthony "Dusty" Miller

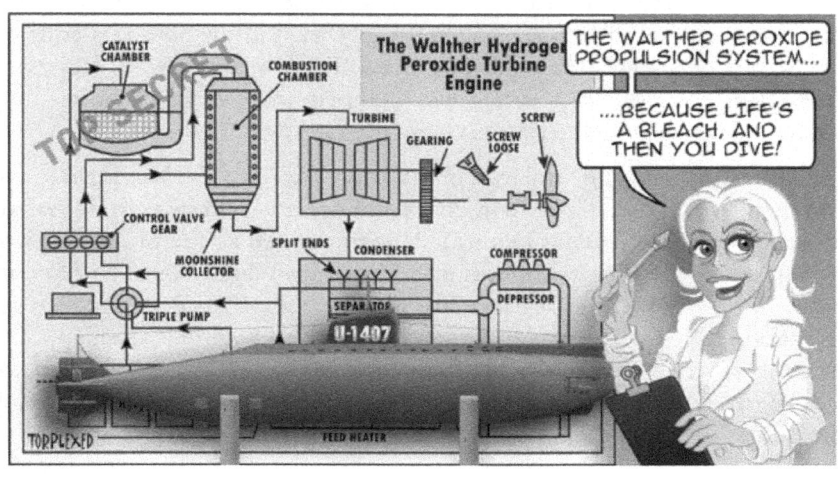

Origin of the Laconia Order

Dr. Maurer Maurer and Lawrence J. Paszek

The *Laconia* affair has long been of interest to students of naval history and international law, but the details concerning one important facet of the story were lost for many years among the voluminous records of U.S. air operations in World War II. In fact, it was not until 1959 that USAF historians began a systematic search for information that would dispel the mystery of the American bombing plane that had been involved in the incident. This search, which indirectly resulted from the publication of Admiral Dönitz's *Memoirs*, extended over four years before the historians were able to locate and obtain all the information required for this fully documented report on the U.S. Air Force role in the *Laconia* affair.

In the war crime trials at Nuremberg following World War II, Admiral Karl Dönitz had to defend the so-called "Laconia Order" in which he had prohibited German submarines from aiding or rescuing survivors from ships they sank. The order was being used by the prosecution to help support a charge that Dönitz had violated the rules of warfare. In his defense, the former commander of the U-boat fleet said that the order was the direct result of an incident which had occurred in the South Atlantic on 16th September, 1942. He explained that, "In spite of flying a large Red Cross flag," one of his U-boats was bombed while engaged in rescuing survivors from a torpedoed British ship, the *Laconia*.

The story of the sinking of the *Laconia* by U-156, Lieutenant-Commander Werner Hartenstein commanding, is well known from various published works, including Dönitz's *Memoirs*, as well as the proceedings of the International Military Tribunal at Nuremberg. With three other U-boats and a submarine tanker, U-156 had sailed from France in August for operations off Cape Town, South Africa. On Saturday evening, 12th September, when U-156 was approximately 900 miles south of Freetown and 250 miles northeast of Ascension Island, Hartenstein sighted and sank the White Star liner *Laconia*, which the British Admiralty was using as a transport. Before she went down, the liner sent two radio signals, one at 2022 hours on the 600-meter international wavelength and the other six minutes later on the 25-meter band, reporting that she had been torpedoed at 04° 34' South latitude, 11°25' West longitude.

Hearing shouts in Italian, Hartenstein picked up some of the people from the *Laconia* and learned that the British ship had been carrying 1,800 Italian prisoners of war. Later, in interrogating the *Laconia's* navigation officer, the Germans learned that in addition to the Italians and a crew of 463 the liner was carrying 268 British Service personnel, 80 women and children, and 160 Poles who had been guarding the Italians. According to the navigator, two torpedoes from U-156 had hit compartments occupied by the Italians, and many of the prisoners had gone down with the ship. Discovering his allies among the survivors, Hartenstein immediately began rescue operations and reported the situation to his superiors. Dönitz ordered the submarines to take aboard as many survivors as possible without interfering with the ability of the U-boats to submerge, for the safety of the submarines was not to be endangered in order to carry out rescue operations.

Later that night, Hartenstein informed Dönitz that U-156 had picked up 193 persons, including 21 British, and that there were hundreds of others in the water. In his radio message, the U-boat captain suggested diplomatic neutralization of the area. A little later, at 0400, he announced, by the means of radio messages sent in English on the 25- and 600-meter bands, that if he were not attacked he would not interfere with any ship coming to the aid of survivors. Meantime, Dönitz had ordered two other submarines, U-506 and U-507, to proceed to the area and help with the rescue work. He also requested the assistance of an Italian submarine, the *Cappellini*, and asked the Vichy French to send warships from Dakar to meet the submarines and take on the survivors.

During Sunday the 13th, U-156 was busy fishing people out of the water and placing them in lifeboats. On the 14th, after learning that the French were sending ships, Dönitz ordered the submarines detailed for the Cape Town operations to go on, but U-156 was to remain. On Tuesday the 5th, U-506 and U-507 joined U-156 in the rescue work, and on the following day the *Cappellini* arrived in the area.

Hartenstein had received a report of an unidentified steamer in the vicinity, and he apparently expected it to respond to his message offering immunity from attack to ships that would help rescue survivors. He also had been informed that there were no airplanes on the British island of Ascension, but that the Allies had planes at Freetown. Thus far, however, the submarines had carried out their rescue operations without assistance or interference from the Allies.

On the morning of Wednesday the 16th, the U-boats were collecting and bringing together lifeboats and rafts in preparation for the rendezvous with the French ships the next day. U-156 had aboard 110 survivors, 55 Italians and 55 British, including five women. Some of the survivors were below, but many were on deck and others were in lifeboats which the U-156 had in tow. At 0925, as he was preparing to pick up another lifeboat, Hartenstein saw a four-engined aircraft with American markings approaching him. He immediately covered the forward gun of U-156 with a large Red Cross flag to indicate the nature of his mission and his "peaceful intentions". As the plane circled overhead, the U-boat captain used light signals to ask where the plane was from and whether it had seen a steamer in the vicinity, but his attempts to communicate with the airplane failed. After some time the plane flew off to the southwest, but about 1030 it returned and bombed the submarine. This was the attack which Dönitz said was responsible for the "Laconia Order".

During the years following World War II, lack of information concerning this unfortunate incident of war gave rise to a number of speculations. Hartenstein had identified the plane as an American Liberator (B-24). In 1956, Samuel Eliot Morison, historian of U.S. naval operations in World War II, stated that the plane had come from the American base on Ascension Island, but in 1959 the U.S. Navy was attempting to identify the plane as one operation from Freetown. A number of naval authors — American, French, British, and German — sought to clarify the incident, but details of the attack by the B-24 on U-156 remained shrouded in mystery. On 4th August, 1963, however, the London *Daily Express* quoted Brigadier General Robert C. Richardson III of the U.S. Air Force as saying that he had commanded the squadron on Ascension where the B-24 was based temporarily on 16th September, 1942, and that he had ordered the attack. Unfortunately, the reporter who interviewed the General played up the sensational side of the story rather than the conditions under which the attack was made.

Richardson, then a captain, arrived on Ascension in August, 1942, when the U.S. Army Air Forces opened an air base on the British island. Construction of the base had been proposed late in 1941 by the AAF's Ferrying (later Air Transport) Command to provide a refueling stop for military aircraft flying the southern ferry route from the United States to Brazil and then across the South Atlantic and Africa to the Middle East. After the United States entered the war in December 1941, the proposal was quickly approved, the plans were drawn, and permission to use the base was obtained from the British. At the end of March, 1942, two Army transports, escorted by two cruisers and four destroyers, arrived at Ascension with men, equipment, and materials for construction. Work began on 13th April, and on 14th August the permanent garrison arrived to take over the operation of the base.

The planning, construction, and operation of the base were veiled in secrecy, but the Army feared that such activity could not be hidden forever from the enemy. Dependence upon shipping to bring in equipment, supplies, fuel, food, and all the other necessary items made the base vulnerable, for the German submarines and surface raiders were active in the South Atlantic and Vichy France had a number of warships at Dakar under the control of officers who were reported to be pro-German. The enemy could put the American base out of commission by sinking tankers and supply ships. Or enemy submarines or surface vessels might shell gasoline storage tanks and other vital installations, such as the plants required to convert seawater into drinking water for the troops. Furthermore the Army was aware that neither Britain nor the United States could spare naval ships to provide a constant guard over the sea-lanes and prevent the enemy from shelling the island. There was also a possibility that the Germans, with the collaboration of Vichy France, might send planes from African bases to bomb Ascension. Or they might land troops to sabotage military facilities or even try to capture the island.

Fearing, then, that the enemy would discover the activity on Ascension and attack the island, the U.S. Army made elaborate plans for defense of the base. The construction job was assigned to combat engineers who were not only equipped to build an air base but were also armed to protect themselves and their site. For added security, the Army sent along an anti-aircraft battery, which went into action for the first time on 15th July, a few days after the runway was ready for emergency landings. With orders to fire on any unannounced airplane that came within range, the battery opened up on a small biplane that flew over the field. The gunners scored three hits as the engineers quickly placed machinery and vehicles on the runway to prevent a landing. Then the plane was identified as friendly, the obstructions were removed, and a Swordfish from H.M.S. *Archer* landed.

Work progressed rapidly, and without interference from the enemy, as the engineers constructed roads, barracks, a hospital, water and electrical plants, gasoline storage tanks, gun emplacements, ammunition dumps, radar sites, and radio towers. In the interest of security, great care was taken in camouflaging the various installations.

While the engineers worked, the units for the permanent garrison were being formed at various camps and bases in the United States. These units sailed from Charleston, South Carolina, aboard the *James Parker* on 26th July and arrived at Ascension on 14th August. Three days later, Richardson arrived from the United States with a flight of medium bombers to be based on the island.

Planes flying the South Atlantic route had begun to use Ascension for refueling late in July, before the Ferrying Command had any personnel based there, but a Ferrying Command detachment of two officers and 22 men arrived in the *James Parker*. To support and defend this detachment and its operations, the Army provided a task force of approximately 1,700 officers and men under the command of Colonel Ross O. Baldwin, an Infantry officer. About one-third of the members of this force were assigned to AAF Composite Force 8012, commanded by an Air Force colonel, James A. Ronin. The largest AAF unit was the 1st Composite Squadron, commanded by Captain Robert C. Richardson III. This squadron, made up of 30 officers and 219 men, had two flights of pursuit planes (18 P-39D's) and the flight of medium bombers (5 B-25C's). Ronin's force also included a signal warning detachment to operate two radar sets, as well as personnel for air base, weather, and communications functions.

The largest unit of the task force was the 3rd Battalion (less two rifle companies) of the 91st Infantry Regiment. The ground forces also included two artillery batteries, a searchlight platoon, and personnel for medical, quartermaster, and ordnance activities. In addition to rifles and side arms, the ground forces had 200 submachine guns, 28 .30-cal. Machine guns, 12 .50-cal. Machine guns, 4 37-mm anti-tank guns, 4 81-mm mortars, 4 155-mm guns, and the 4 37-mm anti-aircraft guns that had been sent in with the engineers, plus 2 5.5-inch naval guns which the American task force took over from the British detachment on the island. The U.S. Navy did not send patrol planes until much later, but 29 Navy men were attached to the Army task force to operate small craft for rescue work and patrolling the harbor.

Baldwin's intelligence officer, Major Walter C. Buethe, believed that by mid-September the enemy was aware of the presence of U.S. Army forces on Ascension. In his opinion, a bombing attack was not likely because of the distance from enemy air bases, and shelling by surface raiders was not probable after the 155-mm guns and radar were in operation. The greatest threat, he believed was from enemy submarines, which might shell the island or land small parties of men to sabotage gasoline storage facilities or other vital installations. Richardson agreed in general with this estimate. He was not concerned about enemy aircraft, but he thought that the enemy might try to destroy the gasoline tanks, which were very vulnerable to attack from the sea. The main function of Richardson's squadron therefore was to patrol the sea around the island, to detect and destroy any enemy submarines and surface raiders in the area, and to protect Allied ships in the vicinity of the island. Although the squadron had been supplied with bombs and depth charges for anti-submarine warfare, it had no special training or equipment for such operations.

By 20th August, 1942, the 1st Composite Squadron was ready for operations. Patrols were scheduled daily, and some aircraft were kept constantly on alert at the base, which had been named Wideawake Field after the wideawakes (sooty terns) that claimed the island in great numbers. In a readiness test conducted on 27th August, a P-39 on alert was able to take off in less than three minutes, and a B-35 got away in ten.

During the first three weeks of operations, through Saturday, 12th September, the day the *Laconia* was torpedoed, the 1st Composite Squadron flew 64 sorties within a radius of 250 miles of the island without sighting a single target. On 3rd September, however, the men of "A" Battery at Southwest Bay, thinking they heard the engines of a U-boat and seeing a craft submerging just off shore, had fired at it. Fortunately, the gunners were either poor shots or were too excited to aim accurately, for the target turned out to be a P-39 which they had not seen land on the water. Returning from a routine patrol, Lieutenant Ben Herbert Smith had ditched his plane when the fuel system failed. The P-38 sank almost immediately, but Smith, who was not injured, got out and floated in his Mae West until he was picked up by a crash boat.

Search and patrol missions from Ascension were entirely under the direction of Air Force personnel, for Baldwin, the task force commander, was not consulted on matters relating to air operations. Whenever the 1st Composite Squadron had planes out, at least one of the key Air Force officers — Ronin, Richardson or Captain Willard W. Wilson, Ronin's operations officer — was already readily available at the base. Through the control tower he could communicate by radio with pilots on search and patrol missions. The Army's radio station on Ascension, WYUC, also received some of the reports of submarine sightings from the Allied ships in the South Atlantic. All the radio equipment had not yet been installed, and WYUC was not in communication with either South America or Africa. The American task force, however, used the British cable for communication with South America and the United States and used the British radio on Ascension, ZBI, for communication with Freetown, where the Royal Navy collected and correlated data on submarine sightings and directed movements of Allied ships in the South Atlantic. A British liaison officer attached to the U.S. task force passed on to the Americans submarine sightings and other intelligence data he received by radio from Freetown.

Neither WYUC, nor ZBI picked up the *Laconia's* signals on Saturday, 12th September, or Hartenstein's messages on the following morning, when he asked ships to come to the rescue of the survivors. In fact, it appears that no British station or ship read the *Laconia's* signals. Freetown, however, evidentially received the messages that Hartenstein sent in English, but the British were suspicious that the Germans might be setting a trap for unwary merchant ships. At any rate, the Americans on Ascension did not learn of the sinking of the British liner until Tuesday the 15th.

The 15th was a busy day at Wideawake Field. At 0700 four A-20s and a Stratoliner, the latter carrying a British admiral and his party, took off for Accra. En route, one of the A-20s sighted two submarines at 04°40' South, 11° West. While circling for a better look, the A-20 was fired on, but not hit, by the boats. The Stratoliner, which was in the vicinity but flying higher, above the clouds, picked up the A-20's radio report of the incident and relayed it to WYUC. Within ten minutes after receiving the message, the 1st Composite Squadron had two B-25's on the way to the area.

Two hours later, at 1210, the British liaison officer gave the Americans a message which either was garbled or indicated that Freetown's information was faulty, for the message said that the *Laconia* had been torpedoed only a few minutes earlier, at 1145 on the 15th, at 05°05' South, 11°30' West, a considerable distance from the spot reported by the British liner. The message indicated that the *Laconia* had carried 700 passengers, but there was no mention of German submarines being engaged in the rescue operations or of Hartenstein's call for the Allied ships to assist with the rescue work.

Neither the B-25's dispatched to the scene of the attack on the A-20 nor the other planes which flew search missions that day had any luck in spotting the enemy. That night, in a message delivered at 2200 to the Americans on Ascension, the British asked the 1st Composite Squadron to assist with rescue efforts being directed from Freetown. There were few Allied ships in that part of the South Atlantic because warships and merchantmen were being assembled for the invasion of North Africa and because shipping had been routed farther west to avoid the greater submarine menace along the African coast. A merchant ship, the *Empire Haven*, was nearby, however, and H.M.S. *Corinthian* was at Takoradi. They were being sent to aid the survivors from the *Laconia*, and Freetown wanted Richardson's squadron to provide air cover for the operation.

The spot where the *Laconia* had gone down was so far from Ascension that a B-25 from Wideawake would be able to remain in the area for less than half an hour. A plane with longer range was needed, and as it happened, there was one at Wideawake on the night of 15th September. It was a B-24D Liberator of the 343rd Bombardment Squadron, which had recently moved across the South Atlantic en route to the Middle East for service in the battle against Rommel's Afrika Korps. On the way over this four-engine bomber had been separated from other planes of the squadron when it was delayed by mechanical trouble. Now it was at Wideawake, and it was pressed into service.

Loaded with depth charges and bombs, the B-24 took off at 0700 on Wednesday, 16th September, and headed north-east. The pilot was Lieutenant James D. Harden, and his crew included Lieutenant Edgar W. Keller, bombardier, and Lieutenant Jerome Permian, navigator. These men, all of whom were flying their first combat mission, were members of the 343rd Squadron, but the co-pilot, Lieutenant Raymond J. Ford, belonged to Richardson's 1st Composite Squadron.

At 0930, Harden spotted a submarine — U-156 — towing two lifeboats and approaching more at 05° South, 11°40' West. While the B-24 circled overhead, its crew saw the U-boat pick up the other two lifeboats and continue on its course. They also saw that the submarine had a white flag with a red cross. Using a signal lamp, the crew challenged the U-boat to show its national flag, but none was displayed. The submarine, however, blinked light signals which could not be read very clearly but which were thought to be "German Sir". After 40 minutes, Harden gave up the effort to communicate and headed southward.

The radio operator on the B-24 soon established contact with WYUC at Wideawake Field, reported the sighting of a submarine towing four lifeboats, and asked what to do next. As Colonel Ronin says, "It was a good question." It could be answered in only one of two ways: return to base, or attack. There were no friendly submarines in that part of the Atlantic, and the Americans, who as yet had no word of the rescue work being conducted by German U-boats, had received no instruction against interference with such operations in that area. Richardson carefully weighted the alternatives. He had responsibility for providing the protection that Freetown had requested for British ships going to the rescue of survivors. If he ordered Harden to come in, he would not only jeopardize the safety of British ships, but would leave the submarine free to continue destruction of Allied shipping. Further, such an order would mean abandoning an important and legitimate military mission that had a chance of successful accomplishment. On the other hand, an order to attack would place in jeopardy the lives of some of the survivors. Harden had to have an answer soon. He could not remain in the area much longer and still have enough fuel to get back to Ascension. After conferring with Ronin, Richardson issued the order: "Sink sub".

Upon receiving the signal, Harden turned back northward and soon found the U-boat. Following is the account of the attack as reported by the pilot of the B-24.

> Upon returning to position, lifeboats had moved away from the sub. One pass dropping three depth charges was made, one hit ten feet astern, and two were about 100 and 200 yards. Made three more runs and bombs failed to fall. This was fixed and a final run was made at 400 feet. Two bombs were dropped one on either side, not more than 15 or 20 feet away. The sub rolled over and was last seen bottom up. Crew had abandoned sub and taken to surrounding lifeboats.

The log of U-156 describes the attack as witnessed from the submarine:

> Aircraft of similar type approached. Flew over, slightly ahead of the submarine, at altitude of 80 meters (about 250 feet). Dropped two bombs about three seconds apart. While four lifeboats in tow were being cast off, the aircraft dropped one bomb in their midst. One boat capsized. Aircraft cruised around for a short time and then dropped a fourth bomb 2-3,000 meters away. Realized that his bomb racks were empty. Another run. Two bombs. One exploded, with a few seconds delayed action, directly under the control room. Conning tower vanished in a tower of black water. Control room and bow compartment reported taking water. All hands ordered to don life jackets. Ordered all British off the boat. Batteries began giving off gas. Italians also ordered off (had no escape gear to give them).

The people in the lifeboats saw the attack from still a different view. One of the boats reached the Liberian coast on 10th October, four weeks after the *Laconia* was sunk. During that time 52 of the 68 persons in the boat had died. The 16 people who reached land safely (15 British and 1 Pole) had suffered terribly. It is no wonder that they were confused as to the chronology of events. Here is their story as reported by the American chargé d'affaires in Monrovia:

About four o'clock on that afternoon (Sunday, 13th September, according to the report) an American liberator bomber appeared, and although the submarine displayed a Red Cross flag, the bomber launched seven depth charges one of which fell near a lifeboat, completely destroying it and drowning all passengers, who were Italian prisoners. The others fell about three yards on either beam of the submarine, the explosion lifting it from the water and obviously caused damage. The submarine continued on the surface for about a mile then submerged, throwing all of the survivors from the deck into the water. Many of these were drowning by suction, but the remaining lifeboats were able to pick up a few.

U-156 had not been sunk, as Harden and his crew believed, but it had sustained considerable damage. Shortly after 1100, Hartenstein returned to the lifeboats and transferred to them the remaining passengers he had aboard. He then submerged and headed westward. So far as he was concerned, the rescue operation was ended. That night, when Dönitz was informed of the attack on U-156, he directed U-506 and U-507 to continue rescue work and hand over the survivors to the French ships that would arrive the next day. Meantime the U-boat captains could retain Italians aboard, but all other survivors were to be transferred to lifeboats. Warning the captains to beware of attack, Dönitz instructed them not to seek protection under the Red Cross flag, but to keep their boats ready to submerge instantly.

While the B-24 was on its way back to Ascension, Lieutenant Richard T. Akins, pilot of a B-25 of the 1st Composite Squadron, reported at 1025 that he had sighted lifeboats and rafts at 05°10' South, 11°10' West, just a few miles south and east of where Harden and his crew had bombed U-156. That afternoon, Richardson flew out to the area and found some lifeboats. He also saw the *Empire Haven,* which he directed towards the boats. An hour later, Captain Virgil D. Holdsworth in another B-25 reported that he had spotted lifeboats at 4° South, 12° West.

That night, a message from Freetown indicated that French warships from Dakar were headed south, but there was nothing in the signal to indicate their mission was to assist in rescuing survivors from the *Laconia*. The men on Ascension were sure that if the enemy had not previously discovered the presence of American forces on the island, he knew it now as a result of the bombing of the submarine. Assuming that the Vichy warships were on their way to Ascension, the men prepared to defend the island. As the historian of the 1st Composite Squadron wrote, "arrangements were made for an American Reputation — *the powder was dry.*" The tension that night was heightened when the radar picked up a surface target 40 miles to the northeast. For an hour and a half the radar followed the track as the target moved to a position 14 miles southeast of the island.

At 0720 on Thursday, 17th September, Harden and his crew were off again in the B-24. They reached the search area at 0905 and began flying a square pattern. At 1030 they sighted a submarine two miles ahead and to the left at 04°51' South, 12°22' West. Increasing his speed to 200 miles per hour Harden went in for the attack. The boat crash-dived, and its conning tower and deck were awash when the B-24 passed over. The bombs failed to release, so Harden went around and made a second pass 45 seconds later. This time two 500-pound demolition bombs and two 350-pound depth bombs fell in train, two landing astern of the submarine and two hitting directly on top. When Harden came back over the spot, the crew saw an oil slick. For 40 minutes, the plane circled the area, but no further results were observed. Harden then headed back to base, the crew believing that they had sunk the submarine or at least badly damaged it. But they were wrong. U-506, which then had more than a hundred survivors aboard, escaped without damage.

In nine other sorties flown by the 1st Composite Squadron on 17th September, only Akins had anything significant to report — at 1500 he saw eight people on a raft at 03°25' South, 13°10' West. When he went back to the same area the next morning he found four empty lifeboats, all in good condition. There were oars in the boats, and Akins thought that there also was food. At the time he had no idea the people might have been removed by the warships from Dakar. One of the French ships, a small vessel which was making 22 knots on a zigzag course northward, was sighted later that morning by Lieutenant J.A. McClellan at 03°45' South, 13°15' West.

That afternoon at 02°56' South, 13°35' West, Lieutenant Philip Main sighted two French ships headed northwest at 17 knots. Reporting by radio from his B-25, Main received instructions to identify the vessels if possible but not to attack unless fired upon. At 1500, Wideawake queried Freetown concerning the status of the French vessels. The British reply, received at 1700, was to shadow but not interfere with them, the message said, "it appears that they are searching for the Italians from *Laconia*." This evidently was the first time that the Americans on Ascension had received any information concerning the rescue operations which had been undertaken by the German U-boats and the Vichy warships. The cruiser *Gloire* and the sloop *Annamite* had arrived the previous day. Now *Gloire* was on her way back to Dakar with more than a thousand survivors taken from U-506 and

U-507 and from lifeboats and rafts the French had found. The following day another French sloop, the *Dumont d'Urville*, met the *Cappellini* and took on 42 survivors, who were subsequently transferred to the Annamite and taken to Dakar.

Meantime, on the 17th, Dönitz had issued the directive that was to become known as the "Laconia Order." Addressed to all commanding officers, it read in part as follows:

> No attempt of any kind must be made at rescuing members of ships sunk and this includes picking up persons in the water and putting them in lifeboats, righting capsized lifeboats and handing over food and water. Rescue runs counter to rudimentary demands of warfare for the destruction of enemy ships and crews.

The "Laconia Order" had a prominent place in the Nuremberg proceedings which resulted in Dönitz being sentenced to prison for ten years. In the trial, however, the order lost most of its value to the prosecution when Fleet Admiral Chester W. Nimitz testified that the war with Japan the U.S. Navy had followed the same general policy and was set forth in the German admiral's directive.

The B-24, which attacked U-156 and U-506, never flew another combat mission. The plane crashed in Palestine on 18th October, 1942, while Harden and his crew were on their way to rejoin the squadron in the Middle East. None of the men were injured, and when Harden returned to duty he turned in a report of his operations from Ascension Island. On the basis of that report, the American commander in the Middle East awarded Air Medals to Harden and other members of the crew for destruction of an enemy submarine on 16th September and for the probable destruction of another on 17th September, 1942.

Hartenstein, having survived the attack by the B-24, continued operations and sank two more ships before returning to base on 16th November, 1942. The U-boat captain perished on another voyage, when U-156 was sunk by U.S. naval aircraft east of Barbados on 8th March, 1943.

A note by Captain S. W. Roskill

It was in about 1954, when I was working on Volume II of *The War at Sea*, that I first came up against the perplexing question of why and by whom an order was given to the Liberator (B.24) aircraft to attack the U-boats which were engaged in rescuing the survivors from the sunken liner *Laconia* on 16th and 17th September, 1942. My chief concern was, of course, to find out to what degree the British authorities were involved, and I therefore made a careful search in the records of the naval bases at Freetown, Sierra Leone, and Cape Town and in the Admiralty's signal files covering the period in question. Though the surviving base records were obviously not complete, my researches left me fairly confident that neither the Flag Officer, West Africa, nor the C-in-C, South Atlantic at Cape Town, nor their associated R.A.F. commands nor the Admiralty was in any way involved in the issue of the order. Though the conclusion from the written records was of course not positive (since there was no evidence regarding who gave the order) it was supported by all the surviving officers who were serving in the British headquarters at the time whom I was able to contact.

In 1958, shortly after Volume II of *The War at Sea*, in which I had dealt briefly with the "*Laconia* Incident," had been published, I was in America where, as always, I was receiving the most cordial co-operation from the U.S. Navy Department. Admiral Dönitz's memoirs had just been published in Germany (the English translation did not actually come out until 1960), and I was aware that he made a very strong attack on the British naval authorities' conduct of operations after the sinking of the *Laconia*. (See *Admiral Dönitz Memoirs*, Weidenfield & Nicolson, 1960, pp. 255-64.) I therefore enquired informally of one of my American friends whether it was not time that the whole question of the issue of the order to bomb the U-boats was cleared up. As I received the very strong impression that further enquiries on the subject would not be welcome, and I was at the time in the position of being treated as an honored guest, I judged it injudicious to press the issue.

When, however, a few months later the *Sunday Times* asked me to prepare a number of serial articles from Admiral Dönitz's memoirs, and to provide a running commentary on them, I took the opportunity to draw attention to the gap in my knowledge (see article "Mystery of U.S. Plane's Attack," *Sunday Times*, 1st February, 1959). This produced no reaction in America beyond a request for a copy from one official in the State Department. The next development was a letter from M. Léonce Peillard, the French naval historian, telling me about his research into the whole story, shortly followed by a copy of his very full and extremely competent study (*L'affaire du Laconia*, Robert Laffont, 1961, translated into English as *U-boats to the Rescue*, Jonathan Cape, 1962). At the end of his book M. Peillard reproduced his correspondence with the U.S. Navy Department's Office of Naval History, and also a report received by that office from the U.S. Air Force Historical Division which had been passed to him. Unfortunately both those documents did contain the implications that the aircraft in question might be, and indeed probably was, British. Thus the

U.S.A.F. Historical Section's report stated that, "Although the attack may have been made by a U.S. B.24, it was probably made by an R.A.F. plane." This unsupported deduction was, in my view, completely incorrect; and it was in fact largely contradicted by the same authority's admission a few sentences later that on the 16th and again on 17th September, 1942, a U.S. Army Air Force B.24 flying from Ascension Island had attacked a U-boat in the approximate position of those engaged in the rescue of the *Laconia's* survivors. In his "Conclusion" (*L'affaire du Laconia* p.282) M. Peillard came emphatically to view that here at last was the truth — with which I cordially agreed. It is, however, an astonishing fact that the whole paragraph giving M. Peillard's conclusions is omitted from the English translation of the book. It was this, and other similar treatment elsewhere of the original text, that caused me to enter a strong protest against the liberties taken by the translator when I reviewed the English edition (*Sunday Telegraph*, 15th September 1963).

Meanwhile I had received, also in 1963, a letter from the U.S. Embassy in London enquiring whether I could produce a copy of my article "Mystery of U.S. Plane's Attack," referred to above for the use of Mr. Maurer who would, I was told, be producing an article on the subject of the *Laconia* incident in the U.S. Air Force Historical Review. That I was very glad to do so, as I realized that the enquiry probably indicated that the uncertainties were at long last to be resolved. Such, then, is the background to the article reproduced above.

The only conclusion that I can draw from the tortuous tale I have recounted is that it shows once again how very unwise it is for a Service Department to try to impose historical censorship — except of course in cases where a good deal more than the commonly pleaded, all-embracing "national interest" is involved. Though the mills of history may grind slowly, they do grind exceeding sure, and all experience shows that the truth is virtually certain to come out in the end — so inevitably producing a stain on the reputation for straight dealing of the department which imposed the original censorship. It is, however, fair to record that the American Service Departments have always been, and still are being, very open-handed towards me; and that their attitude towards the historian is in general far more liberal than that of their British counterparts. Thus I am all the more glad that the U.S. Air Force has at last swept away the cobwebs which have too long obscured the truth regarding the *Laconia* incident.

Contributors

Robert Dexter Armstrong was born in Rome, Georgia on the day after Christmas, 1940. He qualified in submarines in 1960 and served six years served aboard USS *Drum* (SS-228), USS *Diablo* (SS-479), and USS *Requin* (SS-481). Attended University of South Carolina, American University, Washington, DC and University of Oslo, Norway. Worked as statistical draftsman at the International Monetary Fund, Washington, DC, senior field representative for the Potomac Electric Power Company, sales engineer for the Haughton Elevator Company. Mr. Armstrong retired at thirty years as the Deputy Director of Space Acquisition and Building Management of the U.S. General Services Administration (GSA). Wife, Solveig Elise Nordvik (Armstrong) died December 13, 2005. Presently retired and loafing and spends time with daughters Kristine Margaret and Catherine and three grandchildren; Andrew, Grace-Solveig, and Calvin a.k.a. "Bulldozer".

Kelly Asay was the President and Producer for Tesseraction Games for five years. Mr. Asay has also seen duty as a senior developer for Dynamix, a television producer for CBS, a professional musician, recording engineer, news cameraman, and corporate trainer. He now works for his own company Deep Six Online and continues to develop products for naval enthusiasts. Mr. Asay currently lives in Eugene, Oregon with his wife Cindy and their two teenage sons.

Florin Boitor worked as a videogame developer for Ubisoft Romania for 12 years since the opening of the studio in Bucharest. He contributed as programmer, lead programmer, studio manager or producer to the development of more then 12 titles on various platforms. His last position at Ubisoft was executive producer of Silent Hunter III. He currently runs his own game development company called ParKel Soft. He is having a lot more time to enjoy watching movies, skiing, football, and rock music.

Joe Buff is a seasoned risk analyst, professional writer on national security, and best-selling author of *Seas of Crisis*, *Straits Of Power*, *Tidal Rip*, *Crush Depth*, *Thunder in the Deep*, and *Deep Sound Channel*. A former partner in a top-ten global management consulting firm, three of his non-fiction articles received annual literary awards from the Naval Submarine League. Mr. Buff's latest novel, *Seas of Crisis*, won the 2006 Admiral Nimitz Award for Outstanding Naval Fiction from the Military Writers Society of America. Mr. Buff holds a master's degree in math from MIT, earned under a National Science Foundation Fellowship. He worked as an intern at the Argonne National Laboratory. Previously a qualified actuary for twenty years, with extensive experience at interpreting policy implications of dire "what if" scenarios, Mr. Buff is now a member of the Society for Risk Analysis, an international scholarly body headquartered in McLean, VA.

Spencer Burnham is a freelance graphic designer/3D artist who lives with his wife Lisa and son Calvin in Barrington, New Hampshire.

Gerard Cuomo, Catskill, NY, was born 17 August 1976. He is a graduate of Embry-Riddle Aeronautical University, B.S. Aeronautics. He earns a living as a commercial pilot and a part-time technical writing consultant for a dental equipment manufacturer. He has been subsimming since the late 1980s with Microprose's "Red Storm Rising".

Brian Danielson works as a programmer for a major hardware company. He is one of the most experienced subsim skippers around, getting in on the genre from the start in the early eighties. He was a contributor of information for Aces of the Deep and a programmer for Command AOD and has been known to keep mission logs, calculate manual firing solutions, and construct his own TDC programs.

Craig "Torplexed" Dinkelman was born May 12, 1962 on the old USAF Itazuki AFB in Fukuoka Japan. Lived on the Karamursel AFB Turkey, Kadena AFB Okinawa, and Fort Meade, Maryland as a kid. Despite being "an incorrigible Air Force brat", he developed a strange affinity for the naval stuff. Mr. Dinkelman attended the Praxis Art School and the Art Institute of Seattle and learned how to draw on paper with ink and pencils. He moved on to running the art department of a small Tacoma firm, now drawing with computers. His hobbies include wargaming, model making, history, watching bad movies, and cartooning.

Paul Farace began working aboard *USS Cod* (SS 224) in 1976 as a college freshman and now is the curator of this National Historic Landmark submarine. He was honored with the Henry Vadnais award for his restoration and curatorial work aboard *Cod* by the Historic Naval Ships Association. He is married to Teresa Lazusky and whenever possible is scouring thrift stores and flea markets for sub-related artifacts.

Timothy Grab was born in Philadelphia, Pennsylvania and grew up in Burlington, New Jersey. Inspired by the events of the first Gulf War, he enlisted in the Navy at age 23 in 1991. He spent six years in the service, with over four of those years assigned to the Improved Los Angeles class submarine USS Annapolis, SSN-760. Mr. Grab now resides in Massachusetts, where he works as a software/hardware support engineer in the telecommunications industry. His current interests include naval simulation games, U.S. naval history with a focus on submarines, computer programming, and photographing historic naval ships and lighthouses.

Clifford J. Hurgin, Jr. has enjoyed playing video and computer games for thirty years, from the earliest home systems and arcades to the graphically rich games of today. Mr. Hurgin enjoys creative photography, sketching and writing. His favorite reading material is usually in regards to ship disasters, survival situations, and Military history. He is a proud son of a WWII Navy veteran and resident of Danbury, CT.

Jonathan Beck Jorgensen is a Project Manager for IBM in Copenhagen, Denmark. Married with one child, his interests include submarines, traveling and scuba diving. Known as "McBeck", he been a part of the Subsim community since 1998 when he first read about SH2.

Jason Lobo (34) was first introduced to submarines when his father bought him a Revelle U-47 model for Christmas. In 1984, they bought their PC and come Christmas, a copy of Gato appeared under the tree. He was hooked. Mr. Lobo lives in Eugene, Oregon, claims to have five career hockey goals, and plays Silent Hunter III with his first-mate and gun captain, his six-year-old son.

Maurer Maurer (1914-2002) received his B.S. from Miami University, Ohio, and his M.A. and PhD from the Ohio State University. From 1955 until his retirement in 1983, Dr. Maurer worked in the U.S. Air Force Historical Research Center. Among his many books are the four-volume *The U.S. Air Service in World War I*; *Air Force Combat Units in World War II*; and *Combat Squadrons of the Air Force, World War II*.

David "Horsa" Millichope was born in 1947 in Chester, England. He trained at the University of Hull (England) as a biologist and subsequently as a teacher of biology/history, completing with a Master's Degree in Educational Research. Despite his somewhat conventional scientific background his first loves have always been history and the visual arts. For 20 years he taught mainly biology in secondary schools, later resigning his teaching post in 1989 to build and run his own photography/video business. He is married and living in West Yorkshire. There are two grown up daughters that he sees regularly, who now work in London. He feels he is now best described as semi-retired — still working but "no longer living under the tyranny of the alarm clock.".

Bill Nichols is a former U.S. Navy submarine officer. He earned his dolphins on the world's first nuclear submarine, USS Nautilus, during her last years of operation. Since leaving the Navy, he has worked on the Trident missile system, the USS Virginia-class submarine program, the

Navy's Surface Ship Concept Formulation (CONFORN) program, and various classified programs for the Defense Advanced Research Projects Agency (DARPA). Mr. Nichols is presently a systems engineer for the Ballistic Missile Defense program.

Lawrence J. Paszek, a WWII combat infantryman, earned a B.S. in Foreign Service and MSFS at Georgetown University in Washington, D.C. As Historian for the American Battle Monuments Commission, Mr. Paszek compiled research and designed commemorative maps displayed at the Hawaii and Manila American Military Cemeteries before accepting a Fullbright Scholarship to the University of Warsaw, Poland, in 1960. During his subsequent career as Historian and Senior Editor in the U.S. Air Force Historical Program, he authored several books, including the Air Force *Guide to Documentary Sources*. Mr. Paszek ushered through design and publication more than 40 volumes on Air Force history. In retirement, he served in 1992 as Senior Editor for the six-volume *Gulf War Air Power Study*, initiated by the Secretary of the Air Force. He resides in Reston, Virginia.

Donald Ross was born in 1922 and raised in New York City. He spent the entire years of WWII in Cambridge, Mass, where he graduated Harvard College in three years, married his step-sister Harriet Murphy, taught physics and did war research. His 45-year career in fluid mechanics, underwater acoustics, and submarine warfare is covered in his two contributions to this book. He retired in 1992 following a three-year tour in La Spezia, Italy as Associate Director of the NATO ASW Research Laboratory. His post-retirement volunteer career included three years as assistant to the Director of the San Diego County Human Relations Commission and five years as staff analyst for the Domestic Violence Unit of the San Diego Police Department. Dr. Ross currently lives with his second wife, Nancy, in La Jolla, CA, where he uses the Internet to conduct historical research and is producing a web page. He has three married sons scattered around the planet, and three grandchildren.

Laura "Sharky" Sands lives with her husband near Chicago. She began a social life on usenets and BBS's in the eighties. Ms. Sands was the "Kitchen Patrol Priestess" of the Wolfpack League and was instrumental in the success of all three Sub Club meetings.

Theodore P. Savas graduated from The University of Iowa College of Law in 1986 (Honors). He practiced law in Silicon Valley for twelve years and co-founded Savas Woodbury Publishers (subsequently Savas Publishing) in 1990. He sold the company to an East Coast publisher in 2000. He is currently Managing Director of Savas Beatie LLC (www.savasbeatie.com) In addition to teaching legal, history, and business-related college classes, Mr. Savas is the author or editor of fourteen books (published in five languages) including *A Guide to the Battles of the American Revolution* (with J. David Dameron, Spellmount and SBLLC, 2006), *Hunt and Kill: U-505 and the U-Boat War in the Atlantic* (Spellmount, SBLLC, 2004), and *Silent Hunters: German U-boat Commanders of World War II* (Campbell, 1997; Naval Institute Press, 2003). He has also written numerous articles for a variety of journals, magazines, and newspapers. He can be reached at teds@savasbeatie.com

Mariano "Marcantilan" Sciaroni (31), graduated as a lawyer in 1999 and specialized in corporate and commercial law. He has also spent two years in the Higher War School, Argentine Army, attending the Master's Degree Program on Strategy and Geopolitics. Mr. Sciaroni lives with his wife in Buenos Aires, Argentina.

Valerie Stevens is a senior at Brazoswood High School and has been accepted at Texas A&M University to study architecture. She is a model student and an accomplished artist. At age nine, she was among the first volunteers to restore the *USS Cavalla* in Galveston, TX. Ms. Stevens has served over 250 hours in Teen Court as a defense attorney and designed the Brazoswood High School Varsity Tennis shirt and a dedication banner.

Dave Stoops served three-and-a-half years aboard the *USS Permit* SSN-594 as a nuclear trained electrician. After his service during the Cold War, Mr. Stoops picked up an electrical engineering degree and worked 28 years for Phillips Petroleum Company, retiring in 2003. He and his wife live in the NASA area of Houston.

Deryck Swetnam, MA, was born in Portsmouth UK in December, 1946 and did his schooling there and in Malta. He joined the Royal Navy qualifying as an Engine Room Artificer, where he served until 1976. After an industrial accident he applied for and was accepted for a Master's Degree course in Maritime History at Portsmouth University. The Flanders U Boat Flotilla was the subject of his dissertation and he read a paper on it at the New Researchers Conference at Portsmouth Royal Naval Museum in March 2005. Mr. Swetnam's researches into the Flanders flotilla continues with the aim of producing a book on the subject. He is married with two grown children and two grandchildren

Capt. Ernest J. "Zeke" Zellmer was an officer aboard *USS Cavalla* when she sank the carrier *Shokaku* June 19, 1944. Capt. Zellmer has been a key figure in the restoration of *Cavalla* in Seawolf Park, Galveston, Texas. He is retired and living in Satellite Beach, Florida with his lovely wife, Beverly.

Acknowledgments

I express my great debt and appreciation for all the people who have helped Subsim and this book find success. In the beginning, my interest and education in subs, history, literature, and computers was aided and influenced by my parents, Charlotte and Abe Stevens, my Uncle Frank, Stanford's Book Exchange, Dicky Gillespie, Rick Range, Karen Gillenwaters, Ron Martini, Frank Morrisette, Jim Atkins, Dave Stoops, Zeke Zellmer, Jim Rankin, Zeb Alford, Kim Castro, Brian Danielson, and Guðmundur Helgason, to name a few.

I also wish to thank Eduardo Perez, Frank Kulick, Jamie Carlson, Shawn Storc, Bill Nichols, "Gizzmoe," Laura Sands, Bram Otto, Dargo, Craig Dinkleman, Tom Chick, "Crow," Florin Boitor, Mike Jones, John Channing, Pat Miller, John Veverka, Troy Heere, Philip Salminen, and many more fine individuals for their time, care, and inspiration.

This book was created by the generous efforts of the contributors: writers, game developers, navy vets, historians, and artists who were willing to share their knowledge and gifts with the community. Special thanks to Vickie Stevens for her invaluable typing skills and the diligence of Charlotte Stevens, Tom Morris, Daryl Carpenter, and Rachael Owens for assistance in proofing the texts. Their sharp eyes and skill kept this book from descending below test depth.

Neal Stevens

Bibliography

The Flanders U Boat Flotilla 1915 – 1918 - D. Swetnam MA

Bacon Sir R. *The Dover Patrol 1915- 1917* (Hutchinson & Co London) c1919

Corbett J.S. *Naval Operations vol 1* (Longmans, Green & Co, London) 1938

Corbett J.S. *Naval Operations vol 2* (Longmans, Green & Co, London) 1921

Corbett J.S. *Naval Operations vol 3* (Longmans, Green & Co, London) 1940

Dittmar F.J. & Colledge J.J. *British Warships 1914 - 1919* (Ian Allan, London) 1972

Evans A.S. *Beneath the Waves* (Kimber, London) 1986

Fürbringer W. *Fips, Legendary U Boat Commander* (Pen & Sword Books Barnsley) 1999

Gibson R. & Prendergast M. *The German Submarine War 1914 –1918* (Constable & Co, London) 1931.

Grant R.M. *U Boats Destroyed* (Periscope Publishing, Penzance) 2002

Grant R.M. *U Boat Intelligence* (Putnam, London) 1969

Grant R.M. *The U Boat Hunters* Periscope (Publishing, Penzance) 2003

Greene W.G. (Ed.) *The Navy List* (various editions) H.M.S.O.

Groner E. *German Warships 1815- 1945* vol II (Bernard & Graefe) 1983

Hallan T.D. *The Spider Web* Arms &Armour Press 1979

H.M.S.O.

British Vessels Lost at Sea 1914 – 1918 (Patrick Stephens Ltd, Cambridge) Reprint 1977

Humphries R. *The Dover Patrol 1914 –1918* (Sutton Publishing Ltd Stroud) 1998

Jameson W. *The Most Formidable Thing* (Rupert Hart-Davis Ltd London) 1965

Lloyds Register 1914 edition

Marder A.J. *From the Dreadnought to Scapa Flow Vol II The War years: To the eve of Jutland 1914-1916* (Oxford University Press) 1965.

Marder A.J. *From the Dreadnought to Scapa Flow Vol III Jutland and after: May 1916-December 1916* (Oxford University Press) 1966.

Scheer R. *Germany's High Sea Fleet in the World War* (Cassel & Co, London) 1920

Primary Sources from the Public Record Office

ADM 137/3874 *Interrogation of the Survivors of German Submarines UB Boats*

ADM 137/3876 *Interrogation of the Survivors of German Submarines UC Boats*

Air 1/2099 *RNAS Reports*

Air 1/2314 *Anti Submarine Reports*

Primary Sources from the Submarine Museum, Gosport

Naval Staff Monograph (Historical) vol xix May 1939

Periodicals

The Gentlemans' Magazine 1794 and 1798

Internet and email sources

Ryheul J. *The Flandern U Boat bases and U Bootflotte Flandern* September 2002 http://www.uboat.net/articles/index.html?article=48

Ryheul J. *The U Boat Bases of Brugge, Zeebrugge and Oostende* August 2002

http://www.uboat.net/articles/index.html?article=46 downloaded 26/7/03

Mike Y UB123 http://uboat.net/forum/read.php?f=23&i=1193&t=1193 downloaded 7/3/04.

Lowrey M & Dufiel Y. Database of vessels sunk by U Boats 1914 –1918 (unpublished)

TV Programme

Paul Griffin *Wreck Detectives* 8 August 2004[Television series] Channel 4 Produced by Paul Griffin.

The Dreadnought Era - David Millichope

Castles of Steel : Robert Massie : pub Jonathan Cape

The Rules of the Game : Andrew Gordon : pub John Murray.

The Grand Scuttle : Dan van der Vat : pub Birlinn

Jutland 1916, Death in the Grey Wastes : Steel and Hart : pub Cassell

---

I was a regular poster at Subsim when I got a message from Neal asking if I'd be interested in being part of a core group to plan and start a new clan which became the Wolfpack League. That was about eight months before the SHII release date which then got pushed back. I remember being the 4th poster at the WPL and the post was just a simple "Reporting in." We debated and voted on things like whether members should be automatically assigned to a Flotilla based on region, and that the 9th should be given to the German players, if we got enough of them to join. We had no idea how big that place would get!

The integrity of this place stood out like a beacon and the help you got from the people here wasn't just good, it often came from guys who had a naval history and that impressed me. The early names I remember were Bill Nichols and Hondo.

Memories... Beta testing SHII - Destroyer Command multiplay with Neal, Old Man, Ddoutel (Duane), and some others. It was hard to go through the necessary business when you really wanted to cut loose on somebody. Somehow, some way, before we ended for the night Old Man would find a way sink Neal's destroyer.

Yes, guys, I showed a longtime friend, Sharky, the WPL and Subsim and she joined soon after. All you potato peelers have me to blame.

-Rick C. Sniper

When surfing the web, sail over to Subsim.com

Web's #1 submarine and naval game resource.

www.subsim.com

www.ingramcontent.com/pod-product-compliance
Lightning Source LLC
Chambersburg PA
CBHW022100150426
43195CB00008B/203